MW00743428

# SOIL CHEMISTRY

# SOIL CHEMISTRY

### SECOND EDITION

## HINRICH L. BOHN
## BRIAN L. McNEAL
## GEORGE A. O'CONNOR

**JOHN WILEY & SONS, INC.**

New York / Chichester / Weinheim / Brisbane / Toronto / Singapore

This book is printed on acid-free paper. ∞

Copyright © 2001 by John Wiley & Sons, Inc. All rights reserved.

Published simultaneously in Canada.

No part of this publication may be reproduced, stored in a retrieval system or transmitted in any form or by any means, electronic, mechanical, photocopying, recording, scanning or otherwise, except as permitted under Sections 107 or 108 of the 1976 United States Copyright Act, without either the prior written permission of the Publisher, or authorization through payment of the appropriate per-copy fee to the Copyright Clearance Center, 222 Rosewood Drive, Danvers, MA 01923, (978) 750-8400, fax (978) 750-4744. Requests to the Publisher for permission should be addressed to the Permissions Department, John Wiley & Sons, Inc., 605 Third Avenue, New York, NY 10158-0012, (212) 850-6011, fax (212) 850-6008. E-Mail: PERMREQ@WILEY.COM.

This publication is designed to provide accurate and authoritative information in regard to the subject matter covered. It is sold with the understanding that the publisher is not engaged in rendering professional services. If professional advice or other expert assistance is required, the services of a competent professional person should be sought.

***Library of Congress Cataloging-in-Publication Data:***

Bohn, Hinrich L., 1934–
    Soil chemistry / Hinrich L. Bohn, Brian L. McNeal, George A. O'Connor.—3rd ed.
       p.   cm.
    Includes bibliographical references (p. ).
    ISBN 0-471-36339-1 (cloth : alk. paper)
    1. Soil chemistry.   I. McNeal, Brian Lester, 1938–   II. O'Connor, George A., 1944–   III. Title.
S592.5 .B63 2001
631.4′1—dc21                                                        2001017914

10 9 8 7 6 5 4 3 2 1

# CONTENTS

# PREFACE

I thank Linda Candelaria, Gavin Gillman, Robert Harter, Mark Noll, Stom Ohno, and Scott Young for their suggestions and encouragement to write a third edition of Soil Chemistry. I am especially grateful to Linda Candelaria for her excellent suggestions on additions and revisions.

This edition tries to emphasize that the soil and soil solution are the center and heart of the environment. The chemical composition of the biosphere, hydrosphere, and atmosphere depends greatly on the chemistry of the soil.

Since the second edition appeared, much of the interest in soil chemistry has been on the fate of so-called toxic chemicals and elements in soils. This edition points out that (1) all of the chemical elements—toxic and beneficial—were always in the soil, (2) the soil is the safest part of the environment in which to deposit our wastes, (3) there are wise and unwise ways to utilize soil for waste disposal, (4) soil chemistry degrades wastes and converts them into benign or useful substances, (5) environmental activists and the popular media usually ignore the dose–response concept that is central to toxicology and to soil fertility, and (6) how much is in the soil, how fast it is changing, and how easily it transfers to plants and water are more important than what is there. Soil chemistry can answer those important questions. A goal for the future is to answer them better.

The use of aluminium for aluminum, occasionally of natrium and kalium for sodium and potassium, and often of essential elements for essential nutrients is not affectation. The intent is to bring more international usage and correctness into the book. Soil chemistry is a global issue.

Brian McNeal and George O'Connor were unable to work on this edition of our book. I retained much of their excellent contributions to the previous editions. Any errors or omissions are my responsibility. I would be grateful if readers would call them to my attention.

HINRICH BOHN

*Tucson, Arizona*

# 1

# INTRODUCTION

*No one regards what is at his feet; we all gaze at the stars.*
                                   —Quintus Ennius (239–169 BC)

*Heaven is beneath our feet as well as above our heads.*
                     —Henry David Thoreau (1817–1862)

*The earth was made so various that the mind of desultory man, studious of change and pleased with novelty, might be indulged.*
                                   —William Cowper (The Task, 1780)

The quotations illustrate how differently humans see the soil that gives them life and feeds them. Those opinions have been held for a long time. Most people are still at the knowledge level of Quintus Ennius who lived more than 2000 years ago. They take for granted the food that the soil produces, the clean water and air that the soil provides. Thoreau's and Cowper's wonder and fascination of soils is rarely expressed or felt. Yet the soil is wondrous if one looks closely. The soil—the solid but porous surface of the earth to about one meter depth, the depth that roots penetrate—has many mysteries. The soil is as mysterious and exciting as any other science and any other part of the universe.

Soil is a mixture of inorganic and organic solids, air, water, microorganisms, and plant roots. All these phases influence each other: Weathering and adsorption by the soil affect air and water quality, air and water weather the soil, microorganisms catalyze many of the reactions, and plant roots absorb and exude inorganic and organic substances. Soil chemistry considers all these reactions but emphasizes the reactions of the *soil solution*, the thin film of water and its solutes (dissolved substances) on the surfaces of soil particles.

## 1.1 THE SOIL SOLUTION

The soil solution is the interface between soil and the other three active environmental compartments—atmosphere, biosphere, and hydrosphere (Fig. 1.1). The boundaries are dashed lines to indicate that matter and energy move actively from one compartment to another; the environmental compartments are closely interactive rather than isolated. The interface between marine sediments and seawater, and between groundwater and subsoils, is chemically much the same as the interface between surface soils and the soil solution. Sediments remove and release ions from the bodies of water they contact by the same processes as the interface between the soil and the soil solution.

The soil solution is the source of mineral nutrients for all terrestrial organisms. As the soil solution percolates below the root zone, it becomes groundwater or drains to streams, lakes, and the oceans, and strongly affects their chemistry. The amounts of matter transferred are much greater and the rates of these reactions are much faster in the soil than in the other environmental compartments. The soil solution is the most important transfer medium for the chemical elements that are essential to life.

The soil solution differs from other aqueous solutions in that it is not electrically neutral and usually contains more cations than anions. The net negative charge of soil clay particles in most soils extends electrically out into the soil solution, and the charge is balanced by an excess of cations in the soil solution. These cations belong to the solid but are present in the soil solution. Soils in old and heavily weathered soils, as in parts of Australia, Africa, and South America, or in soils of volcanic origin, as in Japan and New Zealand, may have a net positive charge. There the soil solution has an excess of anions.

The interactions of ions and electrical charge at the soil particle–soil solution interface happens at all particle interfaces. In cases outside of the soil, this interaction

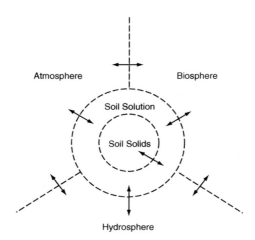

Atmosphere    Biosphere

Soil Solution

Soil Solids

Hydrosphere

**FIGURE 1.1.** The soil and the soil solution are the heart of the environment.

is generally negligibly small. The soil is unique because the soil's surface area is so large that this interaction becomes so extensive. Because of this interaction, the boundary between soil solids and the soil solution is diffuse. The water and ions at the interface belong to both the aqueous phase and to the soil solids.

The diffuse layer extends out as much as 50 nm into the aqueous solution from the particle surface. For clay (colloidal)-size (<2 $\mu$m) particles with their large surface area, this interaction is great enough to significantly affect the composition of the soil solution next to the colloidal particles. Because soils contain considerable clay, a large part of the soil solution is affected by colloids. At the so-called field capacity water content, most of the soil solution is in the < 10 $\mu$m contacts and pores between sand (50–2000 $\mu$m) and silt (2–50 $\mu$m) particles. Clay particles and microbes congregate at these contacts so the soil solution interacts closely with these reactive bodies in these contact zones. The soil solution on open sand and silt surfaces is only 10–100 nm thick, so much of this water is also affected by the particle's charge. The portion of soil solution affected by soil colloids increases as the soil dries.

Most soil reactions occur at the soil solution/soil interface. Ions in water can move and react fast enough to measure easily. Slower but still measurable reactions occur in the weathered surfaces of soil particles. These poorly understood surfaces contain considerable water. Reaction rates in the truly solid phase at soil temperatures, however, are too slow to be measured in our lifetimes.

Because the mass and reactivity of soils are great, the chemistry of the atmosphere and fresh water are largely controlled by the chemistry of the soil solution. Reactions that require days and years in air, and hours in water, require only seconds and minutes in soils. The compositions of the air, water, and biomass compartments in the environment evolved from, and still respond to, the chemistry of the soil. The soil came first and as it changed, it changed the others. The change in the others also changed the soil, but to a lesser extent because of the soil's mass.

The soil solution contains a wide variety of solutes, including probably every element in the periodic table. This book discusses those solutes that are active in the environment—the solutes that affect plant and animal life, or have been of concern in pollution.

Figure 2.1 shows the chemical elements that are essential to living organisms, those that are reactive, and those that are in significant amounts in each environmental compartment. Chapter 2 discusses the life-essential and other important elements in more detail. All of the major and minor essential elements are stored and available for transfer in soils; the amounts in the other compartments, although important, are generally much, much smaller. Major and minor refer only to the amounts needed by organisms; all are essential and all are needed in their proper amounts. Too little is deficiency, too much is toxicity. The essential elements are also called essential nutrients, but elements is a better term because nutrients implies energy content, such as in the carbohydrates and fats of food.

With few exceptions, the soil supplies these elements to living organisms in about the right amounts. Although not perfect, the soil's supplying power is remarkably effective. This is not simply a happy coincidence; life evolved in response to the availability of these elements in soils.

For most people the major reason to study soil chemistry is to insure and to increase production of food and fiber crops. The soil is and will be the main source of human nutrition. The oceans supplement our food supply but their productivity is limited by the osmotic potential of the water, the limited availability of essential elements, the low temperature of ocean water, and the long food chain between photosynthetic organisms and those large enough to harvest. The oceans can provide only a small amount of society's needs and wants. The soil's productivity per unit area is many times that of the oceans. Terrestrial plants remain the cheapest and best means of converting solar energy into life support for this planet. The growth of plants on soils is the basis of most of the world's economy and of a nation's well being.

Soil chemistry is only one of many factors that affect plant growth. In contrast to climate and other uncontrollable factors, however, agriculturalists can influence and modify soil chemistry to considerable extent. The amounts of essential elements needed by plants over a season are small enough that supplementing the soil supply is feasible. Increasing the efficiency of that fertilization is a continuing soil chemistry challenge. The toxicity of materials that harm plant growth can also be controlled by soil chemistry.

People are now learning to appreciate the soil's large role in the biogeochemical cycling of the elements. Soils can mitigate many undesirable human-caused changes (pollution) of the environment. Safe removal of wastes from the environment has been recognized to be as important for continued civilization as food production. The retention, exchange, oxidation, and precipitation of waste in soils make them unequaled as recycling media.

In earlier times when the population was less dense and industries were few and small, wastes were distributed widely on and in the soil and could readily return to their natural biogeochemical cycles. By concentrating wastes in urban areas, industrial facilities, landfills, feedlots, and sewer outfalls—releasing wastes to the air and water rather than allowing them to react in soils—by fertilization, and by creating synthetic chemicals that react slowly, humanity has occasionally and in local areas exceeded the rate at which these materials can return to their biogeochemical cycles. "Advanced" societies sometimes overlook the degradative functions of soil and look instead for expensive, and only partially satisfactory, technological methods of waste disposal. Humanity creates pollution, which has awakened a new awareness of the importance of soil chemistry.

Soil chemistry is closely related to colloid (surface) chemistry, geochemistry, soil fertility, soil mineralogy, and soil microbiology or biochemistry. Soil fertility considers soil as a medium for plant growth. Soil mineralogy examines the structural chemistry of the solid phase. Soil microbiology studies soil biochemical reactions. Such subdivision is necessary to study the soil thoroughly, but these subdivisions sometimes obscure the interaction between soil components, and this interaction is often as important as the properties of the components alone.

Soil chemistry traditionally has had two branches: inorganic and organic, but strict separation of the two fields is difficult and pointless in many cases. The direction of biochemical soil reactions is largely based on the inorganic phase. Soil organic processes affect primarily the rate of soil chemical reactions. Biochemical reactions are carried out by soil microorganisms, whose vast numbers in soils influence many reactions. For several elements, notably carbon, nitrogen, and sulfur, the microbial role almost totally determines soil reaction rates. Biochemical and microbial reactions are primarily catalytic processes affected by the independent variables of soil mineral composition, climate, gas exchange with the atmosphere, and energy from photosynthesis. Despite the importance of biochemical reactions, research in soil chemistry historically has been more oriented toward inorganic processes.

## 1.2  BACKGROUND

Food and fiber production were already important before agriculture began. After fear, food is the dominant concern of every animal. The senate of ancient Athens debated soil productivity 2500 years ago and voiced the same worries about sustaining and increasing soil productivity that are heard today: Can this productivity continue or is soil productivity being exhausted?

In 1790, Malthus noticed that the human population was increasing exponentially and that food production was increasing arithmetically. He predicted that by 1850 the demand for food due to population growth would overtake food production, and people would be starving and fighting like rats for morsels of food. Similar apocalyptic predictions continue to crop up and cannot be disregarded. It is encouraging, however, that productivity has increased since the Greek senate debates and faster than Malthus predicted. In recent history food productivity has been increasing faster than ever. The earth now feeds the largest human population ever and a larger fraction of that population is better fed than ever before. Whether this can continue and at what price to the environment and other organisms is an open question. One encouraging part of the answer is the rapidly declining human birth rate on most continents in recent decades, thus putting less stress on the soil's resources in the future. Another part of the answer lies in soil chemistry, and much progress is still to be made in our understanding of the soil and its potential.

Agricultural practices that increase crop growth—planting legumes, manuring with animal dung and with litter from forests, rotating crops, and liming—were known to the Chinese 3000 years ago. These practices had also been learned by the Greeks and Romans and appeared in the writings of Varro, Cato, Columella, and Pliny. The reasons for their effectiveness, however, were unknown. Little or no further progress was made in the Western world for almost 1500 years because of ignorance and *deductive reasoning*. Deduction is applying preconceived ideas, broad generalities, and accepted truths to particular problems, without testing if the preconceived ideas and accepted truths are valid. One accepted truth, derived from the Greeks, was that matter was composed of earth, air, fire, and water—a weak basis, as we later learned, on which to increase knowledge.

In the early 1500s, Sir Francis Bacon pointed that *inductive reasoning*, the *scientific method*, is a much more productive approach to gaining new knowledge—

observe and measure, derive broader ideas from the data, and test these ideas again. The scientific method brought progress, but the progress in soil chemistry was slow.

Palissy (1563) thought that plant ash came from the soil and when added back to the soil could be reabsorbed by plants. Plat (1590) thought that salts from decomposing organic matter dissolved in water and absorbed by plants were responsible for plant growth. Glauber (1650) thought that saltpeter (Na, K nitrates) was the key to plant nutrition by the soil. Kuelbel (ca. 1700) believed that humus was the principle of vegetation. Boerhoeve (ca. 1700) believed that plants absorbed the "juices of the earth." Others have found this idea in Pliny's writings. None of them, however, had experimental proof.

Van Helmont (1592) tried to test these ideas. He planted a willow shoot in a pail of soil and covered the pail so that dust could not enter. He carefully measured the amount of water added. After five years the tree had gained 75.4 kilograms. The weight of soil in the pot was still "200 lb (90.8 kg) less about two ounces (56 g)." Van Helmont disregarded the 56 grams as what we would today call experimental error. He concluded that the soil contributed nothing to the nutrition of the plant and that plants needed only water for their sustenance. Although he followed Bacon's suggestions and the scientific method as well as he could, he unfortunately came to a wrong conclusion. Many experiments in nature still go afoul because of incomplete control and measurement of all the experimental variables.

John Woodruff's (1699) experimental design was much better. He grew plants in rainwater, river water, sewage water, and in sewage water plus garden mould. The more solutes and solids in the growth medium—the "dirtier" the water—the better the plants grew, implying that something in soil improved plant growth. The idea developed, but without further testing, that the organic fraction of the soil supplied the plant's needs. That idea persists to this day. Organic substances absorbed by plants from the soil may affect plant growth, but this has been difficult to prove.

In 1840 Justus von Liebig persuasively advanced the idea that inorganic chemicals were key to plant nutrition and that an input–output chemical budget should be maintained in the soil. Liebig's theory was most probably based on Carl Sprengel's work in 1820–1830 that showed that mineral salts, rather than humus or soil organic matter, were the source of plant growth. Liebig's influence was so strong that when Boussingault (1865) measured more nitrogen appearing in plants than he put into the soil, his work was disregarded for many years. Microbial nitrogen fixation did not fit into the Sprengel–Liebig model.

Soil chemistry was first recognized as distinct from soil fertility in 1850 when Way and Lawes, at Rothamsted, England, discovered cation exchange. Their work suggested that soils could be studied apart from plants; yet the results would still have implications for soil fertility.

The Information Age is upon us and the information is both accurate and inaccurate. From the information available about nutrition and health, for example, one might deduce that life is becoming riskier. In reality we are living longer and healthier lives than ever before. To make sense of what we hear and read, we sometimes still resort to deductive reasoning. Science no longer has the certainty that it once seemed to have and can be very complicated. People look for answers and ideas they can understand. Many current ideas about health, nutrition, and organic farming and gardening, for example, are popular not because they have been tested, but because their simplicity is easily understandable and because not-always-sound logic makes them look good. Although science cannot provide all the answers, our lives are healthier and longer because we have broken away from deductive reasoning. The spurious logic, oversimplification, and rejection of careful testing that can be part of deductive reasoning are steps backward rather than forward.

## 1.3  SOIL–ION INTERACTIONS

Solutes, electrolytes, and nonelectrolytes in the soil solution are the immediate sources of the elements required by plants for growth. This supply can be continuously renewed by the many mechanisms of ion–soil interaction that remove and add ions in the soil solution: (1) mineral weathering, (2) organic matter decay, (3) rain, (4) irrigation waters containing salts, (5) fertilization, and (6) release of ions retained by the colloid or clay fraction of soils.

Solutes in the soil solution and ions retained on soil particle surfaces are generally the largest fraction of the elements available to plants. Weathering of ions from soil particles is slow compared to plant needs. Organic decay releases ions much faster than weathering, but most of the ions released react with the soil's solid phase before they can be absorbed by plants or microorganisms. When retaining ions, soils strike a delicate balance between preventing losses by leaching and supplying plants and microorganisms. Ion retention by soils does not completely prevent leaching losses but is sufficiently strong that ions can recycle many times through soils, plants, and animals before they are finally lost to groundwaters, rivers, and the sea.

Ions and molecules are retained in soils by cation and anion exchange, precipitation, weak electrostatic attraction, reactions with soil organic matter (SOM), and retention within microbial cells. If each ion were retained by only one such mechanism and by only one soil component, soil chemistry would be relatively simple. Instead, each ion reacts by several mechanisms, and to varying degrees, with many different solid phases. For simplification, soil chemists generally either measure a single parameter that reflects most soil interactions with a given ion (e.g., its overall availability to plants), or fractionate the soil and measure the ion's interactions with each soil component separately. Neither approach has proved totally satisfactory. Single-parameter measurements fail to account for variations from one soil to the next. Fractionation procedures may neglect the important interactions between soil components.

## 1.4  COLLOIDS AND THE SOIL SOLUTION

The complexity of ion interactions with the soil's solid phase is greatly increased by the colloidal properties of the soil's clay and organic fractions. *Colloids* are substances of about 1- to 1000-nm particle size that form unique mixtures when suspended in air or water. The components in colloidal mixtures tend to lose their individual identities so that the mixtures are like new substances. Colloidal mixtures are so ubiquitous and so distinctive that they have their own names: fog, smoke, smog, aerosol, foam, emulsion, gel, soil, and clay. All are small particles suspended in a liquid or gaseous fluid. Other colloidal mixtures include metal alloys, pearls, butter, and fine-grained rocks. The particles do not settle out of the suspending fluids, but also do not mix homogeneously with them. The physical properties of many colloidal mixtures tend to be similar regardless of the chemical composition of the colloid, because the size and the interaction are so important. Colloidal suspensions of starch, soap, salad dressing, and clay in water, for example, look the same, are similarly viscous, similarly colored, opaque, and stable (do not separate into separate phases).

Solutions are also mixtures but, in contrast to colloidal mixtures, solutions retain many properties of the major component—the solvent—while the minor component—the solute—loses its identity. Salts disappear when then dissolve in water. Aqueous solutions containing the same amounts of matter are as fluid and transparent as water. A mixture of 5% Na bentonite in water is a thick white gel; a solution of 5% NaCl is quite fluid and is transparent. When particles larger than ca. 2 $\mu$m are suspended in air or water, they settle out of suspension into separate phases. The properties of such mixtures are the sum of the properties of the separate components.

Colloidal particles interact strongly with the fluid, but the individual particles have some structural integrity, so they cannot be said to dissolve homogeneously. The colloidal mixture behaves so distinctively because of the large surface area of interaction between the particles and water or air. The ions at the boundary interact with the ions and molecules of both phases. This is true at any surface or phase boundary, but the interaction of colloidal phases is large because their surface areas are so large. A 1-mm sand particle has a surface area/mass ratio of about 0.002 $m^2g^{-1}$; a 1-$\mu$m clay particle, 2 $m^2g^{-1}$; and a 1-nm particle, 2000 $m^2g^{-1}$.

The important colloidal properties that clays impart to soils include ion and molecular retention and exchange and water and gas adsorption. The colloidal properties of clay create the intimate mixture of solids, liquids, and gases in the soil that is essential to life.

## 1.5 COMPOSITIONS OF SOILS AND PLANTS

Except for carbon as $CO_2$, H as $H_2O$, and oxygen as $O_2$, plants derive their essential elements from the soil. The minor exceptions are nitrogen and sulfur gases ($NO_x$, $NH_3$, and $SO_2$) absorbed directly from the atmosphere by leaves, plus ions absorbed from dust and foliar sprays on the leaves. Foliar absorption can be significant in polluted atmospheres or where agricultural sprays are purposely applied. Under natural conditions, the major factors affecting ion availability to plants are (1) the ion's concentration in the soil solution; (2) the degree of ion interaction with, and rate of release from, the soil's solid phases; (3) the activity of soil microorganisms; and (4) discrimination by the plant root during ion uptake. This book is concerned primarily with the first two factors: the soil solution and ion interaction with the solid phase.

Table 1.1 shows representative contents of important elements in soils. Soil contents vary and the values in Table 1.1 are averages. The composition of plants is less variable, partly because soil development tends to narrow the range of element availability compared to the range of the elemental composition of rocks. The soil is an O–Si–Al–Fe matrix containing relatively small amounts of the essential elements. The matrix is virtually inert in terms of plant nutrition, but the small amounts of ions held by that matrix are vital.

**Table 1.1. Typical concentrations of essential elements in soils, ratios of plant ash to soil content, annual plant uptake, and the ratios of soil content (to 1-meter depth) compared to annual plant uptake**

| Element | Soil Content (Weight%) | Plant Ash Content/Soil Content | Annual Plant Uptake (kg ha$^{-1}$ yr$^{-1}$) | Soil Content/ Annual Plant Uptake (yrs) |
|---|---|---|---|---|
| Oxygen | 49 | — | — | — |
| Hydrogen | — | — | — | — |
| Silicon | 33 | 0.3 | 20 | 21 000 |
| Aluminum | 7 | 0.03 | 0.5 | 180 000 |
| Iron | 4 | 0.1 | 0.4 | 100 000 |
| Carbon | 1 | — | — | — |
| Calcium | 1 | 25 | 50 | 260 |
| Potassium | 1 | 15 | 30 | 430 |
| Sodium | 0.7 | 1 | 2 | 4 600 |
| Magnesium | 0.6 | 3 | 4 | 2 000 |
| Titanium | 0.5 | 0.08 | 0.08 | 62 000 |
| Nitrogen | 0.1 | 15 | 30 | 40 |
| Phosphorus | 0.08 | 4 | 7 | 150 |
| Manganese | 0.08 | 0.6 | 0.4 | 3 000 |
| Sulfur | 0.05 | 70 | 2 | 320 |
| Fluorine | 0.02 | 1 | 0.01 | 26 000 |
| Chlorine | 0.01 | 10 | 0.06 | 200 |
| Zinc | 0.005 | 5 | 0.3 | 2 000 |
| Copper | 0.002 | 5 | 0.1 | 1 000 |
| Boron | 0.001 | 50 | 0.003 | 400 |
| Tin | 0.001 | 2 | 0.001 | ~ 10 000 |
| Iodine | 0.0005 | 0.1 | 0.00003 | 22 000 |
| Molybdenum | 0.0003 | ~ 2 | 0.003 | 1 000 |
| Cobalt | 0.0008 | 1 | 0.0006 | 17 000 |
| Selenium | 0.000001 | ~ 500 | 0.003 | 40 |

From Vinogradov's data in N. F. Ermolenko. 1972. *Trace Elements and Colloids in Soils*. Israel Program for Scientific Translations, Jerusalem.

Column 2 of Table 1.1 shows the ratio of plant content to soil content of important ions. The hydrogen, carbon, and oxygen ratios are omitted because these ions are not derived directly from soils. The ratios are crude indices of the relative availability of soil components to plants. Calcium, sulfur, nitrogen, and potassium in soils are more available than iron and manganese. One goal of soil chemistry is to explain why ions in soils vary widely in their degree of plant availability.

Column 3 shows the approximate annual plant uptake per hectare (ha) of the elements, assuming an annual dry matter production of $10\,000$ kg ha$^{-1}$. The amounts of calcium, potassium, and nitrogen absorbed greatly exceed plant absorption of the other elements.

Column 5 is the annual plant uptake of the elements divided by their total amounts in the soil. The result is the length of time the soil could supply that element to plants, if the plants were totally removed at harvest and nothing were added to the soil. The weakness of this assumption is that only a small fraction of plant matter is removed from soils during harvesting; most of the plant dies and decays where it grew. The assumption also ignores atmospheric and fertilizer inputs. Nonetheless, the ratios in column 5 roughly illustrate the relative size of the soil's store of essential elements. The soil's reserves of nitrogen, chlorine, and sulfur are low but are continually replenished by microbial nitrogen fixation, gas absorption, and rain. Despite this replenishment, nitrogen and sulfur concentrations are often less than optimal for plant growth. Nitrogen is the element that most commonly limits crop productivity. The low selenium ratio may be questionable because of limited data. The soil's supplies of potassium and calcium are also relatively low but are not cause for alarm. Calcium and potassium are constantly replenished by weathering of lower soil horizons and by decay of plant material, and they are easily replenished by liming or fertilization. The soil/plant conditions indicated in column 5 have existed for several thousand million years and are unlikely to change very much in the near future.

The data in Table 1.1 are illustrative rather than quantitative. Soil has supplied the essential elements to living organisms since terrestrial life began. The Exhaustion Plot at Rothamsted Experiment Station in England, for example, has operated continuously since 1845 and has shown that the worst agricultural practice—removing all plant material at harvest from the soil each year and no fertilization to make up the losses—reduces but does not stop plant growth or crop yields.

## 1.6   NONAGRICULTURAL SOIL CHEMISTRY

Soil chemistry is also important to the nonagricultural uses of soil. Soil is a building material for earthfill dams and roads and is being rediscovered in industrial nations as a building material for homes. Brick and cement are, after all, only baked soil material. Lesser-developed nations have always recognized the livability of mud huts and adobe houses. The physical stability of soil structures depends in part on their soil chemical status. The longevity of mud and wattle construction in medieval European homes and of the adobe buildings of Southwestern Native Americans depends on high calcium and low salt concentrations of the soils used to build those walls.

The persistence through many centuries of the temples at Angkor Wat in Southeast Asia depends on the high concentrations of iron and aluminium oxides in the soils used to form the building blocks. Such *laterite* (now more properly called *plinthite*) materials from soil can dry irreversibly and thus resist slaking and weathering even in a humid tropical climate.

The use of clay suspensions as drilling muds for lubrication and for clay liners for sealing landfills and lakes depends on the ability of $Na^+$ on colloidal surfaces to keep the clay from settling out and aggregating. The strong interaction of clay colloids and water is also useful in overcoming diarrhea, as a diluent for drugs, and as a drying agent.

Many iron and aluminium ore deposits are the end result of long and extreme soil chemical weathering at the end of the plinthite stage. The time scale puts their formation into the category of geochemistry, but the mode of formation involves the same chemical reactions as soil formation.

At the opposite end of the reaction time scale, soil clays are being investigated as catalysts to speed industrial chemical reactions. The "cracking" of petroleum into gasoline and other organic reactions is catalyzed by certain clays. Adsorption to clay surfaces imparts catalysis by holding reactive molecules in positions that encourage reaction and polymerization of organic chemicals. The $Cu^{2+}$, $Al^{3+}$, $N(CH_3)_4^+$, and rhenium phosphine and rhenium phospho-organic complex ions adsorbed on clays have such catalytic properties. The large surface area of the clays produces such an intimate reactant–catalyst mixture as to be almost a homogeneous single phase. At the end of the reaction, the clay catalyst is easily removed by filtration.

## 1.7  BIOGEOCHEMICAL CYCLES AND POLLUTION

The soil is a major part of the cycling of elements at the earth's surface. The elements that humans release as wastes are derived from the soil and the earth. Ions that are weathered at the earth's surface and released to the atmosphere or leached to the seas eventually circulate back to the land. They return to soils as gases absorbed from the atmosphere, as wastes removed from waste water, as solids buried in soils, as solutes in rain, as uplifted marine sediments, and as igneous rock uplifted to the land surface. Carbon, nitrogen, and sulfur cycle rapidly among the atmosphere, oceans, and soils. Other elements cycle more slowly between rocks, soils, and oceans but their movement is still rapid on a geologic time scale. The removal rate of elements from soils is slowed significantly by adsorption, precipitation, pH buffering, and plant uptake.

Chemical pollution is the diversion of chemical elements from the natural biogeochemical cycles. The carbon, nitrogen, and phosphate in municipal wastes released to streams and lakes are removed from the soil–plant cycle, which is the source of the nitrogen and much of the phosphate. If those substances were instead put back directly into the soils from whence they came, much less pollution would result. Air and water only slowly convert their wastes back into their natural sites in plants and soils. Soil, on the other hand, has enormous surface area and microbial catalytic activity plus oxygen and water with which to deactivate pollutants. Soil degrades most wastes quickly and returns the components to their natural cycles, thereby minimizing environmental disturbance.

If one considers pollution to be the rendering of soils to be unfit for plant growth, then the greatest contribution to pollution is the salinization and urbanization of soils. Careless irrigation slowly adds increasing amounts of salts to soil, which can reduce and stop plant growth. Covering soils with asphalt and concrete also renders soil unfit for plant production.

Human consumption of food, water, wood, metals, and fuel diverts substances from their natural cycles, if one assumes that humans are unnatural. Humans are trying to take care of themselves, just as other species do. The difference is that we concentrate wastes on a much larger scale and are better at using water and wind to carry them away. Humans, however, presumably have the capacity to analyze and improve their activities. This is hopeful and seems superior to ants, for example, which denude their environment and leave behind a totally barren land before moving to a new location. Even the worst of too-frequent shifting cultivation, monoculture cropping, clearcutting of forests, industrial pollution, and suburban sprawl are no worse than what ants do, and we have the ability to change and improve the situation.

*Consumption* is the transformation of matter and energy into less useful forms, including dilution to concentrations less than those recoverable by our current technology. Fertilizers, for example, are made from concentrated sources and diluted by spreading on agricultural lands. Over the short term, only a fraction is recovered by plants. The remainder is consumed by the soil when the fertilizer is converted into slowly recoverable or nonrecoverable forms. Not one atom is lost by this consumption, but the availability, chemical states, concentrations, and locations of the atoms change.

Consumption of water includes discarding wastewater to the sea, where marine salts contaminate the water. It is usually much easier and cheaper to remove the small amounts of pollutants in wastewater than to remove the large amount of salt added to the water when it is dumped into the sea. Artificial desalination of seawater is costly because of the energy required. Water is recovered naturally from the sea by solar evaporation. The resulting rain is unevenly distributed over the land so one-time use of water, in arid regions particularly, seems very wasteful. Water can often be used consecutively in the home, industry, and agriculture, thus decreasing water consumption. Wastewater spread on land can be agriculturally beneficial and simultaneously renovated for reuse.

## 1.8   SOIL AND THE HYDROSPHERE

The amount of water in soils is only a tiny fraction of the earth's total water supply. Table 1.2 shows two estimates of the distribution of water at the earth's surface. The two estimates differ widely but agree that the fraction present as soil moisture is small, 0.001 to 0.0005%. Either of these fractions of soil water seems perilously small to supply all terrestrial life and to help moderate climate. The periodic droughts around the world also emphasize that the amount of soil moisture is small. Assuming

**Table 1.2. Two estimates of the distribution of the earth's water**

| | | |
|---|---|---|
| Ocean Water | $80\%^a$ | $97\%^b$ |
| Ice | 19 | — |
| Groundwater | — | 0.06 |
| Pore water in rocks | 1 | 2 |
| Lakes and rivers | 0.002 | — |
|     Fresh | — | 0.007 |
|     Saline | — | 0.007 |
| Soil moisture | 0.001 | 0.0005 |
| | $(2 \times 10^{16}$ kg$)$ | $(7 \times 10^{16}$ kg$)$ |
| Atmospheric water | 0.0006 | 0.0001 |
| Biota | — | 0.0001 |
| Total | $1.7 \times 10^{21}$ kg | $1.35 \times 10^{21}$ kg |

[a] From R. M. Garrels, F. T. Mackenzie, and C. Hunt. 1975. *Chemical Cycles and the Global Environment.* W. Kauggman, Los Altos, CA.

[b] From D. A. Speidel and A. F. Agnew. 1982. *The Natural Geochemistry of Our Environment.* Westview, Boulder, CO.

that the world's soils contain 10% water by volume, the $1.2 \times 10^{14}$ m$^2$ of terrestrial soils contain $1.2 \times 10^{13}$ m$^3$ of water to a meter depth. This imposing amount is the water that supports plants, weathers rocks, forms soil, and is the medium in which most soil chemical reactions occur.

Terrestrial water receives most of its dissolved solutes from the soil, where rain first reaches the earth's surface and where weathering is strongest. The composition of water is less affected when it percolates to greater depths because the water already contains the salts obtained from the soil above. The composition and concentration of dissolved solutes can change at depths if the water contacts subsurface $CaCO_3$ or if it is stored for long periods in underground basins. In most cases, however, percolating waters retain the character of solute composition initially conferred on them by the surface soil.

Stream water is soil drainage plus surface runoff. Natural drainage waters contain relatively low concentrations of the essential ions. This slow and steady input from drainage waters supports what is generally regarded as a natural and desirable aquatic population in streams and lakes.

Runoff water is richer in sediments, nutrients, and organic matter. The fraction of surface runoff in temperate and humid regions is relatively small compared to drainage water under natural conditions. Dense plant cover prevents erosion and the soils are relatively permeable. Agriculture can increase the nitrate and phosphate concentrations, in runoff particularly, by removing the natural plant cover. Proper management can minimize runoff from agricultural lands, but some changes in water composition due to agriculture may be inevitable. Urbanization also increases runoff. The velocity of runoff increases as the overall permeability of the land surface is decreased, and runoff increased, by paving, compaction and destruction of natural contours.

In arid regions, surface runoff is a considerable fraction of the stream flow. The intense storms, sparse plant cover, and relatively low soil permeability create intermittent streams "too thick to drink and too thin to plow." The invisible solute concentration in such waters is also high and can be as important to the downstream ecology as the sediment.

## 1.9   SOIL AND THE ATMOSPHERE

The interaction of gases with soils is much less obvious than soil–water interaction, but is important to maintaining the low amounts of carbon, nitrogen and sulfur gases in the atmosphere. Gases that are foreign to the atmosphere are adsorbed by soils and plants and degraded to the natural and nongaseous forms of these elements. Soils also release gases, such as $H_2O$ and $CO_2$ from organic decay and $N_2$ and $N_2O$ from natural soil nitrogen compounds and from fertilizers. Soil has affected environmental chemistry since the earth began, and the soil in turn has been affected by other components of the environment.

Soil is a prominent part of the natural cycles of carbon, nitrogen, and sulfur. In the nitrogen cycle, nitrate and ammonium ions in rainwater are absorbed by soil, plant roots, and soil microorganisms and converted to amino acids or to gaseous $N_2$ and $N_2O$, which diffuse back to the atmosphere. Ammonia is also emitted and absorbed by soils. Under natural conditions the gaseous nitrogen loss is approximately balanced by $N_2$ uptake and conversion to amino acids by symbiotic and free-living soil microorganisms.

Soil absorbs sulfur dioxide, hydrogen sulfide, hydrocarbon, carbon monoxide, nitrogen oxide, and ozone gases from the air. The reactions are subtle and are often forgotten in considerations of the composition of the atmosphere. Direct soil absorption is perhaps most obvious in the case of the rapid disappearance of atmospheric sulfur dioxide in arid regions. The basicity of arid soils makes them an active sink for acidic compounds from the atmosphere. The relative amount of direct soil absorption of atmospheric gases, inappropriately termed dry fallout by atmospheric scientists, is less in humid regions where plant absorption and rain washout of the gases are substantial.

The soil's role in the carbon cycle is very large. Table 1.2 shows the estimated amounts in the active carbon reservoirs at the earth's surface: the atmosphere, living biomass, freshwater, and ocean water above the thermocline (a temperature inversion that isolates the surface 50 m from the deeper ocean). The amount of soil carbon far exceeds the others. The emission of $CO_2$ by organic decay in soils is the largest $CO_2$ input to the atmosphere. The change in amount of organic soil carbon resulting from climate changes must have affected atmospheric $CO_2$ concentration. Varying organic decay rates in soils due to climate and atmospheric $CO_2$ changes have not been considered adequately in most discussions of the carbon cycle. A slower decay rate would have the same net effect on atmospheric $CO_2$ that direct soil absorption of $CO_2$ from the atmosphere would have. *Peat* (organic-rich soil) has been accumulating in Canada and elsewhere since the glaciers retreated at an average rate of about 1 mm $yr^{-1}$. Virgin soils, when cleared and cultivated, on the other hand, lose one-third to one-half of their original organic carbon content by oxidation to $CO_2$. This amount of $CO_2$ production over the last 100 years equals or exceeds the highly publicized amount released by fossil fuel combustion.

Great changes in soil organic carbon levels have occurred during geologic time. The enormous carbon accumulation by soils during the Carboniferous Era must have affected the $CO_2$ content of the atmosphere greatly. Carbonate accumulation and loss in soils is probably a smaller buffer of atmospheric $CO_2$ than is organic soil carbon. The mass of soil carbon as carbonate is less, and its turnover rate is slower.

The atmosphere, in turn, affects soil development by providing oxygen and by wind erosion and deposition. Sand dunes are only the most obvious example. Loess soils are deposits of silt-sized particles carried by winds from riverbeds and glacial outwash. A large fraction of the clay content of the soils along the eastern shores of the Mediterranean Sea has been carried by winds several thousand kilometers from the Sahara Desert and Atlas Mountains of North Africa. Trade winds carry Saharan clay particles several thousand kilometers out into the Atlantic Ocean.

## 1.10  SOILS AND THE DEVELOPMENT OF LIFE

Soils have had a large role in the development of life and probably in the origin of life. Plants obviously have adapted to the physical and chemical characteristics of soils. These characteristics could also have aided the first steps of chemical organization that preceded life.

The earth's early atmosphere was rich in $CH_4$, $H_2$, $CO_2$, CO, $NH_3$, $N_2$, and $H_2S$, and contained no $O_2$. In the late 1940s Miller and Urey showed that lightning discharges in such a gas mixture could produce small amino acid-like and other organic molecules that could be the organic precursors needed for life. The next question is how do these molecules polymerize to larger molecules?

In the late 1920s a Soviet scientist noticed that hydrophobic substances in seawater cluster together, like the tar balls familiar to beachgoers. He suggested that this could have been a step in the organization of simple organic molecules into the more complex ones of living organisms. One problem with this hypothesis is that the molecules formed in lightning discharge are hydrophilic. They dissolve in water and disperse in the oceans rather than accumulate as do hydrophobic hydrocarbon globules.

In the 1930s the British biologist Haldane, inspired by this idea and rather ignorant of soils, thought that the initial polymerization of small molecules might have happened in an organic-rich ocean, which he called the "primordial (pre-life) soup." This phrase caught the public's attention even though the source of the simple organic molecules was then unknown. To make even a thin 1% soup in the approximately $1.4 \times 10^{21}$ kg of ocean water requires about $1 \times 10^{19}$ kg of organic carbon compounds. The total amount of organic carbon on earth is estimated to be $8 \times 10^{18}$ kg, but these estimates were unavailable in Haldane's time. All of the carbon existing as methane and carbon dioxide in the atmosphere would have had to combine into small organic molecules to create this primordial soup. The oceans would disperse these organic molecules rather than concentrate and orient them into structural polymers.

Bernal pointed out soon after that the adsorption by clays in soils and sediments would be important to concentrate and polymerize simple molecules. Although water neither concentrates nor catalyzes polymerization, his ideas were incorporated into the marine origin of life by marine workers' suggestions that the polymerization occurred in drying tidal pools. The drying also concentrates NaCl, which is inimicable to life. The marine biologists were eager to attribute as much importance to their discipline as possible. Evolution is opportunistic and would take advantage of the best chemical and physical conditions available for retaining, accumulating, and polymerizing organic compounds. These conditions are much more better realized in soils than in ocean water.

Clay particles formed before life began, so primordial soils had essentially the same chemical and physical characteristics as today's soils. Clays could adsorb and concentrate simple organic molecules as they fell in rain, or adsorb them directly from the atmosphere. Amino acids, for example, have been shown to polymerize when adsorbed on clay surfaces; benzene and phenol polymerize spontaneously on Fe(III)- and Cu(II)-coated clays. Whether such reactions actually led to the origin of life is speculation, but these reactions are much more likely in soils than in tidal pools.

In addition to concentrating organic molecules and catalyzing their polymerization, soils have several other properties that might be important in the origin and development of life: (1) protection of organic molecules from breakdown by ultraviolet light; (2) periodic drought, which encourages organic condensation (reactions that form water); (3) the relatively constant chemical composition of soil surfaces and the soil solution, much more so than that of the parent rocks; (4) higher availability of phosphate and microelements than in water; (5) the much lower osmotic potential of soil solution than marine water; and (6) a cation-rich solution. Living organisms produce mostly organic anions, which require cations for balancing and a great amount and variety of cations than anions. These factors suggest that life could have begun more easily in the soil than in the oceans.

Furthermore, the order of cation composition in plants and animals is Ca > K > Na = Mg and is close to the order of availability in soils, Ca > Mg > K = Na. The order in living organisms is quite different from the composition of seawater, Na > Mg > Ca = K. The amounts of solutes in body fluids of plants and animals is also closer to that of soil solutions than of seawater. The osmotic potential of body fluids is lower than that of seawater. Terrestrial plants and animals therefore die by ingesting seawater; marine plants and animals survive by excluding NaCl at a considerable expense of energy. As a result, life is more active in and on terrestrial soils than in the sea. Indeed, the open sea is barren.

The concept of life originating in marine or tidal pools nonetheless persists and is supported by the rather complete fossil record in marine sediments. The fossil record on land is erratic, but this should be expected because of the rapid turnover of land surfaces compared to marine sediments. Soils have eroded, weathered, and dissolved into soluble salts into the sea many times since life began. A complete fossil record under these circumstances would be highly unlikely. Much of the ocean bottom has remained intact, so fossils can accumulate readily and sequentially.

## 1.11 THE ROLE OF SOIL IN THE ENVIRONMENT AND THE MAINTENANCE OF LIFE

The role of soil in the origins of life is controversial. Its role in maintaining life and the environment is clear, but often unrecognized and taken for granted. Soil scientists are partly responsible because they have been too modest and too reticent about soil's control of the atmosphere and hydrosphere. The atmosphere, biosphere, and hydrosphere are weakly buffered against change and can fluctuate wildly when perturbed. Soils strongly resist chemical change and are a steadying influence on the other three environmental compartments. Changes in the hydrosphere, atmosphere, and biosphere due to human activities are often the result of bypassing the soil. The soil is the most robust environmental compartment. Treating solid, liquid, and gaseous wastes in soils before release to the hydrosphere or atmosphere can minimize environmental change. The soil is the source of most human wastes and ought to be site of their disposal.

The environment is large and very complex, too complex for anyone yet to understand fully. Scientific training necessarily tends to specialize—learning more and more about less and less. Scientists try to expand that myopic background to the whole environment, with mixed results. Among other things, they bring along biases, of which one is usually that their background field is the most important. Atmospheric scientists, for example, naturally believe that the atmosphere is the most important part of the environment. The authors of this book are no different. We argue without apology that the soil plays the central and dominant role in the environment (Fig. 1.1).

Table 1.3 shows which of the four environmental compartments store significant amounts of the essential elements. Table 1.4 shows the estimated quantities of the elements that circulate rapidly through the environment—C, fixed N, P, S, and water. The amounts shown for the oceans are those above the thermocline, a temperature inversion at about 50 m depth, which separates the surface water from the deeper water. The soil is the largest reservoir of almost all the essential elements and is the only reservoir for most of the essential elements.

The general chemical and related physical properties of the four active environmental compartments are summarized as follows:

*Atmosphere.* Poorly buffered against chemical changes; high amounts of readily available C as $CO_2$; high $O_2$; $H_2O$ variable; otherwise chemically uniform worldwide; entrained dust is the only reactive surface; high mass transfer rates.

*Biosphere.* High C, N, and P concentrations; energy rich; high energy and mass transfer rates; high reactivity; ratio of biomass/soil is variable and climate dependent; complex and highly ordered structures.

**Table 1.3. Distribution of carbon, nitrogen, sulfur, and oxygen at the earth's surface**

| Reservoir | Carbon ($\times10^{12}$ kg) | | Nitrogen ($\times10^{12}$ kg) | | Sulfur ($\times10^{12}$ kg) | | Oxygen ($\times10^{12}$ kg) | |
|---|---|---|---|---|---|---|---|---|
| Atmosphere | 700 | ($CO_2$) | 3 800 000 | ($N_2$) | 4 | | 1 000 000 | ($O_2$) |
| Surface ocean (above the thermocline at 50-m depth) | 800 | | 1 | (organic) | 0.1 | (organic) | 10 000 000 | ($H_2O$) |
| Living biomass | 500 | | 14 | (organic) | 2 | (organic) | 500 | (carbohydrate) |
| Soils, to 1-m depth | 1 800 | (organic) | 180 | (organic) | 20 | (organic) | 20 000 | (soil water) |

**Table 1.4. Estimated annual gains and losses from soils worldwide. [a](rates before human influence, if significantly different, are in parentheses)**

| Process | Carbon $(\text{Tg yr}^{-1})$ | Nitrogen $(\text{Tg yr}^{-1})$ | Sulfur $(\text{Tg yr}^{-1})$ |
|---|---|---|---|
| Soil organic matter decay | −35 000 | — | — |
| Cultivation of virgin soils | −1000 to −2000 | −100 to −200(0) | −10 to −20(0) |
| Peat accumulation | 200 | — | — |
| Soil nitrogen fixation | — | 200 | — |
| Atmospheric $N_2$ fixation | — | 4 | — |
| $NO_x$ (mostly $N_2O$), net | — | −8 to −25 | — |
| Ammonia, net | — | −20 to −60 | — |
| Nitrogen fertilization (1970) | — | 30(0) | — |
| Sulfur emission from flooded soils, net | — | — | −3 |
| Soluble or eroded to the sea | 100 to 200 | −0 to −20 | −60 |
| Sea spray and solutes | 100 to 200 | 20 to 20 | 20 |
| Weathering | — | — | 40 |

[a] Largely from B. H. Svensson and R. Soderlung, eds. 1976. *Nitrogen, Phosphorus and Sulphur—Global Cycles.* SCOPE Report 7, NFR, Stockholm.

[b] For comparison, the carbon released by fossil fuel combustion in 1975 was 3000 Tg yr$^{-1}$.

[c] The sulfur released by fossil fuel combustion in 1975 was 130 Tg yr$^{-1}$.

*Hydrosphere.* Poorly buffered against chemical changes; moderate $O_2$ availability; low water availability because of high osmotic potential of seawater; low ratios of biomass/unit area and photosynthesis/unit area (because of little light below the surface); low temperatures; low essential elements (especially P and Fe); high NaCl.

*Soil.* Chemically stable; strong buffering of elemental availability; variable $H_2O$ availability; soil solution is cation-rich; active microbial and inorganic catalysis; source and sink of all the elements essential to living organisms; high P but low availability; ratios of biomass/unit area and photosynthesis/unit area vary widely; photosynthesis is seasonal; high mass and energy transfer rates.

## 1.12  CHEMICAL UNITS

The number of atoms, not their mass or volume, determines the extent of a chemical reaction. The unit of numbers is the *mole*, Avogadro's number ($6.02 \times 10^{23}$) of molecules, ions, electrons, and so on. The *atomic weight* and *molecular weight* are the mass in grams (g) of 1 mole of that substance. For example, one mole of $H_2$ has a mass of 2 g and contains $6.02 \times 10^{23}$ $H_2$ molecules and $12.04 \times 10^{23}$ hydrogen atoms. Ions have the same mass as their atoms, because the mass of electrons is insignificant. The numbers of moles of reactants and products can change during a chemical reaction, but the number of ions and the mass remain constant.

The *atomic weight* of a chemical element is usually not a whole number, because it is the weighted average of the natural isotopes of the elements. Hydrogen, for example, has three isotopes: normal hydrogen ($_1^1H$), deuterium ($_1^2H$), and tritium ($_1^3H$). The superscript is the atomic weight of the isotope, the number of protons plus neutrons in the nucleus. The subscript is the *atomic number*, or number of protons, which chemically distinguishes each element. Deuterium and tritium are rare, so the distribution of isotopes in natural hydrogen yields an average atomic weight of 1.008 g. The atomic weights of the elements come closest to whole numbers when the natural mixture of carbon isotopes is assigned a mass of 12. The atomic weights and atomic numbers of the elements are shown in the periodic table (Table 1.5).

Gravimetry, measuring the mass of products and reactants, is the most accurate form of chemical measurement. Aqueous solutions are more conveniently measured by volume, as *molarity* (M), the number of moles per liter of the solution. Sometimes *molality*, moles per 1000 g of solvent, is preferred for concentrated solutions, such as seawater or when temperature is far from 25° C.

A 1 M (one molar) solution of $CaCl_2$ is 1 mole (147.03 g) of $CaCl_2 \cdot 2H_2O$ dissolved in water and the mixture made up to 1 liter (L). The mixture is made up to volume after adding the solute, because salts and miscible liquids change the volume of water when added to it. Mixing 50 mL of ethanol with 50 mL of water, for example, yields about 80 mL of mixture. Salts also change the volume of water, but less dramatically.

The numbers of moles of products and reactants may be unequal, as in

$$H_2SO_4 = 2H^+ + SO_4^{2-} \tag{1.1}$$

The numbers of *moles of ion charge* (formerly chemical *equivalents*), however, are equal in the equation. Equation 1.1 means, "one mole of sulfuric acid yields two moles of hydrogen charge and two moles of sulfate charge." Moles of ion charge are the number of moles multiplied by (1) the number of moles of $H^+$ or $OH^-$ that react with 1 mole of the substance, or (2) the number of "moles" or Faradays of electrons that 1 mole of the substance accepts or donates. The volumetric unit is *molarity of ion charge*, or moles of ion charge per liter.

Low concentrations are sometimes expressed as negative logarithms, the "p" scale. This is most familiar as "pH," the negative logarithm of the $H^+$ ion concentration:

$$pH = -\log(H^+) \tag{1.2}$$

The p scale is also used for other ions—pNa, $pH_2PO_4$, and pCa—and for the negative logarithm of equilibrium constants (pK) and solubility products ($pK_{sp}$). The negative logarithm increases as the concentration or value decreases.

High concentrations of components in solid, liquid, or gaseous mixtures can be more conveniently expressed as *mole fractions*. The mole fraction is the ratio of moles of a substance to the total moles of all substances in the mixture.

Gas concentrations are usually expressed as mole fractions or *partial pressures*. Partial pressure is a volume/volume ratio, and is equal to the mole fraction, at low

# Table 1.5. Periodic table of the elements (the data for each element are, from top to bottom, atomic number, atomic weight, chemical symbol, and common oxidation states in soils and plants)

Each cell below lists: atomic number — atomic weight — symbol — common oxidation states.

| PERIOD | IA | IIA | IIIB | IVB | VB | VIB | VIIB | VIII | VIII | VIII | IB | IIB | IIIA | IVA | VA | VIA | VIIA | 0 |
|---|---|---|---|---|---|---|---|---|---|---|---|---|---|---|---|---|---|---|
| 1 | 1 · 1.008 · H · 1 | | | | | | | | | | | | | | | | | 2 · 4.003 · He · 0 |
| 2 | 3 · 6.94 · Li · 1 | 4 · 9.01 · Be · 2 | | | | | | | | | | | 5 · 10.81 · B · 3 | 6 · 12.01 · C · ±4,2,0 | 7 · 14.01 · N · 5,±3,0 | 8 · 16.00 · O · -2,0 | 9 · 19.00 · F · -1 | 10 · 20.18 · Ne · 0 |
| 3 | 11 · 22.94 · Na · 1 | 12 · 24.31 · Mg · 2 | | | | | | | | | | | 13 · 26.98 · Al · 3 | 14 · 28.09 · Si · 4 | 15 · 30.97 · P · 5 | 16 · 32.06 · S · 6,-2,0 | 17 · 35.43 · Cl · -1 | 18 · 39.95 · Ar · 0 |
| 4 | 19 · 39.10 · K · 1 | 20 · 40.03 · Ca · 2 | 21 · 44.96 · Sc · 3 | 22 · 47.90 · Ti · 4 | 23 · 50.94 · V · 5,4 | 24 · 52.00 · Cr · 6,3 | 25 · 54.94 · Mn · 4,3,2 | 26 · 55.85 · Fe · 3,2 | 27 · 58.93 · Co · 2 | 28 · 58.71 · Ni · 2 | 29 · 63.54 · Cu · 2 | 30 · 65.37 · Zn · 2 | 31 · 69.72 · Ga · 3 | 32 · 72.59 · Ge · 4 | 33 · 74.92 · As · 3,5 | 34 · 78.96 · Se · 6,4,0 | 35 · 79.91 · Br · -1 | 36 · 83.80 · Kr · 0 |
| 5 | 37 · 85.47 · Rb · 1 | 38 · 87.62 · Sr · 2 | 39 · 88.91 · Y · 3 | 40 · 91.22 · Zr · 4 | 41 · 92.91 · Nb · 5,3 | 42 · 95.94 · Mo · 6,3 | 43 · (99) · Tc · 7 | 44 · 101.0 · Ru · 4 | 45 · 102.9 · Rh · 2 | 46 · 106.4 · Pd · 2,0 | 47 · 107.9 · Ag · 1,0 | 48 · 112.4 · Cd · 2 | 49 · 114.8 · In · 3 | 50 · 118.7 · Sn · 4,2 | 51 · 121.8 · Sb · 3,5 | 52 · 127.6 · Te · 4 | 53 · 126.9 · I · -1 | 54 · 131.3 · Xe · 0 |
| 6 | 55 · 132.9 · Cs · 1 | 56 · 137.3 · Ba · 2 | 57 · 138.9 · La · 3 | 72 · 178.5 · Hf · 4 | 73 · 181.0 · Ta · 5 | 74 · 183.0 · W · 6 | 75 · 186.2 · Re · 7 | 76 · 190.2 · Os · 4 | 77 · 192.2 · Ir · 4 | 78 · 195.1 · Pt · 4,0 | 79 · 197.0 · Au · 3,0 | 80 · 200.6 · Hg · 2,1,0 | 81 · 204.4 · Tl · 1 | 82 · 207.2 · Pb · 4,2 | 83 · 209.0 · Bi · 3 | 84 · (210) · Po · 2 | 85 · (210) · At · -1 | 86 · (222) · Rn · 0 |
| 7 | 87 · (223) · Fr · 1 | 88 · (226) · Ra · 2 | 89 · (227) · Ac · 3 | | | | | | | | | | | | | | | |

**Lanthanide series (Period 6):**

| | | | | | | | | | | | | | | |
|---|---|---|---|---|---|---|---|---|---|---|---|---|---|---|
| 58 · 140.1 · Ce · 3,4 | 59 · 140.9 · Pr · 3,4 | 60 · 144.2 · Nd · 3 | 61 · (147) · Pm · 3 | 62 · 150.4 · Sm · 3 | 63 · 152.0 · Eu · 3 | 64 · 157.2 · Gd · 3 | 65 · 158.9 · Tb · 3 | 66 · 162.5 · Dy · 3 | 67 · 164.9 · Ho · 3 | 68 · 167.3 · Er · 3 | 69 · 168.9 · Tm · 3 | 70 · 173.0 · Yb · 3 | 71 · 175.0 · Lu · 3 |

**Actinide series (Period 7):**

| | | | | | | |
|---|---|---|---|---|---|---|
| 90 · 232 · Th · 4 | 91 · (231) · Pa · 5 | 92 · 238 · U · 6 | 93 · (237) · Np · 5 | 94 · (242) · Pu · 4 | 95 · (243) · Am · 3 | 96 · (247) · Cm · 3 |

**Table 1.6. Some SI units adopted by the Soil Science Society of America**

| Quantity | Unit | Symbol | Definition | Former or English Units |
|---|---|---|---|---|
| Length | Meter | m | | = 3.281 feet, 39.87 inches, $10^6$ microns, $10^{10}$ angstroms |
| Area | Square meter | $m^2$ | | |
| | Hektar | ha | $10^4$ $m^2$ | = 2.471 acres |
| Volume | Cubic meter | $m^3$ | | |
| | Liter | L | $10^{-3}$ $m^2$ | = 1.057 U.S. quarts |
| Mass | Kilogram | kg | | = 2.204 pounds |
| Time | Second | s | | |
| Energy | Joule | J | $kg\ m^2\ s^{-2}$ | = 0.2390 calories |
| Force | Newton | N | $kg\ m\ s^{-2}$ | = $10^5$ dynes |
| Pressure | Pascal | Pa | $N\ m^{-2}$ | = $10^{-5}$ bars, 9.87 × $10^{-6}$ atm |
| Power | Watt | W | $J\ s^{-1}$ | = 1.341 × $10^{-3}$ horsepower |
| Electric Current | Ampere | A | | = coulombs/second |
| Potential difference | Volt | V | $J\ A^{-1}\ s^{-1}$ | |
| Resistance | Ohm | Ω | $V\ A^{-1}$ | |
| Conductance | Siemens | S | Ω | = 1 mho (1 mmho/cm = 1 dS $m^{-1}$) |
| Charge | Coulumb | C | A s | = 1/(96 516) faradays |
| Amount of substance | Mole | mol | $6.03 \times 10^{23}$ entities | Avogadro's number |
| | Moles of ion charge | $mol^{(+)}$ or $mol^{(-)}$ | mol × ion charge | = equivalent |
| Temperature | Kelvin | K | | |
| | Celsius | °C | | 0°C = 273.13 K |
| Radioactivity | Becquerel | Bq | $s^{-1}$ | = 2.7 × $10^{-11}$ curies |
| Absorbed dose | Gray | Gy | $J\ kg^{-1}$ | = 100 rad |
| Concentration | Moles per unit | | $mol\ m^{-3}$ | |
| Liquid | volume | M | $mol\ L^{-1}$ | = molarity |
| | | | $mol^{(+)}\ L^{-1}$ | = normality of cations |
| | | | $mol^{(-)}\ L^{-1}$ | = normality of anions |
| | molecular weight unknown | | $kg^{-3}$ | = 1000 ppm |
| Gas | Moles per unit volume | $mol\ m^{-3}$ | | = $10^6$ mg/$m^3$ |
| | molecular weight unknown | $kg\ m^{-3}$ | | = $10^6$ mg/$m^3$ |
| | mole fraction, or volume per volume | $m^3\ m^3$ or $mol\ mol^{-1}$ | | = $10^6$ ppm (v/v) |

pressures, where gases behave ideally. The volume of a gas is independent of other gases and is 22.4 L mol$^{-1}$ at 25° C and 1 atmosphere pressure. The volume of each gas in a mixture is proportional to the number of molecules or moles in the gas. Low gas concentrations are often given as parts per million (ppmv), meaning the number of molecules of the gas per 1 million molecules of the gas mixture. Ppmv is a volume/volume and mole/mole unit for gases. (Ppm is a mass/mass unit for solids and liquids.) Ppmv in gases can be converted to mass/volume units by

$$\text{mg m}^{-3} = \frac{\text{molecular weight}}{24.06} \times \text{ppmv} \qquad (1.3)$$

The metric system was modified as Systeme International (SI) units (Table 1.6) to prevent some confusion. The SI is based on seven fundamental units—including the mole, meter, kilogram, and second—from which the others are derived. The significant changes for soil chemistry are mole of ion charge for equivalent, siemens for mho, joule for calorie, and pascal for pressure. Table 1.4 summarizes the SI units most frequently encountered in soil chemistry. SI allows easier conversion and communication between disciplines, but unfortunately discards some useful and familiar units, such as angstrom and equivalent.

SI also recommends changes in the writing of numbers and units: avoiding the solidus (/), using the seven basic units in the denominator, using prefixes for large and small numbers (Table 1.7), and standardization of the decimal point. The solidus is removed by giving units in the denominator a negative exponent, so m/sec$^2$ becomes m sec$^{-2}$. Large numbers should be written without punctuation and with a space for each thousand—twenty five thousand is 25 000. Periods and commas in numbers are

**Table 1.7. SI prefixes**

| Prefix | | Multiple | Symbol |
|---|---|---|---|
| Exa | | $10^{18}$ | E |
| Peta | | $10^{15}$ | P |
| Tera | | $10^{12}$ | T |
| Giga | | $10^{9}$ | G |
| Mega | | $10^{6}$ | M |
| Kilo | | $10^{3}$ | k |
| Hekto | Avoid when possible | $10^{2}$ | h |
| Deka | | $10^{1}$ | da |
| Deci | | $10^{-1}$ | d |
| Centi | | $10^{-2}$ | c |
| Milli | | $10^{-3}$ | m |
| Micro | | $10^{-6}$ | $\mu$ |
| Nano | | $10^{-9}$ | n |
| Pico | | $10^{-12}$ | p |
| Femto | | $10^{-15}$ | f |
| Atto | | $10^{-18}$ | a |

confusing: the English and American 2,500 means $2\frac{1}{2}$ to Europeans. In SI the period and comma both denote the decimal point; 2.5 and 2,500 mean $2\frac{1}{2}$ in SI.

For soils in the field, soil area is usually more convenient and more meaningful than soil mass, but the square meter is too small and the square kilometer too large. The hektar or hectare (ha $= 10^4 \text{m}^2$) has been adopted as an area unit. To convert soil mass to area, the common factor for mineral soils is 2 million kg ha$^{-1}$ 15 cm$^{-1}$ of soil, assuming a soil bulk density of 1300 kg m$^{-3}$. Cultivated organic soils have bulk densities of about 300 kg m$^{-3}$.

## ADDITIONAL READING

Bolt, G. H. (ed.). 1990. *Interaction at the Soil Colloid–Soil Solution Interface*. Kluwer, Dordrecht.

Cresser, M. S., K. Killham, and T. Edwards. 1993. *Soil Chemistry and Its Applications*. Cambridge University Press, Cambridge, UK.

Huang, P. M. (ed.). 1998. *Future Prospects for Soil Chemistry*. SSSA Spec. Pub. 55, American Society of Agronomy, Madison, WI.

McBride, M. B. 1994. *Environmental Chemistry of Soils*. Oxford University Press, New York.

Russell, E.W. 1987. *Soil Conditions and Plant Growth*, 11th ed. A. Wild, ed. Longman, London.

Sposito, G. 1994. *Chemical Equilibria and Kinetics in Soils*. Oxford University Press, New York.

Tan, K. H. 1998. *Principles of Soil Chemistry*, 2nd ed. Dekker, New York.

## QUESTIONS AND PROBLEMS

1. Starting with a cube 1 m on a side, calculate the change in surface area by subdividing it successively into cubes 1 mm, 20 $\mu$m, or 1 $\mu$m on a side. How many particles would be in each size group?

2. What is the mass of each of the above size particles? The specific gravity of aluminosilicates is 2.65. Calculate the surface area/mass ratios of the above cubes, assuming that they are aluminosilicates.

3. What were the shortcomings of van Helmont's experiment and how could it have been improved?

4. Compare the lists of macroelements for plants and for animals. Do the same for the microelements. Which are derived directly from the soil? What is the soil's role, if any, in providing the essential elements not directly supplied by the soil? Which elements are influenced significantly by the atmosphere and by humans?

5. Which of the essential elements are metals? Nonmetals? Cations? Anions?

6. Why was it necessary that the macroelements be of low atomic weight for life to evolve as we know it?

7. Early in the earth's lifetime, the atmosphere was apparently rich in methane and hydrogen and low in oxygen. Lightning can create simple amino acids in such a gas mixture. Describe the sequence of reactions and the soil properties that might aid polymerization of these amino acids in soils.

8. Discuss the differences between an element's extractability, "availability," and total content in soils.

9. Discuss why $CO_2$ exchanges between soils and the atmosphere.

10. Calculate the amount of water held in the world's soils if the average soil moisture content is 20% by mass (the land area is $1.2 \times 10^8$ km$^2$, the soil's specific gravity is 1.3, and the soil is 1 m deep). Compare this value with the soil water data in Table 1.2.

# 2

# IMPORTANT IONS

This chapter discusses the individual chemical elements in soils. Most questions about soil chemistry are about specific ions and many people are satisfied with a quick, albeit incomplete, answer. The processes that govern these ions in soils and can yield a more complete answer are discussed in the later chapters. If the chemical symbols and nomenclature used for conciseness are unfamiliar, please refer to Chapter 3, where they are explained.

As the earth formed and cooled during its origin, the lighter chemical elements tended to float to the surface. The earth's center is thought to be iron-rich and has a density of $>6000$ kg m$^{-3}$. The density of the outer crust, or mantle, is about 2800 kg m$^{-3}$. The density of the rock minerals at the earth's surface is about 2650 kg m$^{-3}$. The elements in the rock minerals at the earth's surface (Table 1.1), are the starting materials for soils and also contain the essential elements from which soil and life evolved.

Soils contain all of the natural chemical elements in the periodic table, including the elements that are essential to living organisms and those labeled as toxic. Only a few elements make up $>95\%$ of the mass and volume of the earth's crust (Table 1.1). The remainder are present in soils in small but important amounts. Even the rarest natural element, the alkali metal francium, $^{87}$Fr, which has never been isolated, occasionally appears as faint lines in the arc spectrum of arid soils where alkali metal salts accumulate.

Only in a few areas, at least as far as we now know, are any elements naturally in a state that can cause harm. In a few other unfortunate areas, soils contain synthetic chemicals, synthetic radioactive isotopes, and elements that have been concentrated by human activities. These areas are a serious problem. In an agricultural sense, the severest concentration problem is the widespread accumulation of salts in irrigated arid lands. The mere presence of a chemical in soils is rather insignificant. What matters is the availability of a substance to plants and to the soil solution; its total concentration is of much less significance. Many soils labeled as contaminated because of their total element content are benign, or can easily be made benign, because their availability to plants and movement to groundwater is minimal. The working definition of a contaminated substance is that it will make you sick if you eat it, a poor definition to apply to soils. Soil processes tend to change the availability of elements added to soils back to that in native soils. The plants and groundwater on copper, lead, and zinc ore deposits are not necessarily contaminated, but the total amounts of those ions in the soils are very high.

For most purposes, the important ions in soil chemistry are those that are essential, or toxic, to living organisms. Although opportunistic, evolution took little or no advantage of the most prevalent ions—O, Si, Al, Ti, Na, and Cl—in soils and oceans. Figure 2.1 shows the essential ions in bold type, and the toxic ions are cross-hatched. The soil is both the source of the essential elements and the safest disposal site to return the elements back to their native biogeochemical cycles. The elements are present as ions in nature because the zero oxidation state is unstable for most elements. The exceptions are $O_2$, $N_2$, the inert gases, and the precious metals Au, Pt, Ag, and so on. In the elemental state atoms bind only to each other. As ions, the elements are active and react with other ions.

In recent decades soil chemistry has increasingly been involved with *pollution*— unnaturally high concentrations of chemicals in the environment caused by industrial activities and by high population density, which concentrates wastes. Much of that harm could be mitigated by reacting those substances with soils rather than releasing the chemicals to the atmosphere or hydrosphere. The soil reactions that control natural concentrations in soils can also buffer the effects of large additions to soils.

The distinction between essential elements and toxic elements is not clearcut. The essential ions have a concentration range in which they are essential for organisms to complete their life cycle. The boundaries of this range vary with the species of the organism. At concentrations below that boundary, organisms suffer deficiency. Above that boundary, increasing amounts do not increase growth, and if the concentration is well above the beneficial range, the essential elements can be toxic. "Toxic" elements are those that are more likely to be found at this excessive concentration, or elements that are toxic at low concentrations. The excessive concentrations are almost always due to human activity. Evolution has accepted the natural concentrations and availabilities of these elements in soils, water, and air. The availability is governed largely by their reactions in soils and sediments with the soil solution and fresh and marine waters.

Some elements have incorrectly been labeled as all good or all bad. Oxygen at 20 volume percent $O_2$ in air is necessary for aerobic organisms but is deadly to anaerobes. Concentrated, 100%, $O_2$ is helpful for humans to breathe only for a short time; longer periods are harmful. Arsenic is known as a toxic element, but low As concentrations are essential to living organisms.

All substances are toxic if their concentrations are too high. The toxic ions in Fig. 2.1 are distinctive because, as far as is yet known, they do not benefit living organisms at any concentration, and they are toxic at low concentrations in the

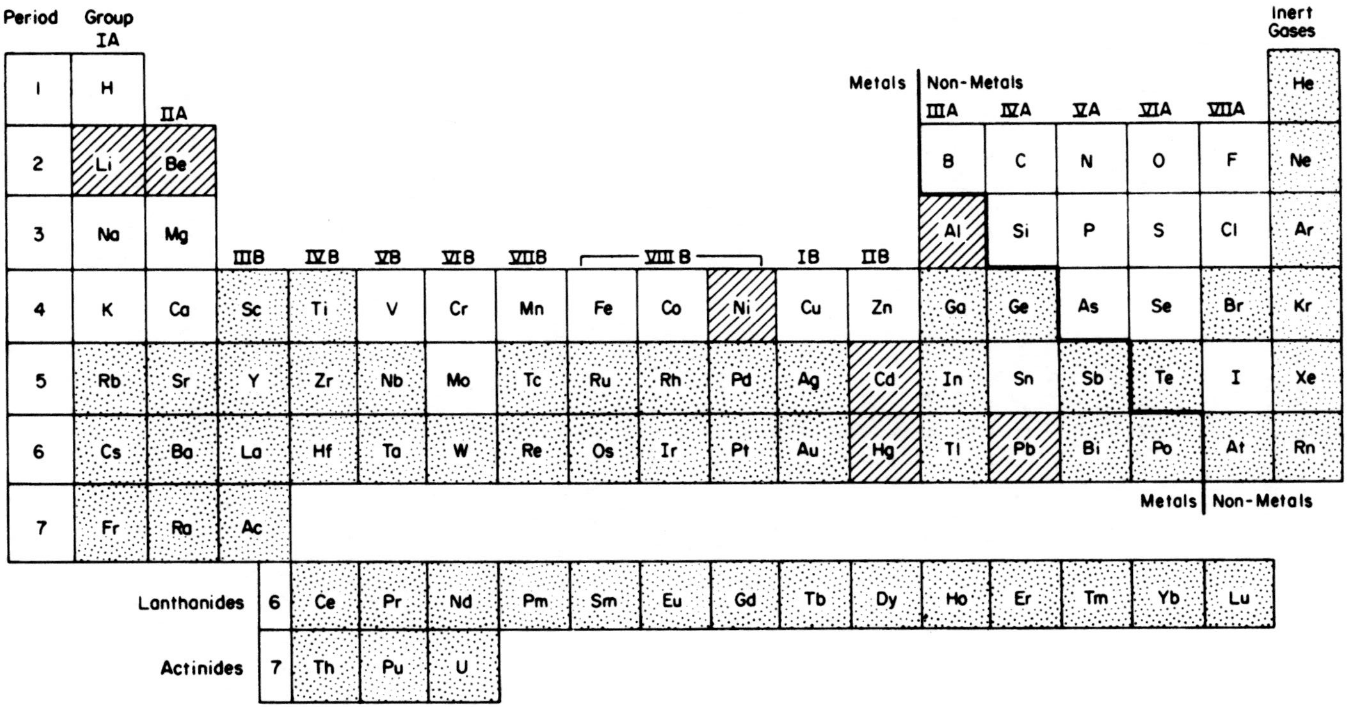

**FIGURE 2.1.** Periodic table of the elements differentiated on the basis of soil chemistry. Essential elements are in white; common toxic elements are in crosshatched areas; elements normally present in inconsequential amounts are in dotted areas.

28

food/water/air web. The list is biased because it refers mainly to humans, but other organisms probably have the same response as humans. Indeed, we assume this when we use laboratory animals for testing toxicity.

With the important exception of plant toxicity by aluminium in acid soils, toxicity in soils appears to be due to human-caused conditions. The number of toxic ions would be much greater except that humans have not been exposed to other elements enough for health and regulatory concern.

The "toxic" ions are present in all soils, plants and water, but the natural concentrations and availabilities are so low that they appear to pose no danger to organisms. Animals have an effective buffer against high concentrations because they ingest plants rather than soils. Plants are more tolerant of toxicity, absorb these ions in small amounts, and tend to retain them in the roots rather than translocate them to plant tops. This plus soil unavailability keeps toxic ions out of the animal food web. Isolating and assessing the health effects of the background concentrations of the toxic ions is difficult. Lead, for example, is ubiquitous in the environment but is retained strongly by soils, so the Pb concentrations in food and water are low. Some people worry about Pb in drinking water, but the amounts added to the human body by poor plumbing, flaking old paint, some folk remedies, and some pottery glazes are far greater.

Although most of the elements in Fig. 2.1 are shown as neither essential nor toxic, our knowledge is incomplete. The number of essential elements is probably, and the number of toxic elements is certainly, larger than shown in Fig. 2.1 Water, plants, and dust contain many ions, so the exposure of organisms to most of the elements in the periodic table is continuous, but in very small amounts. Proving their essentiality or toxicity under these conditions is difficult. To show the essentiality of chlorine to plants, for example, the plants were grown in greenhouses with carefully filtered air to remove dust and water droplets, the plants were grown hydroponically because even acid-washed quartz sand contains sufficient Cl to satisfy the plant's small Cl requirement, the water was doubly distilled and could not contact glass to avoid leaching Cl out of the glass, and the plants had to be second generation grown in this environment. Normal seeds contained enough Cl to satisfy this annual plant's life cycle requirement of Cl.

This experiment may have been of more than just academic interest. Because the Cl input to most regions is marine Cl entrained in rain, and because soils retain Cl weakly, continental regions far from the oceans could have low Cl in the soil–plant–water system. Chlorine deficiencies may occur, just as has been shown for Cl's chemical relative, iodine. The human disease goiter, due to I deficiency, was endemic in the interiors of the continents until the 20th century when I supplements in table salt and wider distribution of food fish became common. Low Cl concentrations in soils may subtly affect plant and animal health. NaCl improves plant growth in some parts of Australia, although NaCl is best known as being harmful to plants at high concentrations. The Australian plants may use the Na to supplement a low K supply rather than to overcome a Cl deficiency, but a suggestion of Cl deficiency remains. The strong desire of animals and humans for more NaCl in their diet may be due to a subtle Na and Cl deficiency as well as due to taste.

## 2.1  ESSENTIAL ELEMENTS

All organisms need the essential macroelements or macronutrients—C, H, O, P, K, N, S, Ca, Fe, and Mg—in relatively large amounts to complete their life cycle. A mnemonic phrase to remember them is "See Hopkins cafe, mighty good." Iron is needed in smaller amounts than the others, but is necessary for the mnemonic. Animals additionally require Na and Cl as macroelements, but since animal biomass is only 1/10000 that of plant biomass, animal requirements are insignificant in the overall picture. Macroelements have been called macronutrients for a long time, but "nutrient" implies energy content, as in carbohydrates.

Organisms need the essential *microelements* or *micronutrients*—B, Si, F, Cl, V, Mn, Fe, Cr, Co, Ni, Cu, Zn, Mo, Ar, Se, Sn, and I—in smaller amounts than the macroelements, but they are just as essential. Whether all the microelements are required by all organisms is unclear. Several have been shown to be essential for only one species. Because proving essentiality is so tedious and expensive, if an element is essential for one or a few species, it is assumed to be essential for other organisms as well. The list of essential microelements will probably grow as experimental techniques become more and more refined.

The soil is involved with all the essential elements. Most are present as ions in the soil solution and flow into the plant as it absorbs water. Plants obtain hydrogen, carbon, and oxygen from water and air, but soils provide water-holding capacity, provide pore space for $O_2$ and $CO_2$ movement between plant roots and the atmosphere, and supply $CO_2$ to the atmosphere through the decay of organic matter by soil microorganisms.

Animals derive their essential elements from plants. The ability of plants to supply these elements to plants and animals depends on a combination of factors: availability of the ions in the soil solution, plant selectivity at the soil solution/root interface, and ion translocation from root to plant top. The system is good but not perfect. The concentrations of essential elements are occasionally too high or too low for animals, because plants can tolerate a much wider range of elemental concentrations than can animals. Plant contents of iron, for example, tend to be lower than ideal for the human diet. Grazing animals in a few semiarid parts of North America may suffer from high selenium, in Australia from low cobalt, and formerly in Norway from low phosphorus availability because of unsatisfactory amounts in soils. The supply of essential elements by plants to animals is generally adequate; cases of too little or too much are noteworthy because of their rarity.

## 2.2  TOXICITY AND DEFICIENCY

The boundaries between toxicity, sufficiency, and deficiency are vague. The amounts vary with the species of plant and animal, vary within the species's growth cycle, and vary with the organism's general health and the supply of the other essential elements. The elements in the crosshatched areas of Fig. 2.1 are the toxic elements of primary concern to government regulatory agencies—Be, Cd, Hg, Pb, and all radioactive elements. As far as is presently known, these elements are harmful to humans and probably other organisms when present above typical natural concentrations, and are of doubtful benefit at any concentration. The nonradioactive elements are present in soils, but at low concentrations that have not yet been shown to affect life. Probably most of the unmarked elements in Fig. 2.1 are "toxic," but exposure at above their natural concentrations is too rare to be a general health concern.

Other elements that can be in possibly toxic concentrations in soils, water, and plants are Ni, Cu, Zn, and As. Toxic elements can be trendy and political as one or more reach popular attention. Regulations generally use total soil concentration, rather than soluble or plant-available concentration, as the criterion of toxicity because defining "soluble" and "available" is difficult. The ash from nerve gas incineration, for example, may have been incorrectly classified as a hazardous waste because it contained 50 mg kg$^{-1}$ lead. The native soils in the area contained 70 mg Pb kg$^{-1}$. The stigma of the ash's origin as nerve gas waste probably had much to do with its being labeled a hazardous waste. Copper mine tailings contain 0.1% Cu—does that make them a toxic waste? The native ore contains 0.5 to 1% Cu and does not contaminate groundwater. Opponents to fluoridation of water would add F to the toxic list because the water additive NaF is a rat poison at high concentrations. Plutonium is very toxic and was feared to be at significant concentrations in tobacco. The possibility of the radioactive gas radon diffusing from soils into poorly ventilated basements recently raised much concern. Some groups are labeling Cl as a dangerous element because of its effects on stratospheric ozone and want to ban Cl$_2$ as an industrial chemical. Since elements can be both beneficial and harmful, banning elements to create a completely risk-free environment is as sensible as banning fire.

The other issue is told in the adage, "The poison is in the dosage." The amount of the substance is as important as the substance. NaF is a rat poison at high concentrations but it prevents tooth decay, without known side effects, at low concentrations. Under natural conditions toxic concentrations of the regulated elements Be, Cd, Hg, and Pb are very rare, but they are present in all soils and plants at low concentrations. They are the most likely elements, however, to be found at toxic concentrations under anthropogenic (human-caused) conditions, such as metal industries, mine spoils, and landfills. Overall, the dangers created by these elements are small compared to organic compounds—pesticides, drugs, allergens, and so on—but in local areas they can be serious.

Toxicities of Be, Cd, Hg, Pb, and others have been reported where people and plants have been exposed directly, without the protection of the soil–plant system. Beryllium dust in industrial air, F from aluminium smelting and phosphate fertilizer manufacture, Pb and Cd from water pipes and from copper and zinc smelting, Se from drainage of high-Se soils, and a recent scare about Hg amalgams in tooth fillings are examples. Although public awareness of chemical pollution is increasing, exposure to toxic elements and substances is decreasing. People are living longer and healthier than ever before. The public media and alarmists to the contrary, part of that is due to decreasing exposure to toxic concentrations of chemicals.

In the "good old days," lead water pipes were common and lead was in pewter tableware and in ceramic glazes. Lead arsenate was a pesticide spray on apples and other crops. Workers in Cu, Pb, and Zn smelters were exposed to extremely high amounts of those and other metals in those ores. People ate antimony salts in the 19th century to make their hair black and glossy. The Mad Hatter in *Alice in Wonderland* may not have been unusual then because Hg compounds were used to make felt and a symptom of Hg poisoning is unbalanced behavior. If these elements are in soils, they are generally retained strongly and kept unavailable to plants. If absorbed by plants, they tend to accumulate in roots and little is translocated to plant tops. Our gastrointestinal tract also insulates us to some extent from these elements in food and water.

All substances are toxic above an ill-defined threshold concentration that varies with each substance and with each species. Under natural conditions and at our current state of knowledge, toxicity is rare. One exception is Se accumulation by legume plants in a few semiarid regions of North America. Se is toxic to grazing animals, but the legumes show no symptoms. Another example is occasional Mn and Fe toxicity to rice plants in acidic Asian rice paddies. The most prevalent natural soil toxicity problem is largely hidden from us. Aluminium toxicity to plants in acid soils is so widespread that we accept it, perhaps appreciate it, in forested areas. Forest plants are more tolerant of Al toxicity than food crops and grasses, which might otherwise compete with the trees in those areas. Because trees are beautiful and valuable, we accept the effects of Al toxicity until we want that land to grow food crops and grasses.

Plants must grow in the environment within reach of their roots and leaves. To survive, plants have had to evolve a broader tolerance to imbalances of essential and toxic elements than have animals. Deficiencies and toxicities in plants are usually evident by reduced yield or reproductivity, abnormal coloration, and plant and fruit deformities. Animals can supply their more stringent nutritional requirements by ingesting food grown on a variety of soils.

Deficiencies of soils to supply the essential elements are more obvious in plants than in animals. The nitrogen content of most soils is low enough that plants benefit from added N. In many soils, plants, especially food crops, also grow better with added P, K, and S. Because the response differs among plant species, fertilization can change the distribution of plant species. Since we usually define the untouched plant community as the ideal state, species changes are considered detrimental.

Soil microbes lessen the changes due to fertilization because they compete more effectively than plants for ion uptake, but microbial changes are rarely noticed. Changes due to fertilization are more obvious in air and water, which cannot resist change as well as soils, or at least the changes are more obvious. Green and turbid waters teem with actively metabolizing algae and bacteria, but not with the larger fish that we appreciate. Hence, we call these waters polluted. Microorganisms degrading the dead algae and bacteria compete too strongly for oxygen in those waters for fish to survive. Only when the essential element and nutrient content in a water body is relatively low can fish compete effectively. When Lake Erie was "dead" years ago, due to high inputs of industrial and agricultural wastes, the total metabolic activity was much greater than it is now. The fish population, however, was low. The microbes consumed the wastes and, as the input decreased because industrial and municipal wastewater was cleaned, the microbes consumed themselves. Now fish again survive; the clear waters that we enjoy are the aquatic equivalent of a rather barren soil.

The ability of soils and soil microbes to neutralize excess fertilizers has limits. Nitrogen and phosphate draining from excessive fertilization of sugar cane fields favors cattail plants over native plants in the Florida Everglades. Nitrogen and phosphate from fertilizers and municipal wastewater are causing water problems worldwide.

Iron anemia (Fe deficiency) in humans is common, but is due to low availability to plants rather than low amounts in the soil. Molybdenum deficiency, which prevents microbial nitrogen fixation, and cobalt deficiency in Australian sheep that prevents rumen bacteria from synthesizing vitamin B12, have been reported in Australia. Phosphate deficiency that led to weak bones in grazing animals was reported in Norway.

Table 1.1 lists the total amounts of the essential ions in soils, including ions within the crystal lattices of clay, silt, sand, and gravel particles, which are unavailable by plants. Total concentrations are poor indicators of plant availability. A more useful value would be the amount of each ion available during the plant's growing season. Measuring this availability has been, and still is, the subject of much research. Assessing availability is difficult because ion uptake by plants varies with plant variety and growth conditions as well as with soil chemistry. The major factor influencing an ion's availability is its soil chemistry, but differences in plant response are also important.

One aid to developing a perspective of soil chemistry is to arrange the elements according to their behavior in soils. The elements of primary interest exhibit a wide range of chemical and biochemical behavior. Their essentiality or toxicity to organisms is evidence that the chemistry of each is unique. Their soil chemistry, however, is less distinctive, and their active fractions can be arranged into behavioral groups. Table 2.1a shows the important chemical elements organized into six behavioral groups: exchangeable cations, water-soluble anions, weakly water-soluble anions, weakly water-soluble transition metals plus aluminium, toxic ions, and redox-active elements.

The major exchangeable cations neutralize and are retained by the negative charge of soil minerals and organic matter. The major anions are at lower concentrations in the soil solution than are the cations in most soils. These anions neutralize cations and the negative charge of soil particles. Weakly water-soluble anions are present at concentrations of $\leq 10^{-5}$ M in most soil solutions. The concentrations of the transition metal and aluminium ions in the soil solutions of typical agricultural soils are also in the $\leq 10^{-5}$ M range. This group includes essentially all of metals in the periodic table except for the alkali and alkaline earth metals. Toxic ions are of current concern in the literature. Aluminium is included in this group because of its widespread plant toxicity in acid soils. Many more transition metals would be in this group except that soil reactions markedly depress their availability to plant uptake and movement to groundwater.

The last group in Table 2.1a contains the elements active in oxidation–reduction reactions in soils. Changes in oxidation state greatly change the chemical properties

**Table 2.1a. Ions of major interest in soil chemistry, grouped according to major behavioral modes and shown as their most common states in soil solutions**

| Ion | Comments |
|---|---|
| *Major Exchangeable Cations* | |
| $Ca^{2+}$ $Mg^{2+}$ $Na^+$ $K^+$ $NH_4^+$ $Al^{3+}(H^+)$ | Occur predominantly as exchangeable cations in soils; these ions are relatively easily manipulated by liming, irrigation, or acidification; exchangeable $Al^{3+}$ is characteristic of, though rarely the predominant exchangeable cation in, acid soils; productive agriculural soils are rich in exchangeable $Ca^{2+}$ |
| *Major Anions* | |
| $NO_3^-$ $SO_4^{2-}$ $Cl^-$ $HCO_3^-$, $CO_3^{2-}$ | Present in considerably lower concentrations than the major cations in all but the most coarse-textured and strongly saline soils, where they are essentially equal; sulfate and $NO_3^-$ are important nutrient sources for plants; sulfate, $Cl^-$, and $HCO_3^-$ salts accumulate in saline soils; carbonate ions are present in appreciable amounts only in soils of pH > 9 |
| *Weakly Soluble Anions* | |
| $H_2PO_4^-$, $HPO_4^{2-}$ $H_2AsO_4^-$, $AsO_2$ $H_3BO_3$, $H_2BO_3^-$ $Si(OH)_4$ $MoO_4^{2-}$ | Strongly retained by soils; borates are the most soluble of the group; retention or fixation by soils is pH dependent; molybdate and silica are more soluble at high pH; phosphate is more soluble at neutral or slightly acid pH |
| *Transition Metals and Aluminum* | |
| $Al^{3+}$, $AlOH^{2+}$, $Al(OH)_2^+$ $TiOOH^+$ (?) $Fe(OH)_2^+$, $Fe^{2+}$ $Mn^{2+}$ | Insoluble hydroxides tending to accumulate in soils as silica and other ions weather; iron and manganese are more soluble in waterlogged or reduced soils |

*(continued)*

of the elements. Table 2.1a shows the elements in their most common oxidation states in soils, including the likely ion species in soil solutions. The higher oxidation states are typical of the soil solutions and secondary minerals of aerobic soils. The lower oxidation states tend to be found in organic matter, flooded soils, and igneous minerals.

Some elements appear in more than one group because they have more than one important function. In addition, chemical properties overlap. The arrangement in Table 2.1a implies only that the elements in each group exhibit similar, rather than identical, behavior in soils. Table 2.1a does little to indicate the relative importance of the chemical elements. Certain elements dominate soil reactions because of their greater abundance, because of the rapidity of their reactions, or because they are a source of energy. For example, all transition metal ions and aluminium produce acid-

**Table 2.1b. (Continued)**

| Ion | Comments |
|---|---|
| $Cu^{2+}$<br>$Zn^{2+}$ | More soluble than the above cations in all but very acidic soils; availability increases with increasing soil acidity; complexed strongly by SOM |
| | *Toxic Ions* |
| $Cd^{2+}$, $Al^{3+}$<br>$Pb^{2+}$<br>$Hg^{2+}$, $Hg$<br>$Be^{2+}$, $AsO_4^{3-}$, $CrO_4^{2-}$ | Soil behavior similar to transition metals; $Al^{3+}$ is a hazard to plants; the others are of more concern to animals; $CD^{2+}$ is relatively soluble, available to plants, and its retention is relatively independent of pH; remaining ions are less available to plants with increasing pH, except perhaps for As; the last three ions have received relatively little study in soils |
| | *Active in Oxidation–Reduction Reactions* |
| $C$ (organic to $HCO_3^-$)<br>$O$ ($O^{2-}$ to $O_2$)<br>$N$ (—$NH_2$ to $NO_3^-$)<br>$S$ (—$SH$ to $SO_4^{2-}$)<br>$Fe$ ($Fe^{2+}$ to $FeOOH$)<br>$Mn$ ($Mn^{2+}$ to $MnO_{1.7}$)<br>$Se$ (organic to $SeO_4^{2-}$)<br>$Hg$ (organic to $Hg^0$ or $Hg^{2+}$) | Soil biochemistry revolves around the oxidation state changes of soil carbon, nitrogen, and sulfur compounds; molecular oxygen is the main electron acceptor; FE(III), Mn(III–IV), nitrate, and sulfate are electron acceptors when the oxygen supply is low |

ity during their hydrolysis during soil weathering. Aluminium, however, dominates because of its abundance.

## 2.3   ALKALI AND ALKALINE EARTH CATIONS

The major alkali and alkaline earth cations—Na, K, Ca, and Mg—are roughly 2% by mass in igneous rocks. They are prominent in soils even though large amounts of these ions are lost during weathering of rocks to soils. The other alkali and alkaline earth metals—Li, Cs, Rb, Fr, Sr, Ba, and Ra—are present in only trace amounts, mg/kg and less, in rocks and soils. Although Be is in the alkaline earth column, it is omitted here because it reacts chemically like aluminium. In sedimentary rocks, the relative amounts are probably Ca > Mg = K ≫ Na because weathering has released many ions before the rocks form. Na is released to the soil solution and leaches from the soil to the sea.

The alkali and alkaline earth cations are the major cations in the soil solution, as aquated (water-surrounded) ions and as the charge-neutralizing cations on the surfaces of soil colloids. The ions associated with soil surfaces can easily be *exchanged* for other cations. If a soil suspension contained initially equal amounts of the four major cations, their distribution would be Na > K > Mg > Ca in the salts in the water phase away from colloidal surfaces (the *bulk solution*) and Ca > Mg > K > Na in the ions associated with clay surfaces and organic matter (the *exchanger phase*). The Ca > Mg > K > Na relation is also the relative strength by which soil colloids retain the cations. In soils in humid and temperate regions, most of the cations are associated with colloids, so the amount of cations exceeds the amount of water-soluble anions.

The cations in productive agricultural soils are present in the order $Ca^{2+}$ > $Mg^{2+}$ > $K^+$ > $Na^+$. Deviations from this order can create ion-imbalance problems for plants. High Mg, for example, can occur in soils formed from basaltic serpentine rocks. The Mg inhibits Ca uptake by plants. High Na occurs in soils where water drainage is poor and evaporation rates exceed rainfall. High Na creates problems of low water flow and availability in soils. Low Ca occurs in acid soils.

The total salt concentration in the bulk solution of well-drained soils from humid and temperate regions is generally in the range of 0.001 to 0.01 M. In irrigated and arid soils, the soluble salt concentration is higher. It may be five to ten times higher than that of the irrigation water applied, because evapotranspiration leaves the salts behind. Where salts accumulate due to improper irrigation, high groundwater tables, or seawater intrusion, salt concentrations (particularly Na salts) can reach 0.1 to 0.5 M.

The trace alkali and alkaline earth cations are present in the following amounts: lithium, 10–300 mg kg$^{-1}$; rubidium, 20–500 mg kg$^{-1}$; beryllium, 0.5–10 mg kg$^{-1}$; strontium, 600–1000 mg kg$^{-1}$; barium, 100–3000 mg kg$^{-1}$; and radium, perhaps $10^{-7}$ mg kg$^{-1}$. Some varieties of fruit trees are sensitive to as little as 1 mg L$^{-1}$ $Li^+$ in irrigation water, but $Li^+$ toxicity is rare. Rubidium, cesium, strontium, and barium have all been studied in the laboratory, but have received little attention in the field. Strontium has been studied because its radioactive isotope $^{90}$Sr (half-life $= 28$ years) is produced by nuclear fission and could cause long-term soil contamination after nuclear explosions or accidents. In soils the toxic $Be^{2+}$ ion behaves more like $Al^{3+}$ than like the other alkaline earth cations.

The strength of cation retention by soil particles increases with increasing ion charge and with decreasing hydrated ion size. For the monovalent cations, the increasing order of retention is

$$Li < Na < NH_4 = K < Rb < Cs < Ag < H(Al)$$

The H(Al) indicates that the $Al^{3+}$ ion is responsible for the acidity in acid soils; the amounts of $H^+$ are quite small. Only Na, NH$_4$, K, and H(Al) are in significant amounts in natural soils.

The increasing order of retention of divalent cations by soils is

$$Mg < Ca < Sr < Ba$$

Only Mg and Ca are common in soils, and Ca dominates the cations associated with clay surfaces and organic matter in most soils.

### 2.3.1 Calcium

Most economic crops yield best in soils when $Ca^{2+}$ dominates the exchangeable cations. High Ca indicates a near-neutral pH, which is desirable for most plants and soil microorganisms. Calcium is an essential element for plants and animals, and the amounts are rarely deficient in soils. Other problems appear before the Ca itself is deficient: Mg in soils derived from Mg-rich serpentine rocks competes with Ca for plant uptake, and the $Al^{3+}$ in strongly acidic soils can severely hinder plant growth even though the amount of available Ca is still adequate. When the exchangeable $Na^+$ concentration exceeds 5–15%, water flow problems in the soil can result.

The goal of reclaiming problem soils is to replace most of the exchangeable Al or Na with Ca so that the surface is dominated by Ca. *Limestone* ($CaCO_3$) neutralizes soil acidity and supplies Ca in acid soils. *Gypsum* ($CaSO_4 \cdot 2H_2O$) and other materials supply Ca to displace the Na in sodic soils. These treatments have to be repeated periodically as weathering leaches Ca from the soil in humid regions and irrigation water adds Na in arid regions. Both maintenance treatments yield long-term rather than immediate economic returns. Since farmers are understandably concerned about the short term, they want to delay these relatively expensive treatments as long as possible. Assessing how much and when treatment is needed is an important concern for soil chemistry.

The process of weathering includes the continual loss of the alkali and alkaline earth elements from soils. The loss of $Ca^{2+}$ following liming of acid soils is about equal to the rate of loss during natural soil weathering. The average rate of Ca loss from natural weathering is approximately $10^4$ moles Ca $ha^{-1}$ $yr^{-1}$, 400 kg Ca $ha^{-1}$ $yr^{-1}$, or 1 tonne limestone $ha^{-1}$ $yr^{-1}$.

In the subsoils of arid and semiarid soils, Ca commonly precipitates as *calcite* ($CaCO_3$) rather than being leached away. It is found as indurated layers (caliche and other local names) in many arid soils and as more diffuse $CaCO_3$ in Aridisols and Mollisols. Precipitation of $CaCO_3$ in soils is affected by the rates of soil water movement, $CO_2$ production by roots and microbes, $CO_2$ diffusion to the atmosphere, and water loss by soil evaporation and plant transpiration. $CaCO_3$ layers are also derived from upward movement and evaporation of Ca-rich waters. Calcium carbonate accumulations can amount to as much as 90% of the mass of affected soil horizons. Gypsum precipitates in some arid soils, despite being about $10\times$ as water soluble as Ca carbonate.

Although Ca is very important in plant nutrition, soils derived from limestone (crude $CaCO_3$) parent material can be unproductive. The weathering of limestone releases only $Ca^{2+}$ and $HCO_3^-$, which leach away, leaving no soil behind. An example is the Terra Rossa soils of Italy. The limestone parent material lacks silicates to form the cation exchange capacity that retains Ca and other cations. The fertility of limestone-derived soils generally increases with the amount of silicate impurities in the parent material. Organic limestone soils, such as the peat soils in central Florida, managed like sand cultures, can be highly productive because managing their nutrient additions is relatively easy. They require high fertilizer inputs and the soils are unable to retain them effectively.

## 2.3.2  Magnesium

Despite being the second most abundant exchangeable cation in soils, $Mg^{2+}$ is the least-studied ion in this group. Excessive or deficient amounts are uncommon. Plant Mg deficiencies have been reported in some acid sandy soils and are a concern in northern Europe as a by-product of the large amounts of acid rain during the past century and continuing. Liming often corrects both the acidity and a $Mg^{2+}$ deficiency, because agricultural limestone usually contains appreciable Mg impurities. Under chronic conditions and with crops having high Mg requirements, *dolomite* $(CaMg(CO_3)_2)$ or limestone containing dolomite is a satisfactory liming material.

High exchangeable Mg is sometimes associated with low water permeability, soil crusting, and high pH, similar to the characteristic conditions of sodic (Na-rich) soils. This is sometimes the result of soil formation under marine conditions, where $Na^+$ and $Mg^{2+}$ predominate. The $Na^+$ may have produced the poor soil structure, leading to low water permeability, and then leached away, leaving a Mg soil with an inherited soil structure. *Serpentine* (an Mg silicate rock)-derived soils have high $Mg^{2+}$ levels that repress Ca availability to plants.

Magnesium is an important constituent of many primary and secondary aluminosilicate minerals (with the exception of the feldspars). Magnesium in mafic $(Mg^{2+}$- and $Fe^{2+}$-rich) minerals often leads to the formation of chlorite and montmorillonite clay minerals in soils.

## 2.3.3  Potassium (Kalium)

Potassium is the third most important fertilizer element, in terms of amounts added as fertilizer, after nitrogen and phosphorus. Many soils of humid and temperate regions are unable to supply sufficient $K^+$ for agronomic crops. Farmers in these areas long ago recognized the benefits of applying wood ash and other liming materials to their acid soils. Both the alkalinity of the ash (to counter $Al^{3+}$ toxicity) and its K and Ca content are beneficial.

Soils retain $K^+$ more strongly than $Na^+$, because the hydrated K ion is smaller than the hydrated Na ion. In addition, K fits well between the sheets of several soil clay minerals, while Na does not, so K is retained strongly in soils containing these clay minerals. The K concentration in soil solutions is low but is replenished by K diffusion from between the sheets of these clay minerals, from the slower weathering of K-containing feldspar minerals, and from the decay of soil organic matter.

The K concentration in the soil solution must be continuously replenished to supply plants. Soil testing methods using extracting solutions try to measure this replenishment rate. The amount of K extracted depends on the ability of the extracting cation to separate the silicate sheets and the ability of the anion to react with $K^+$. Sodium tetraphenylborate is a strong $K^+$ extractant because $K^+$ forms a complex ion with the tetraphenylborate ion and because $Na^+$ tends to separate the silicate sheets. The exchangeable plus soluble K content determined by this single-extraction soil test is often closely correlated to the amount of K that the soil will supply to plants over a growing season.

The ability of some soils to supply $K^+$ for plants is remarkable. Tropical soils derived from easily weathered basaltic rocks have supplied up to 250 kg ha$^{-1}$ yr$^{-1}$ of K to banana plants for many years without noticeable soil impoverishment. Such rocks typically contain high concentrations of K-containing minerals, which weather rapidly because of their small crystal size and the climate. Soils in temperate regions that supply adequate K for crop needs often contain considerable K-containing mica in their clay fractions.

Continual K fertilization is necessary for K-deficient soils. In addition to inherently low total K in such soils, certain layer silicates strongly retain, or fix, much of the added fertilizer K. Fixed K is released back to the soil solution too slowly to satisfy plant needs. Soils containing appreciable sand- and silt-sized vermiculite are particularly troublesome in this regard. Saturating the K-fixation sites is uneconomical because of the large amounts of K required. Potassium losses by leaching are small compared to the amounts of K fixation, except in very sandy soils. Estimates for sources of K fertilizer indicate a 800-yr lifetime at present rates of use.

### 2.3.4  Sodium (Natrium)

Sodium, in contrast to $K^+$, is a soil chemical concern when it occurs in excess, more than 5–15% of the exchangeable cations. Sodium can accumulate in these amounts in areas inundated by seawater, in arid areas where salts naturally accumulate from the evaporation of incoming surface or ground water, and in irrigated soils because irrigation water often contains high Na. The 5–15% exchangeable $Na^+$ can inhibit water movement into and through many soils. The lower value applies to fine-textured soils, especially those containing high contents of swelling clays, and soils wetted by rainwater. A high exchangeable Na content can work to the benefit of water infiltration, on the other hand, for strongly swelling soils if the extensive cracks that develop upon drying allow water penetration. Also, sandy soils can be irrigated at 20 or 30% exchangeable sodium. The high Na status can slow water infiltration to more manageable rates during irrigation. Chapter 11 is largely devoted to Na problems in soils.

Sodium is not required by plants but can replace part of the K requirement of some plant species. The attraction of some animals to NaCl and the long history of NaCl as an important article of commerce suggest that soils in humid regions may provide inadequate $Na^+$ and $Cl^-$ for animal diets. Humans ingest more NaCl than they require, because of taste preferences and because NaCl is used in food preservation, but a suggestion of insufficiency in animal diets remains. Of the essential elements, $Na^+$, $Cl^-$, $I^-$, and $F^-$ are unique in that much of their supply to humans comes from additions to our diet rather than from natural foodstuffs.

Saline soils are a problem for plants because the high osmotic potential of the soil solution makes it unavailable for plants. The plant has to expend so much energy to take up water that little energy is left for growth and crop yield. This is similar to the problem of organisms in marine water—"all that water and none of it fit to drink." Sodium is toxic to some plants at high concentrations, but for most plants this is a relatively minor problem compared with the restricted water uptake and movement that normally precede Na toxicity. Fruit and nut trees and berries are sensitive to Na and may show toxicity symptoms before water deprivation. The high soil pH that accompanies Na accumulation in arid soils is generally of secondary importance compared to water problems and the microelement deficiencies induced by the high pH. In cold regions, the amounts of NaCl added to roads for snow and ice clearance seem to be too small to cause significant plant damage, because of dilution by snow and rain.

Most plant roots repel most of the NaCl in the soil solution. *Halophytes* are distinctive because they allow NaCl to enter, which is a great advantage in high-salt soils and waters. Research to adapt these plants to agricultural use is under way. The problems are that the seeds are too small to harvest easily and the plants are too salty to be edible forage for grazing animals.

The alkali and alkaline earth cations are the major exchangeable cations, and they determine the pH of most soils. The relation of the dominant exchangeable cations to soil pH can be summarized by the following reactions:

$$\text{soil—Al}^{3+} + H_2O = \text{soil—AlOH}^{2+} + H^+ \qquad pH \sim 5 \qquad (2.1)$$

$$\text{soil—}(Ca^{2+}, Mg^{2+}) + H_2O = \text{no net reaction} \qquad pH \sim 7 \qquad (2.2)$$

$$\text{soil—Na}^+ + H_2O = \text{soil—H}^+ + Na^+ + OH^- \qquad pH \sim 9 \qquad (2.3)$$

The $Ca^{2+}$- and $Mg^{2+}$-saturated soils hydrolyze water very little, so these soils have a neutral pH. Exchangeable $Na^+$, particularly at low salt concentrations, is weakly adsorbed by soil clay. This causes some water molecules to dissociate into the more strongly adsorbed $H^+$ and leave $OH^-$ in the soil solution. Acid soils characteristically have significant amounts of exchangeable $Al^{3+}$. The amounts of $H^+$ in acid soils are very small. Neutralizing acid soils means neutralizing the large amounts of available Al, the *reserve acidity*, in the soil. Some soils are more acid than pH 5 due to organic acids from decaying organic matter or due to the microbial oxidation of sulfur and sulfides to sulfuric acid. Other sources of acidity are phosphate and nitrogen fertilizers and acid rain.

Although most plants and many soil microbes are hampered at soil acidities as low as pH 5, plants in water culture grow well down to pH 3. The effects of soil acidity are due to toxicity by $Al^{3+}$ and transition metal ions that are more soluble at low pH and coincidentally release $H^+$ by reacting strongly with water. The pH is an indicator of ion status in soils rather than being harmful per se.

## 2.4  MAJOR SOLUBLE ANIONS

The anion concentration in the aqueous phase of most nonsaline soils is less than the cation concentration. Much of the negative charge in these soils is due to soil colloidal particles. Most soil colloids have a net negative charge, which is balanced by cations in the water surrounding the colloidal particles. Those conditions apply to North American and European soils, which are relatively young and weakly weathered. Other continents, where weathering has been intensive, have soil clays with both appreciable positive and negative charge. Positive charge increases anion retention, particularly sulfate, in the same way as the negative charge retains cations.

The major anions in the soil solution are $Cl^-$, $HCO_3^-$, $SO_4^{2-}$, and $NO_3^-$. The soil solution concentrations can indicate the sulfur and nitrogen availability for plants in those soils. In humid region soils, the anion sum rarely exceeds 0.01 M in the soil solution; in arid regions the concentration can reach 0.1 M. The relative amounts of these anions vary with fertilizer and management practices, mineralogy, microbial and higher-plant activity, saltwater encroachment, irrigation water composition, and atmospheric fallout.

In saline soils, anion (and cation) concentrations are higher. A typical distribution in the soil solution is $Cl^- > SO_4^- > HCO_3^- > NO_3^-$. At high pH (pH > 8.5), the distribution might be $(HCO_3^- + CO_3^{2-}) > Cl^- > SO_4^{2-} > NO_3^-$.

The major soluble anions are retained weakly by most soils. Nitrate and $Cl^-$ move through soils at virtually the same rate as the water. Sulfate and $HCO_3^-$ lag slightly behind the wetting front because they interact with $Ca^{2+}$, $Mg^{2+}$, and $Al^{3+}$ and with positively charged sites on clay particles. This interaction is weak, however, compared to the strong retention of anions such as phosphate. In strongly weathered soils rich in Al and Fe(III) hydroxyoxides, anion retention of sulfate and phosphate increases greatly. These soils can develop significant positive charge, so their anion retention (exchange) capacity can exceed their cation retention (exchange) capacity.

The nitrate concentration in aerobic soil solutions is an index of, but only a small fraction of, total soil nitrogen. The nitrate concentration reflects a steady state representing nitrogen turnover in the soil and plant availability of nitrogen in the soil. Fertilization can temporarily change this steady state until denitrification, leaching, and nitrogen uptake by plants and microbes restore the nitrogen balance.

Nitrate is actively taken up and reduced by soil organisms, but is chemically inert in the absence of soil microbes. If leached below the surface soil horizons and root zone, nitrate tends to move unhindered and unchanged through the subsoil. Unwise fertilization and organic waste disposal can therefore increase the $NO_3^-$ concentrations in groundwaters and drainage waters. If downward water flow is very slow and the groundwater table is deep, as in arid regions, the time may allow nitrate to slowly disappear by microbial transformations.

The contribution of agricultural practices to nitrate pollution of streams and well waters is a hot issue. Agriculture increases the nitrogen concentrations of these waters, but in some cases land clearing and leaching of nitrogen from geologic strata in arid regions where nitrate had accumulated before cropping contribute to nitrate in the groundwater. Nitrogen in soil organic matter eroded from soils subsequently oxidizes to $NO_3^-$ in aerated streams.

Nitrate is the major anion of pollution concern because of its effects on the ecology of streams and lakes and because of potential harm to infants. The current upper limit of $NO_3^-$ in drinking water in the United States is 10 mg $L^{-1}$ nitrate as N (45 mg $L^{-1}$ as $NO_3$) or $7 \times 10^{-5}$ M. The nitrate is reduced to nitrite ($NO_2^-$) in the body, but infants lack the enzymes to reduce $NO_2^-$ further. Infants can accumulate toxic $NO_2^-$

concentrations if their water or diet contains high $NO_3^-$ concentrations. The nitrate concentration in the soil solution is greatest near the soil surface, where the ion is produced by microbial decay of organic matter, but it can leach rapidly. Root and microbial uptake reduce the nitrate concentration in the root zone.

The major sources of soil $NO_3^-$ are nitrate fertilizers and the oxidation of organic and fertilizer nitrogen. Rainfall contributes an additional small amount, perhaps 1 to 2 $kg\ ha^{-1}\ yr^{-1}$, or about 1% of the nitrogen turnover. The first raindrops in a storm front contain $NO_3^-$ from natural and anthropogenic $NO_x$ washed out of the air, plus $HNO_3$ formed by lightning. Direct absorption of $NO_x$ (NO plus $NO_2$–$N_2O_4$) gases from the atmosphere in industrial regions may add an additional 1 to $3 \times 10^{10}$ kg $yr^{-1}$ (2 to 4 $kg\ ha^{-1}\ yr^{-1}$) of N to soils.

Nitrogen and, to a slower extent, sulfur cycle between the soil and the atmosphere. Acid rain in the atmosphere contributes considerable sulfur to soils. Depending on the proximity to industrial sources, the sulfur fallout ranges from 0 to 100 $kg\ ha^{-1}\ yr^{-1}$. In humid regions the highest rates are accompanied by plant damage from $SO_2$ and $H_2SO_4$, and perhaps by higher soil acidity, which increases Al and transition metal toxicity to roots. The lower rates of sulfur fallout can be beneficial to plants in regions where the native soil sulfur content is marginal to low. The benefits are enhanced if the atmospheric acidity has already been neutralized by dust and $NH_4^+$ before the sulfur contacts plants and the soil.

The sulfate concentrations in soil solutions are good indicators of sulfur availability to plants. Sulfur is being increasingly recognized as in short supply in humid region soils. The traditional NPK fertilizer scheme is being broadened to NPKS. The extent of sulfur deficiency in soils is increasing as superphosphate fertilizer is being replaced by triple superphosphate. Superphosphate contains 0.5 mol fraction sulfate, while triple superphosphate contains none. The sulfate supply in arid soils is adequate for plant needs.

Sulfur readily changes oxidation state in soils between sulfate ($SO_4^{2-}$), elemental S, and sulfide ($S^{2-}$) as oxidizing–reduction conditions change. Several species of sulfur bacteria catalyze these changes. Sulfide reacts strongly with Fe and other transition metals in soils so $H_2S$ evolution from soils is minimal. Any $H_2S$ formed during decay of organic matter diffuses through soil pores before reaching the atmosphere. Diffusion is slow, leaving ample time to react with the soil's solid phase.

Sulfate ions are retained rather strongly in acid soils, in soils having appreciable positively charged clays, and in soils having considerable Al and Fe hydroxyoxides. The sulfate behaves somewhat as if $AlOHSO_4$ is formed, although this compound has not been identified in soils.

## 2.4.1 Halides

The monovalent anions $F^-$, $Cl^-$, $Br^-$, and $I^-$ are the only oxidation states of the halogens in soils. Chloride is essential for plants in trace amounts and for animals in larger amounts. Although F is held to some degree in soils, as a group these ions are retained weakly and are in low amounts in well-drained soils. Plants are much more tolerant of high Cl concentrations than of high concentrations of other micronutrient ions. Except for the specific Cl sensitivity of fruit and nut trees and berries, the effect of excess Cl in soils is mainly to increase the osmotic pressure of soil water and thereby to reduce water availability to plants. Plants absorb $Cl^-$ from the soil solution and the biomass is a significant Cl reservoir in terrestrial environments. As mentioned for $Na^+$, the natural rate of Cl supply to animals from soils through plants may be less than optimal.

The loss of Cl from the continents to the sea is about $3 \times 10^{12}$ mol $yr^{-1}$ (or about 200 mol $ha^{-1}$ $yr^{-1}$ or 8 kg $ha^{-1}$ $yr^{-1}$). Assuming that the oceans are a steady state, this equals the input rate to soils from the atmosphere and from parent material weathering. Chloride is only a minor constituent of igneous rocks, although apatite contains up to 0.01 mol fraction of Cl and micas can contain slight amounts. Most of the Cl input to soils is from rain, marine aerosols, salts trapped in soil parent materials of marine origin, and volcanic emissions. Chloride fallout at seacoasts can be as much as 100 kg $ha^{-1}$ $yr^{-1}$, but this rate decreases rapidly with distance to 1 or 2 kg $ha^{-1}$ $yr^{-1}$ in the continental interiors. This is apparently adequate for plants, since natural Cl deficiencies in plants have not yet been found. In recent years water softening, industrial brines, and road deicing have contributed Cl to local areas.

Except for much greater Cl accumulation in soils of arid regions, the soil chemistry of I and Br resembles that of Cl, except that I and Br are retained more strongly, especially by acid soils. The major input of I to soils appears to be atmospheric. Endemic iodine deficiency (goiter in humans) occurs in mountainous and continental areas isolated from the sea. Fortunately, supplementing NaCl with small amounts of I effectively supplies the I required in animal diets. Iodide and Br are both potentially toxic, but no natural cases have been reported. Bromide reactions in soils have been investigated as a tracer for the movement of water, nitrate, and soil solutions in soils.

Fluoride is the most unique halide chemically and is the most common halide in igneous rocks. The igneous minerals fluorspar ($CaF_2$) and apatite ($Ca_5(F,OH)(PO_4)_3$) are both insoluble in water. Fluoride can substitute for $OH^-$ to some extent in soil minerals. This mechanism is probably also responsible for F retention by aluminium and iron hydroxides in acid soils. Fluoride also associates strongly with $H^+$. HF is a weak acid, $pK = 3.45$.

Fluoride concentrations in soil solutions, groundwaters, and surface waters of humid and temperate regions are generally less than $5 \times 10^{-6}$ M (0.1 mg $L^{-1}$). In arid regions, the $F^-$ concentrations in groundwaters can reach $10^{-4}$ M (2 mg $L^{-1}$) and higher. Some well waters near Pecos, Texas and Phoenix, Arizona contain as much as $10^{-3}$ M (20 mg $L^{-1}$) F. United States public health agencies recommend that drinking water contain about $2-5 \times 10^{-5}$ M F (0.5–1 mg $L^{-1}$) to reduce dental caries. The amount of F in our diet has increased due to increased phosphate fertilization. The F impurities in the fertilizer are absorbed to some extent by plants.

Possible fluoride air pollution has aroused concern near phosphate fertilizer plants and aluminium and iron smelters. The $F_2$, HF, and $SiF_4$ gaseous by-products of these industries are potentially toxic to plants and animals. Once in the soil, however, fluoride is considerably less hazardous. It is retained somewhat by all soils. Soils receiving high amounts of phosphate fertilizer may acquire as much as 20 kg $ha^{-1}$ $yr^{-1}$ of fluorine impurities from the fertilizer. This input is apparently harmless.

**Table 2.2. Average content of oxyanions in the lithosphere and ranges in soils**

| Oxyanion | Average in Lithosphere $(mg\ kg^{-1})$ | Range in Soils $(mg\ kg^{-1})$ |
|---|---|---|
| $B^{3+}$ $(BO_3^{3-})$ | 10 | 2–100 |
| $N^{5+}$ $(NO_3^-)$ | Negligible | 100–1 000 |
| $C^{4+}$ $(CO_3^{2-})$ | — | 100–50 000 |
| $Si^{4+}$ $(SiO_4^{4-})$ | 250 000 | 100 000–400 000 |
| $P^{5+}$ $(PO_4^{3-})$ | 1 000 | 100–1 000 |
| $S^{6+}$ $(SO_4^{2-})$ | 500 | 200–10 000 |
| $As^{3+}$ $(AsO_4^{3-})$ | 2 | 0.1–40 |
| $Se^{4+}$ $(SeO_3^{2-})$ | 0.1 | 0.03–2 |

## 2.5 WEAKLY SOLUBLE ANIONS

The weakly soluble anions are primarily oxyanions ($H_3BO_3$, $H_4SiO_4$, $H_2PO_4^-$, $HMoO_4^-$, $SeO_4^{2-}$, and $H_2AsO^{4-}$ are the forms at pH 6–7)—small, highly charged cations surrounded by strongly associated oxygen or hydroxyl ions. The average contents in soils and in the lithosphere are shown in Table 2.2. Selenium behaves in some respects like sulfate but is included in this group because of its low concentrations in soils. The more soluble sulfate oxyanions are discussed above.

The weakly soluble oxyanions are weak acids; in aqueous solution they gain and lose $H^+$ as the acidity increases and decreases. In solids, the number of $H^+$ ions depends on the crystal structure and the need for charge neutralization in the structure. The oxyanions tend to be weakly soluble in the soil solution because they associate strongly with $Fe^{3+}$, $Al^{3+}$, $Ca^{2+}$, $Mg^{2+}$, and transition metal ions to form uncharged, and therefore insoluble, molecules. Phosphate is the weakly soluble anion of greatest economic, agricultural, and political interest, so it is discussed in a separate section.

### 2.5.1 Boron, Silicon, Molybdenum, Arsenic, and Selenium

The major soil species by far is silicate, which is the backbone of soil and rock mineral structures. The soluble silica ($H_4SiO_4$, or better described as $Si(OH)_4$) concentration in soil solutions ranges from about 6–10 mg $L^{-1}$ ($2 \times 10^{-4}$ M) in temperate region soils to 100 mg $L^{-1}$ ($2 \times 10^{-3}$ M) in high pH soils and soils containing amorphous silica and opal. Silicon is necessary to plants and animals in trace amounts and is very beneficial to plants. Silica lodges in cell walls and provides physical strength and resistance to insect and fungal attack. Silica may also aid in overcoming plant injury from soil salinity. These benefits are more evident in hydroponics because many growth solutions lack silicon.

$Si(OH)_4$ is a very weak acid, $pK_1 = 9.1$. It polymerizes and precipitates as amorphous silica when concentrations reach $10^{-3}$ M. Such high soluble silica concentrations probably occur only in basic soil solutions, in the interstitial solution between expanded 2:1 layer lattice silicates, and perhaps on the surfaces of actively weathering minerals. Concentrations much below $10^{-4}$ M in soil solutions are unusual in soils.

Evolution has taken little advantage of silicon's ubiquity and relatively constant solubility. Silicon is a useful strengthening agent of the plant's cell wall and forms a cast of the cell wall's morphology. These can remain intact in soil as "phytoliths" after the plant decays. Silicon is essential for animals and only in trace amounts.

Silicon reactions are central to rock weathering and soil development. Silicon is the soil component lost in greatest amount from rock minerals during weathering, and the transformations of silica into secondary minerals are the major reactions of soil development. The sand fraction of soils is usually >90% *quartz* ($SiO_2$), the most prevalent form of Si in soils. Highly weathered soils may contain as little as 20% Si (Table 2.1a). Al and Fe ore deposits are essentially highly weathered soils from which most of the Si has been lost.

Secondary silicates form as clay minerals in soils after weathering of the primary silicates in igneous minerals. The secondary silicates include amorphous silica (opal) at high soluble silica concentrations and the very important aluminosilicate clay minerals: kaolinite, smectite (montmorillonite), vermiculite, hydrous mica (illite), and others. Kaolinite tends to form at the low silicate concentrations of humid soils, whereas smectite forms at the higher silicate and Ca concentrations of arid and semiarid soils. The clay fraction of soils usually contains a mixture of these clay minerals, plus considerable amorphous silicate material, such as allophane and imogolite, which may not be identifiable by x-ray diffraction.

Borates, molybdates, selenates, and arsenates occur in such trace quantities in soils that they probably exist only as impurities in major soil particles and on particle surfaces rather than as separate minerals. Soils in humid regions sometimes benefit from borate and, less often, molybdate additions. The range between deficient and excess is narrow, so spreading a few kg ha$^{-1}$ (a few lbs/acre) uniformly over the soil is difficult unless the borate and molybdate salts are mixed in with other fertilizers or inert materials.

Boron is widely and rather uniformly distributed in rocks and sediments. One mineral is tourmaline (($Mg,Fe,Ca,Na,Li,K)_{4-8}B_2Al_3Si_4O_{20}(OH)_2$), but more commonly B is a minor impurity in other minerals. In soils B is diffuse and is not in identifiable B minerals. Marine sediments were once thought to be characteristically higher in boron than terrestrial sediments, but this view is no longer so widely accepted. Boron released to solution by weathering interacts primarily with Fe and Al hydroxyoxides, with maximum adsorption at pH 7 to 9. Aluminosilicates adsorb B only weakly.

Boron exists in solution as boric acid ($H_3BO_3$ or $B(OH)_3$, $pK = 9.2$). Higher borate polymers such as borax ($Na_2B_4O_7 \cdot 10H_2O$), the common boron fertilizer, dissociate to monomers in dilute solutions. Plant deficiencies of boron are well documented in highly weathered soils, and boron toxicity is known in arid and irrigated soils. The range between deficiency and toxicity in soils is narrower for boron than for any other essential element. Boron concentrations greater than a few milligrams per liter in the soil solution can be toxic to sensitive plants, and concentrations less than several tenths of a milligram per liter may indicate deficiency.

Borates are slightly volatile. Evaporation and condensation from the atmosphere may circulate boron geochemically and may have contributed to a rather uniform distribution of B in soils worldwide. Boron's ubiquitous nature and buffering of boron concentrations by soils has apparently allowed life to evolve with a narrow range of boron sufficiency.

Molybdenum is present in soil solutions as molybdate ($H_2MoO_4$, $pK \approx 5$). Molybdate can reduce to $MoO_2^-$ under reducing conditions. As is true for boron, molybdate forms complex polymeric ions at high concentrations in water. In soil solutions, however, only the monomer exists.

Molybdate reacts strongly with Fe hydroxyoxides. Its solubility and plant availability increase with increasing pH. The soluble Mo concentration may be quite low in acid soils but reaches an upper limit as pH increases. Soil solutions at pH 6.5 may contain about $3 \times 10^{-8}$ M Mo (3 $\mu$g Mo $L^{-1}$). Groundwater concentrations greater than $5 \times 10^{-8}$ M (5 $\mu$g molybdenum $L^{-1}$) are unusual.

Molybdenum is essential for the symbiotic nitrogen-fixing microorganisms growing on the root nodules of legumes and some other plants. Mo deficiency occurred in the Victoria province of Australia. Mo fertilization at a rate of only 70 g $ha^{-1}$ increased forage production 12–16 times by stimulating nitrogen fixation in legumes. In other areas of the world, Mo deficiency in some acid soils can result from adequate soil content but inadequate availability. Liming corrects the acidity and the low Mo availability. Direct Mo fertilization is also effective, although Mo salts, as with boron, must be diluted and spread carefully to prevent Mo toxicity.

Arsenic is an essential microelement for animals, but As is known mainly as a toxic element. Arsenic problems in soils are primarily the result of anthropogenic activities. Lead arsenate and copper acetate–arsenate ("Paris green") were once common insecticides. Spraying apple trees in Washington with Paris green for many years led to As concentrations high enough to harm the trees and hinder replanting. The soils have recovered after several decades as the Pb and As in the soils have become less available. The recovery is not due to Pb and As leaching below the root zone. Movement of As through soils is minimal unless large quantities are concentrated in a small or unreactive soil volume, such as might be the case for industrial waste disposal on coarse-textured soils.

The soil chemistry of arsenate ($AsO_4^{3-}$) resembles that of phosphate. Arsenate, however, can be reduced to arsenite. Arsenate is considered to be the state existing in soils, but As concentrations of soil solutions increase under reducing conditions, suggesting reduction to arsenite ($AsO_2^-$, $Eh^\circ = +0.56$ v). Both $H_3AsO_4$ and $HAsO_2$ are weak acids and are strongly retained by soils. Elemental As, arsine ($AsH_3$), and $As_2S_3$ are stable under strongly reducing conditions, but whether they form to an appreciable extent in soils is unknown. As with other cations and anions that are weakly soluble in soils, arsenic added to the soil solution decreases in concentration with time. The decrease is due to increased strength of retention rather than loss by volatilization of $AsH_3$. Arsenic is retained strongly in soil under oxidizing and reducing conditions.

Selenium and sulfur chemistry are similar, except that selenates ($SeO_4^{2-}$) and selenites ($SeO_3^{2-}$) are weaker acids than their sulfur counterparts. Because of their weak-acid character, $SeO_4^{2-}$ and $SeO_3^{2-}$ should be retained more strongly by soils than the corresponding sulfur anions. The potential of the selenate–selenite couple, $Eh° = 1.15$ v, indicates that selenite is the stable oxidation state over most of the range of soil conditions, but selenate is the form most often reported. Selenate is stable under strongly oxidizing conditions, and elemental selenium is stable under reducing conditions ($Eh = 0.74 - 0.059 \log(SeO_3^{2-})$ for the selenite–selenium couple). Selenium redox reactions are apparently rather reversible and may not require microbial catalysis. Some reports indicated that Se may be lost from soils by reduction to $H_2Se$, a gas. The amounts of $H_2Se$ evolved from soil, like $H_2S$, are small.

Selenium is essential for animals; natural Se deficiencies as well as toxicities have been recognized in grazing animals and fowls. Selenium salts may accumulate in poorly drained areas of arid regions, particularly in the northern high plains of North America. Legumes growing in these soils can accumulate toxic quantities of Se. Such plants are among the "loco weeds," which cause locomotion problems and eventual death of grazing animals.

Plants from some areas in northern California have been found to be low in Se and animals benefit from Se supplements. This may be the basis for the current popularity of Se as a diet supplement for humans. Selenium gained notoriety in central California when drainage water from a large irrigation project evaporated in a reservoir that is also a waterfowl nesting area. The evaporation concentrated Se and it was blamed for abnormalities in ducklings hatched there. The drainage water was originally supposed to drain to San Francisco Bay, where the Se might have been beneficial or at least be diluted to insignificance. Funding difficulties instead stopped the drainage line at the reservoir, which has no external drainage. The fear arose that Se might also be accumulating excessively in other poorly drained areas in the western United States and Canada, but research indicates that the California case is unique because the water originated from a region of high-Se soils.

### 2.5.2  Phosphate

Crops growing in large areas of the world's agricultural soils respond to phosphate fertilization. Phosphate is in the shortest supply of the major (NPKS) fertilizer elements. One estimate is that present ore deposits can supply phosphate for no more than 200 years at current rates of use. Other phosphate deposits exist but they are of lower quality.

Phosphate chemistry in soils has been studied more intensively than that of any other element save nitrogen. Phosphate added to soils is first adsorbed quickly and is later "fixed" into increasingly less soluble states as time increases. Despite this great effort, quantitative predictions of phosphate concentrations in soil solutions are poor and no techniques have been devised to release the large amounts of unavailable phosphate in soils, nor to prevent fixation of fertilizer phosphate by soils. The uncertainties about soil phosphate chemistry and the difficulty of increasing phosphate availability are due to phosphate's strong interaction with many inorganic and organic soil components. The uptake by plants and microorganisms, continual return from organic decay, and slow phosphate release rates in soils continuously release phosphate but at rates that are often below those needed for optimal plant growth. Phosphate chemistry is of greatest interest in acid soils. Phosphate is increasingly unavailable as soil acidity increases, due to retention by Al and Fe. The decay of phosphate in organic matter bypasses the inorganic soil fraction to some extent and maintains some phosphate availability to plants over the whole range of soil pH.

Phosphate is P's only oxidation state in soils and plants. Phosphite ($PO_3^{3-}$, phosphine ($PH_3$), or organophosphorus compounds of lower phosphorus ($< V$) oxidation states have not been found in soils. The $H_2PO_4^-$ and $HPO_4^-$ ions are predominant at the pH of soil solutions. The pH distribution of the $H_3PO_4$–$PO_4^{3-}$ series in solution is shown in Fig. 2.2. The phosphate mineral apatite ($Ca_5(OH,F)(PO_4)_3$) is common in rock minerals. In acid soils, most solid-phase phosphate is associated with Fe and Al and their hydroxyoxides. In basic soils, phosphate is associated with Ca in apatite-like forms. Whether phosphate truly adsorbs on surfaces or precipitates within the weathered layer on particle surfaces is uncertain. Phosphate fixation is appreciable in all but very coarse-textured soils and is particularly strong in soils rich in amorphous Fe and Al hydroxyoxides or allophane (in volcanic soils). Cycling of phosphate by reduction of Fe(III) phosphate during flooding and reoxidation after drainage is thought to account for the continued phosphate availability for centuries in paddy soils.

Soil solution concentrations of phosphate are of the order $10^{-6}$ to $10^{-7}$ M (0.01 to 0.1 mg $L^{-1}$). For the purpose of soil testing, many workers have tried to devise extraction procedures yielding phosphate values related to crop response of phosphate

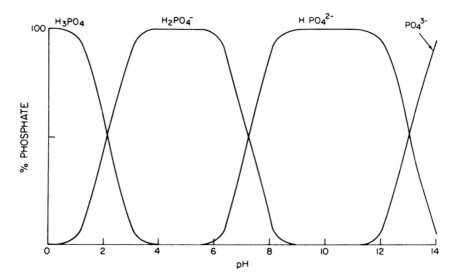

**FIGURE 2.2.** The pH distribution of the phosphate series.

fertilizer. Results from one area and one crop are often inapplicable elsewhere and to other crops. Radioactive $^{32}P$ enabled researchers to follow phosphate uptake and inactivation in soils. Soil phosphate was called labile or nonlabile, according to the rate at which it would exchange with radioactive phosphate. The labile fraction was considered to be the amount available to plants. This attractive concept oversimplified the situation, however, and has lost favor.

Later, researchers assumed equilibrium with various phosphate minerals to calculate phosphate ion concentrations in soil solutions. Minerals such as variscite ($AlPO_4 \cdot 2H_2O$), strengite ($FePO_4 \cdot 2H_2O$), and the series between pure variscite and strengite have solubilities in the range of phosphate concentrations of acid soils. Soluble phosphate concentrations in basic soils are in the range of solubilities of octocalcium phosphate ($Ca_4H(PO_4)_3$). The correspondence between soil concentrations and mineral solubility is unfortunately too crude to accurately predict the soil phosphate status for plants.

Phosphate behavior is also described by Langmuir and Freundlich adsorption equations, although these models may be too simple to accurately explain soil phosphate behavior. Kinetic models of phosphate retention by soils are also being employed. Although kinetics can suggest retention mechanisms, the complexity of soil-phosphate behavior makes this prospect difficult to achieve.

The calcium phosphate series ranges from the most soluble $Ca(HPO_4)_2$ through $CaHPO_4$ and $Ca_4H(PO_4)_3$ (octocalcium phosphate) to the least soluble $Ca_5(OH,F)(PO_4)_3$ (apatite). Phosphate fertilizers are made by treating rock phosphate, mostly apatite, with sulfuric acid to make superphosphate (nominally $CaHPO_4$) or with phosphoric acid to make triple superphosphate (nominally $Ca(HPO_4)_2$). The marketing hyperbole comes from the early use of rock phosphate as a fertilizer on acid soils. During the transformation of the fertilizers to Fe and Al phosphate in acid soils, or to octocalcium phosphate and apatite in alkaline soils, perhaps one-third of the phosphate becomes available to plants. The acid-treated phosphate is more soluble and hence more immediately available than rock phosphate. In the transition from superphosphate to the "improved" triple superphosphate, crop yields in some cases declined. This was because superphosphate contains 50 mole percent sulfate, which overcame a sulfur deficiency in these soils. Triple superphosphate lacks sulfur.

The phosphate concentrations of waters draining from soils usually are about $10^{-7}$ M. Worldwide, this amounts to a phosphate loss of $17 \times 10^{10}$ moles $yr^{-1}$ (10 mol $ha^{-1}$ $yr^{-1}$ or 1 kg $ha^{-1}$ $yr^{-1}$). Phosphate in eroded soil particles reaching the sea is estimated to be an additional $13 \times 10^{10}$ mol $yr^{-1}$. Fertilization affects the phosphate content of sediments eroded from surface soils, and increases the phosphate concentrations of drainage waters and groundwaters. Preventing erosion has the added benefit of reducing phosphate inputs to streams and lakes.

The rate of phosphate loss from soils by weathering is about the same as the overall weathering rate, so the total amount of phosphate in soils tends to remain constant throughout soil development. The availability of phosphate to plants, however, decreases as soils become more acid and the proportion of phosphorus as aluminium and iron phosphate increases.

Human activities affect the global phosphorus cycle to an ever-increasing degree. The rate of phosphate mining equals or exceeds the rate of phosphate lost naturally from the continents. About $35 \times 10^{10}$ mol yr$^{-1}$ are used as phosphate fertilizer, while $6 \times 10^{10}$ mol yr$^{-1}$ are used by other industries and in the home. The industrial forms are particularly soluble, so the phosphate in sewer discharge, as well as the lesser-available forms from farm runoff and erosion, is already affecting freshwater and marine ecology in North America, Europe, and Asia.

Our ignorance of the state of phosphate in soils and our inability to increase the availability of soil and fertilizer phosphate ranks as one of the great frustrations and challenges of soil chemistry. The unsuccessful attempts to overcome soil fixation of phosphate fertilizer include (1) adding soluble silica to soils to replace adsorbed phosphate, (2) using ammonium phosphate fertilizers to avoid the strong acidity produced when calcium phosphates dissolve, (3) creating polymeric phosphates, which are more soluble than monomeric orthophosphates, in the hope of forming less readily fixed phosphate ions in soil solutions (soil microorganisms, however, rapidly break the phosphate–phosphate bonds of such polymers), and (4) using elemental phosphorous, nitrogen–phosphorus, and organophosphate compounds. As yet, no economically or chemically feasible alternative to orthophosphate fertilization of soils has appeared, except for the centuries-old practice of increasing the pH of acid soils by liming. Phosphate availability from organic manures seems to be greater per unit of P, but their P contents are low.

## 2.6   ALUMINIUM, HYDROGEN, AND TRANSITION METALS

Aluminium in soils is closely connected to soil acidity and is also discussed in the chapters on acid soils and ion–water reactions. The acidity of acid soils is due to the reactions of water with exchangeable $Al^{3+}$ on the surface of soil particles. The strong Al–water reaction repels $H^+$ from the water molecules into the soil solution. This can create soil acidities as low as pH 4.5. Stronger acidity means other $H^+$-yielding reactions—organic acids from soil organic matter decay, sulfur and sulfide oxidation, phosphate fertilizers, ammonia oxidation, acid rain, and Fe– and Mn–water reactions—are active.

The Al–water reaction continuously and slowly liberates $H^+$ as Al in aluminosilicate minerals weather to *gibbsite* ($Al(OH)_3$), the Al state in the most highly weathered soils. The Al–water reaction itself is fast, but the weathering of soil minerals to create surface $Al^{3+}$ is slow. Exchangeable $Al^{3+}$ and its hydrolyzed–polymerized forms ($Al(OH)_x(H_2O)_{6-x}^{(3-x)+}$) produce the acidity of most soils as they hydrolyze further toward $Al(OH)_3$. The $Al^{3+}$ is tightly bound to clay surfaces; the exchangeability of $Al^{3+}$ depends on the concentration and cation charge of the salts in the extracting solution. Most workers attribute low exchangeability to strong adsorption of the polymeric ions on clay surfaces. Only $Al(H_2O)_6^{3+}$ is considered truly exchangeable, and it is present in appreciable amounts in soils only at pH $< 5.5$. While other major exchangeable cations are leached from soils during weathering, Al is retained in soils as solid-phase $Al(OH)_3$. The large amounts of limestone to neutralize ("sweeten") acid soils is due to the amounts of $Al^{3+}$ that have to convert to $Al(OH)_3$.

The amount of $H^+$ is an appreciable fraction of the total cations in the soil solution only in extremely acid (pH < 4) soils. Such high concentrations are generally caused by active oxidation of sulfides and elemental sulfur, in the immediate vicinity of dissolving phosphate fertilizer granules, or from organic acids from decay of organic matter. High sulfide contents are typical of some mine wastes and recently drained coastal sediments, particularly in the tropics and subtropics. When the soils are drained and oxygen can enter the soil, sulfide microbially oxidizes to $H_2SO_4$, creating acidities as great as pH 1.5 or 2. When the supply of sulfur is exhausted, the $H_2SO_4$ leaches away and the soil pH reverts to near-neutral values. When the soil is reflooded, the oxidation stops and the pH also returns to near neutrality as the sulfate is reduced to sulfur and sulfide.

In mine spoils of humid regions, complete sulfur oxidation may require several years. During this time, plant growth is inhibited and erosion can be severe. The resulting soil damage can be evident for decades after acid production has ceased and the acidity has been leached away. In more arid regions, the natural reclamation of mine spoils, in terms of sulfide oxidation, proceeds even more slowly. Limited moisture inhibits sulfur oxidation and acidity production, as well as plant establishment. Plants are usually unaffected by the acidity because the basicity of arid region soils is adequate to neutralize the acidity produced. The salinity resulting from neutralization, however, can reduce plant growth.

Although the $H^+$ concentration in soils is normally small, it is extremely important. The chemistry of many ions and the activity of soil microorganisms is so closely tied to pH that no other single soil measurement conveys as much information. The measurement of soil pH in acid soils has received considerable study. The measurement is easy and seemingly straightforward but the simplicity is only apparent. Changes of the soil's electrical double layer, salt concentration, partial pressure of $CO_2$, products of microbial reactions, and extraneous electrical potentials at the soil–electrode interface can cause significant aberrations in soil pH measurements. The pH measurement is closely tied to predictions of the lime requirement of acid soils. The pH of basic soils has less economic importance and has received correspondingly less attention. In addition, the pH of calcareous soils is often controlled by $CaCO_3$ dissolution. Laboratory pH measurements of calcareous soils, and field measurements at similar $CO_2$ concentrations, tend to be monotonously similar, in the range from pH 7.5 to 8.5. A soil pH > 8.5 often indicates appreciable exchangeable $Na^+$. High exchangeable sodium can also exist at lower pH under marine conditions.

Transition metals (groups IB through VIIB and group VIII of the periodic table, Fig. 2.1) are distinct from the elements at either end of the periodic table in that electrons are added to and removed from inner electron orbitals. The chemistry of the transition metals therefore changes more subtly from element to element than elements having electron changes only in the outer, $s$ and $p$, orbitals. In addition, many transition can have more than one oxidation state in soil.

Under aerobic conditions, almost all of the transition metals plus aluminium and beryllium, essentially all of the metals in the periodic table except the alkali and alkaline earth metals, exhibit one major reaction in soils. They associate strongly with $O^{2-}$ and $OH^-$ ions and tend to precipitate as insoluble hydroxyoxides or as minor components of insoluble aluminosilicates. The precipitation may be on the surface and be called adsorption, but the effect on water solubility and plant availability is the same. The exceptions are Cd, Hg, Pb, Zn, and Cu, which also react extensively with organic matter and sulfide, if available.

In anaerobic soils, the individual chemistry of the ions is more distinctive. The transition metal ions in the middle of each period of the periodic table—chromium, manganese, iron, nickel, cobalt, and copper—can reduce to lower oxidation states, while the end members—scandium, titanium, and zinc—have only one oxidation state. The lower oxidation states are more water soluble but still tend to precipitate as carbonates and sulfides, or associate with organic matter, thus reducing their movement but increasing their plant availability.

Small amounts of transition metal and Al ions are associated with clay surfaces and soil organic matter. The amounts of exchangeable and water soluble ions increase with soil acidity. The approximate order of trivalent cation retention is

$$Al < Fe < Sc < Y < Eu < Sm < Nd < Pr < Ce < La$$

Only $Al^{3+}$ is a common exchangeable cation in soils and is significant only in moderately to strongly acid soils, pH $< 5.5$. Al and Fe are usually the second and third most common cations in total soil content, but only tiny fractions are exchangeable on soil particle surfaces and dissolved in the soil solution.

The important soil chemical elements of this group are aluminium and all (except scandium) of the first-row transition metals: titanium, vanadium, chromium, manganese, iron, cobalt, nickel, copper, and zinc. They are important either because their amounts in soils are large (Al, Fe, Ti) and therefore important to soil development, or because they are essential elements to living organisms. A few other transition metals outside of this first row—molybdenum, cadmium, tin, mercury, and lead—are also of interest because of their essentiality (Mo, Sn) or toxicity (Cd, Hg, Pb) to organisms. The total amounts in soils range from as much as 30% Fe and Al in highly weathered soils to less than 1 mg $kg^{-1}$ (Table 2.3). Those not listed are normally present in concentrations of $<1$ mg $kg^{-1}$. Trace contents in soils have been defined as $<100$ mg $kg^{-1}$. Plant uptake and decay, plus the strong retention of transition metals by inorganic and organic soil components, lead to slight accumulation of the trace metal ions near the soil surface. Total concentrations in the surface centimeters of untilled soil can be several times the concentrations shown in Table 2.3, and several times those in their subsoils.

The ions of this group precipitate from pure solutions (in the absence of soils) as hydroxyoxides of varying degrees of hydration, such as $Al(OH)_3$, $AlOOH$, $Al_2O_3$, $FeOOH$, $Fe_2O_3$, and $MnO_x$. Their solubility products can all be expressed as $(M)(OH)^x$. The ion concentrations in soil solutions and soil extracts of all ions in this group, except Fe(III) and aluminium, are usually several orders of magnitude less than those calculated from their solubility products. This indicates that soils retain these ions much more strongly than do pure hydroxyoxides. The relative soil retention of these cations, however, roughly follows the order of the solubil-

**Table 2.3. Total contents of transition and related metal ions in the lithosphere and in soils**[a]

| Element | Average in Lithosphere $(mg\ kg^{-1})$ | Soil content $(mg\ kg^{-1})$ |
|---|---|---|
| Beryllium | 6 | 1–40 |
| Aluminum | 70 000 | 10 000–200 000 |
| Titanium | 4 400 | 1 000–10 000 |
| Vanadium | 150 | 20–500 |
| Chromium | 100 | 5–3000 |
| Manganese | 1 000 | 200–3000 |
| Iron | 50 000 | 10 000–300 000 |
| Cobalt | 40 | 1–50 |
| Nickel | 100 | 10–1000 |
| Copper | 70 | 2–100 |
| Zinc | 80 | 10–300 |
| Yttrium | 30 | 20–200 |
| Zirconium | 220 | 60–2000 |
| Molybdenum | 2 | 0.2–5 |
| Cadmium | 0.2 | 0.01–7 |
| Tin | 40 | 5–10 |
| Lanthanides, total | — | 10–500 |
| Mercury | 40 | 0.005–0.1 |
| Lead | 10 | 2–200 |

[a] After D. J. Swaine. 1955. Commonwealth Bureau of Soil Science Technical Communications 48, Farnham Royal, Buckinghamshire, England; W. H. Fuller. 1977. EPA-600/2-77-020; and J. A. McKeague and M. S. Wolynetz. 1980. *Geoderma* **24**:299.

ity products of the oxidized cations. Reducing conditions increase the $Fe^{2+}$ and $Mn^{2+}$ concentrations in the soil solution, and complexing by soluble organic anions increases the concentrations of $Cu^{2+}$ and $Zn^{2+}$.

Because these cations are multivalent, their hydroxyoxide ion products involve the second, third, or fourth power of the $OH^-$ concentration. Their concentration changes in soil solutions, however, tend to be proportional to only the first power of $OH^-$ or $H^+$. This is partly explainable by the soluble ions being hydrolyzed. For example, the mechanism of Fe(III) dissolution and precipitation at the pH of normal soils is probably

$$FeOOH + H^+ = Fe(OH)_2^+ \tag{2.4}$$

Hydrolysis explains only part of the disagreement between metal ion concentrations in soil solutions and corresponding solubility products. Soils apparently retain trace metal cations by other mechanisms in addition to precipitation as pure phases. Such mechanisms include coprecipitation as minor constituents in Fe, Al, Ti, Mn, and Si hydroxyoxides; adsorption on soil surfaces; complexation with organic matter; and incorporation into plant tissues and decay products. These mechanisms reduce cation concentrations to well below those predicted by hydroxyoxide solubility products.

An exception to that extra-strong attraction may be Fe(III). If constant aeration minimizes $Fe^{2+}$ production, and if biological reactions reach some sort of steady state, Fe(III) concentrations generally agree closely with those calculated from the solubility product of FeOOH. Competing reactions under these conditions appear to be overshadowed by the relatively rapid dissolution/precipitation of FeOOH. The large amounts of FeOOH in soils, compared to the amount of competing reactions, permits the solubility product to control the Fe(III) concentration in soil solutions. Aluminium is also prevalent in soils, but the $Al(OH)_3$ solubility product controls $Al^{3+}$ concentrations in soil solutions only after the cation exchange capacity has been saturated with $Al^{3+}$ (at uncommonly low pH values).

Trace metal ions in soil particles are probably contaminants in crystals or in amorphous gels of more abundant ions. The solubilities of such contaminant ions might still be definable by solubility products if the cations formed an ideal solid solution within the crystal. The solubility should then be a function of their mole fractions and solubility products of the minor and major hydroxyoxides. Ideality in such cases seems doubtful.

The solid hydroxyoxides adsorb $H^+$ in acid soil solutions and adsorb $OH^-$ (or release $H^+$) in basic soil solutions. The surface charge of such solids changes from positive in acid solutions to negative in basic solutions. This behavior is part of the reason that soil cation exchange capacities increase with pH, although the contribution of hydroxyoxides to soil exchange capacities is small except in highly weathered soils.

The manganese, iron, cobalt, copper, zinc, and molybdenum ions of this group are all essential for plants and animals. In addition, vanadium, chromium, nickel, and tin are essential for animals. The soil solution concentrations and plant availabilities of these ions generally decline with increasing pH. Molybdenum is an exception and becomes more available with increasing pH. Reducing conditions dissolve $Mn^{2+}$ and $Fe^{2+}$.

Compared to the major fertilizer elements N, P, K, and S, plant deficiencies of microelements are infrequent. Plant variety and growth rate seem to be at least as important as soil factors in determining microelement deficiencies. Common examples are Fe and Zn deficiencies in irrigated crops, especially fruit trees, grown on basic soils. Such deficiencies are caused in part by high growth rates and consequently high plant demands for these elements. Native plants apparently have adapted to the natural rates of trace metal recycling. Plant growth rates without irrigation in arid regions are of course considerably slower.

Deficiencies of zinc, copper, cobalt, and molybdenum have also been recognized in some acid soils. The Zn, Cu, and Co deficiencies seem to be attributable to the low contents remaining in highly weathered soils. Weathering appears to remove these ions faster than Fe, Al, Mn, and other trace metal ions. Zinc and Cu also complex with soil organic matter, which does much to control their concentrations in the soil solution. Soluble organic anions from decomposing organic matter can dissolve Fe and Al and move them downward where they reprecipitate in the soil profile. This is characteristic of Spodosols, or Podzols, and the movement there is restricted to 10–20 cm downward in the soil profile.

Attempts to estimate the amounts of microelements available to plants have been somewhat successful. Soils to be tested are usually extracted with solutions containing chelates, such as DTPA or EDTA. The amount of metal extracted is then correlated with plant response to fertilization with the element. The solubilities of trace metal chelates is much greater than the solubilities of the hydrated (aquated) cations, so chelates added to soils can increase the solubility and plant availability of some trace metals. The Fe, Cu, and Zn complexes are stronger and form preferentially in soils of pH < 7. In basic soils $Ca^{2+}$ and $Mg^{2+}$, because of their high concentrations, compete with the transition metals for the chelating anions. Some chelates form stronger complexes with iron and zinc than EDTA. Hence, they can maintain higher Fe, Cu, and Zn solution concentrations in basic soils, where Fe and Zn deficiencies occur. The chelates are effective until soil microbes degrade the chelates. Fertilization with chelates has been limited by their high cost and by their biodegradability in soils.

## 2.7  TOXIC ELEMENTS IN SOILS

Although ion retention by soils can cause elemental deficiencies, it also prevents excessive or toxic concentrations in most soil solutions. The evolution of life took advantage of the naturally low concentrations in water and low plant availability; higher concentrations evolved as toxic. Table 2.4 shows the natural soil contents of ions that are generally harmless. The upper values are conservative estimates of soil contents that might lead to toxicity. All soils contain the "toxic elements," even in amounts that are mined as ores, and these concentrations do not necessarily harm plants or groundwater.

This section deals with the elements known as *heavy metals*. Heavy metal is defined as any metal having a specific gravity greater than iron (s.g. = 5.5) and is often used nowadays to mean toxic metal ions. Toxic organic (carbon-containing) compounds, such as pesticides and many industrial and municipal wastes, are much more prevalent and present a greater hazard to plants and animals than heavy metals; they are discussed elsewhere. Heavy metals may be toxic when their ions are available to plants or are in groundwater at concentrations higher than in native soils. *Trace metals* have been defined as being less than 100 mg kg$^{-1}$ of soil.

"Heavy metal-contaminated soils" is a common phrase. Regulatory agencies were forced to make criteria for contamination and understandably were very conservative. The criterion is basically, "if you eat it and it makes you sick, it is contaminated." By that standard, every soil is contaminated. The real issue is whether soil components affect groundwater composition or can be taken up by plants and soil fauna. Mixed adequately with soil, metal ions quickly react with the soil and are adsorbed/precipitated and tend to revert to their native states, and native availability, in soils. In the laboratory, Pb added to a soil suspension was adsorbed by a roughly two-step reaction. The half-life of the first step of retention was 2 minutes and accounted

**Table 2.4. Natural, and apparently safe, soil and plant concentrations of elements that have been designated as being toxic**[a]

| | Total Soil | | | |
|---|---|---|---|---|
| Element | Typical Value (mg kg$^{-1}$) | Range (mg kg$^{-1}$) | Soil Solution (mg L$^{-1}$) | Plants Range (mg kg$^{-1}$) |
| Aluminum | 50 000 | 10 000–200 000 | 0.1–0.6 | — |
| Arsenic | 5 | 1–50 | 0.1 | — |
| Beryllium | 1 | 0.2–10 | 0.001[b] | — |
| Cadmium | 0.06 | 0.01–7 | 0.001[b] | 0.1–0.8 |
| Chromium | 20 | 5–1000 | 0.001[b] | — |
| Cobalt | 8 | 1–40 | 0.01[b] | 0.05–0.5 |
| Copper | 20 | 2–100 | 0.03–0.3 | 4–15 |
| Lead | 10 | 2–100 | 0.001[b] | 0.1–10 |
| Manganese | 850 | 100–4000 | 0.1–10 | 15–100 |
| Mercury | 0.05 | 0.02–0.2 | 0.001 | — |
| Nickel | 40 | 10–1000 | 0.05[b] | ~ 1 |
| Selenium | 0.5 | 0.1–2.0 | 0.001–0.01 | — |
| Zinc | 50 | 10–300 | <0.005 | 8–15 |

[a]From W. H. Allaway. 1968. *Adv. Agron.* **20**:235; R. P. Murrman and F. R. Koutz. 1972. Special Report No. 171, U.S. Army Cold Regions Research and Engineering Lab, Hanover, NH; and G. R. Bradford et al. 1975. *J. Environ. Qual.* **4**:123.

[b]Estimated as 30 times its concentration in seawater.

for 78% of the Pb removed from the water. The half-life of the second step was 58 hours. Compared to rates of ion uptake by plants and water movement in soils, both rates are fast. In the field, soil retention rates are slower because of less mixing, but are fast enough to prevent Pb movement and plant uptake.

Nonetheless, soil contamination is a concern because mixing with the soil can be inadequate. For example, the surface soil next to older wooden buildings can contain high Pb because of Pb-containing paint flaking off the walls and because Pb moves downward very slowly. Children frequently play next to homes and soil gets in their mouths. Industrial waste dumps may contain high concentrations of heavy metals and acidic components that increase the metal's solubility in the water leaching from the dumps.

In many cases, soil "contamination" can be overcome by covering the soil with an unaffected layer or by thoroughly mixing the contaminated layer with "clean" soil beneath. Excavating and moving the material to a landfill does nothing to alter its chemical state and can expose many more people than leaving the soil in place. "Hazardous waste landfills" can be hazardous because they concentrate the wastes and they rely on suspect technology to isolate the wastes.

Landfill leachate is a potential source of heavy metal contamination and a popular horror story in newspapers. Landfills concentrate wastes, the wastes are not always mixed thoroughly with soil, and soils have limits in their capacity to react. Landfills have strong reducing conditions and the reduced oxidation states of metals tend to be more water soluble. These conditions, however, also form organic matter and sulfide, which react with and inactivate the ions.

A popular method of minimizing leachate pollution may be counterproductive in the long run. Thick liners of polyethylene sheets or clay below the landfill prevent water from leaching out, until cracks or leaks occur. Then the water flows down through a small soil volume below the leak and may oversaturate the soil. Liners then accentuate the pollution hazard. Wise water management and treatment would appear to be a better answer than liners. Evidence for landfill contamination of groundwater is meager anyway. Liners that encapsulate landfills are an overreaction to a problem that is less serious than was perceived. Entombing wastes is an ostrich-in-the-sand approach to waste management. Sealing off oxygen input allows only fermentation of organic wastes rather than oxidizing them. The goal should be oxidation, that is, destruction of waste compounds, and returning their constituent elements to their oxidized and most benign states.

The total amount of a heavy metal or any substance in the soil is a poor indicator of its availability to plants or ability to move in soils. Defining the amounts that are toxic to plants or to animals subsisting on those plants is very difficult. Plant and soil concentrations that are harmful to plants are unknown for most ions and for most plants. In addition, the harmful concentrations vary with the species, age, health, and general nutrition of the organization. The soil solution column of Table 2.5 gives estimates of the concentrations that are found in soils and that are immediately available

**Table 2.5. Representative ion concentrations in soil solutions of temperate region soils[a]**

| Ion | $mg\,L^{-1}$ | Molar, $\times 10^{-6}$ |
|---|---|---|
| $Cl^-$ | 60–600 | 2000–20 000 |
| S (as $SO_4^{2-}$) | 50–500 | 500–5000 |
| $Ca^{2+}$ | 30–300 | 800–8000 |
| $Mg^{2+}$ | 5–50 | 200–2000 |
| Si (as $Si(OH)_4$) | 10–50 | 400–2000 |
| $K^+$ | 1–10 | 20–200 |
| $Na^+$ | 0.5–5 | 20–200 |
| $F^-$ | 0.1–0.5 | 5–20 |
| Mo (as $HMoO_4^-$) | 0.001–0.01 | 0.001–0.1 |
| $Mn^{2+}$ | 0.1–10 | 2–20 |
| $Cu^{2+}$ | 0.03–0.3 | 0.5–5 |
| P (as $H_2PO_4^-$) | 0.002–0.03 | 0.006–1 |
| Al (as $AlOH^{2+}$) | <0.01 | <0.4 |
| $Fe^{2+}$ + $Fe(OH)_2^+$ | <0.005 | <0.01 |
| $Zn^{2+}$ | <0.005 | <0.01 |

[a]From R. P. Murrman and F. R. Koutz. 1972. Special Report No. 171, U.S. Cold Regions Research and Engineering Lab, Hanover, NH, pp. 48–74.

**Table 2.6. Maximum recommended concentrations of toxic ions in drinking water for livestock[a]**

| Ion | Upper Limit $(\mathrm{mg\ L^{-1}})$ |
|---|---|
| Aluminum | 5 |
| Arsenic | 0.2 |
| Beryllium | No data |
| Boron | 5.0 |
| Cadmium | 0.05 |
| Chromium | 1.0 |
| Cobalt | 1.0 |
| Copper | 0.5 |
| Fluoride | 2.0 |
| Iron | No data |
| Lead | 0.1 |
| Manganese | No data |
| Mercury | 0.01 |
| Molybdenum | No data |
| Nitrate + nitrite | 100 |
| Selenium | 0.05 |
| Vanadium | 0.10 |
| Zinc | 24 |
| Total dissolved solids | 10 000 |

[a]From *Water Quality Criteria*, Environmental Studies Board, National Academy of Science, National Academy of Engineering, 1972.

to plants. The recommended maxima for livestock drinking water (Table 2.6) are very conservative guides to the desirable maxima in soil solutions, because plants are much more tolerant of high concentrations than are animals. Plants had to evolve a greater tolerance to both toxicity and deficiency because of the limited soil volume within reach of their roots. Animals can range over a much wider area. Soil retention, exclusion by plant roots, and limited translocation to the plant top all exclude soil ions from the animal food web.

Soil solution concentrations of most trace metals are largely unknown because of difficulties in measuring small concentrations. The values in Table 2.4 marked with *b* are only rough estimates derived from the composition of seawater. Reported Mn and Cu concentrations in soil solutions are about 30 times greater than their concentrations in seawater. This factor was applied to the remaining ions of the table as well.

All of the essential microelements and most, if not all, of the trace elements are toxic at soil concentrations much above normal. Naturally occurring high concentrations of toxic elements are rare in soils, except for widespread $Al^{3+}$ phytotoxicity in acid soils. Soil contamination by toxic elements generally is a result of human activities. Anthropogenic pollutant elements and their important oxidation states that have received attention include, in order of atomic number rather than importance: Be(II), F(−I), Cr(III–VI), Ni(II–III), Zn(II), As(III–V), Cd(II), Hg(O–I–II), and Pb(II–IV).

Soil contamination from smelting, metal plating, manufacturing, municipal and industrial wastes, and automobile traffic can increase soil concentrations of these ions
to possibly toxic levels. Even so, animal problems generally occur only when contaminated plants are the sole food supply. Grazing animals confined to contaminated
areas show the most serious effects of toxic metal accumulations in soils.

Figure 2.3 shows the release of six transition metals from ten widely different soils
of central Europe that had been affected by smelting and other industrial pollution.
Despite the differences in soils and amounts of exposure, the percent desorbed from
the soils at constant pH is quite consistent for each ion. In this group Cd is retained
least strongly and this has been generally true in soils; Ni and Cr are retained the
strongest in this experiment. The relative positions of the ions change slightly with
the method of soil extraction. Lead is retained strongly by many soils and chromium
as Cr(VI) can be more mobile and more available to plants than Fig. 2.3 suggests.
Ni, Zn, Hg, and Pb availability are generally less responsive to pH than Al and Be.

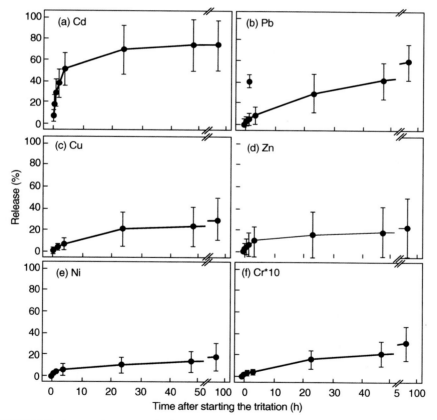

**FIGURE 2.3.** Average heavy metal release during the $pH_{stat}$ experiment in percentage of the total concentration ($n = 10$). Standard deviations are plotted as error bars and show the variability
between the samples.

The availability of the selenate, vanadate, arsenate, and chromate anions decreases with the soil's Fe and Al hydroxyoxide content and increases with pH, since anion retention decreases as pH increases. The pH response for anions, however, is generally less dramatic than for cations, which adsorb/precipitate as hydroxyoxides in soils.

The metals Hg(0–I–II) and Cd(II) are retained less strongly by soils than the other toxic cations, and hence pose a more serious problem of movement and plant availability. Cadmium is a rather soluble transition metal that behaves somewhat like $Ca^{2+}$ except that $Cd^{2+}$ reacts more strongly with organic matter and with sulfides. By absolute standards, however, Cd movement and plant availability in soils are small. Because of its extreme toxicity, $Cd^{2+}$ is a serious concern in soils used as waste-disposal sites. As might be surmised from the insolubility of $Hg(OH)_2$, Hg(II) is retained rather strongly by soil. Hg(II) can be reduced to Hg(I), $Eh = 0.92$ V, and to Hg(0), $Eh^\circ = 0.79 - 0.059 \log(Hg^+)/4$. The Hg(II) oxidation state probably predominates, but reduction to elemental Hg(0) occurs. Elemental Hg is slightly volatile and diffuses as a gas through soil pores. Compared to the other toxic metals in soils, Hg is relatively mobile. The toxic compound dimethyl mercury ($Hg(CH_3)_2$), formed in contaminated and highly reduced aquatic sediments, seems to be rare in soils.

Time decreases the availability of ions added to soils. Time allows ions to diffuse to the strongest sorptive sites, including incorporation into the surfaces of weathered soil particles. Time also leads to the aging of soil solids, with smaller and more reactive phases transforming into larger, less reactive, and less plant-available and water-soluble particles and organic matter. Leaching of the toxic cations and anions (except $NO_3^-$ and perhaps $CrO_4^-$) from soils is generally negligible. Toxic elements tend to remain within a few centimeters of where they first contact the soil unless the soil is stirred by cultivation. If retained at the immediate soil surface, they are above the most active portion of the root zone.

The chromate ($CrO_4^{2-}$) anion moves through well-aerated soils of moderate to high pH. Although chromate is a strong oxidizing agent and hence easily reduced at high concentrations, $CrO_4^{2-}$ stability increases at increasing pH and at the dilute ($10^{-6}$ M) Cr concentrations in soil solutions.

Despite concerns about Cd, Hg, and Cr in soils, the soil is still a far safer medium for the disposal of pollutant wastes than any other part of the environment. Soils are better able to oxidize (the oxidized state is generally the least toxic state), to retain pollutants safely, and to remove them from the food web, than air or water. For example, the increase in toxic metal concentration in plants is about one-half to one-tenth of the corresponding concentration increase in soils. Assuming a total dry matter crop production of 10 000 kg ha$^{-1}$ and a soil bulk density of 1.3, adding the high amount of 130 kg ha$^{-1}$ of a heavy metal in wastes to the surface meter of soil (a mass of 1 300 000 kg ha$^{-1}$ m$^{-1}$) increases the element's concentration in the first crop by about 1 mg kg$^{-1}$ or 10 g ha$^{-1}$, and that concentration and amount decreases with time. The amount in the harvested portion of the plant is much less than that in the total plant.

Municipal waste could be a valuable source of nitrogen, phosphate, and water for crops. Under careless management, however, waste application can lead to trace metal pollution of soils. Part of the answer is to prevent initial toxic metal contamination of municipal wastes by the relevant industries. Where such contamination is unavoidable, the safest procedure might be to dispose the wastes on forested land, land dedicated to waste disposal, or where the plants grown are far from a food web, and to insure that the wastes are actively mixed and diluted into the soil.

## 2.8   CARBON, NITROGEN, AND SULFUR

These three elements are grouped together because (1) their geochemical cycles are rapid and include the atmosphere, (2) they change oxidation states rapidly and cyclically in soils and the environment, (3) the changes in oxidation state provide the energy for life, and (4) C, N, and S processes in soils are closely interrelated. Photosynthesis of carbon and its oxidation back to $CO_2$ provides the energy that drives life and the N and S reactions. Nitrogen and sulfur oxidation–reduction also involve energy changes but the amounts are small in comparison and usually involve a net consumption of energy from carbon sources. The changes and turnover rates of C, N, and S compounds are sometimes of the order of hours and days. Turnover rate is the time required for one complete change of that substance from one phase to another. The other elements in the periodic table cycle annually between plants and soil and between soils and rocks in geologic time scales.

In humid soils, the C, N, and S are predominantly in organic compounds, resulting in the C/N/S mass ratio of about 100/10/1. With increasing aridity, the amount of soil organic matter decreases and the amounts of carbonate, sulfate, and nitrate anions in the soil solution tend to increase. In all soils, the N and S cycle through many soil/plant/soil cycles before some $N_2$ is formed and lost to the atmosphere. The $N_2$ later returns to the soil by nitrogen fixation by soil microorganisms, and as nitrate formed by lightning and washed out of the air by rain. The exchange of S between soil and the atmosphere is much slower. Although $H_2S$ may form in soils, it reacts rapidly with Fe and other transition metals in soils before being released to the atmosphere. The fallout of industrially formed sulfate and nitrate from the atmosphere as part of *acid rain* is a modern phenomenon. The nitrate in acid rain is more than that formed naturally in the atmosphere by lightning strokes.

Oxidation state changes cause great changes in the physical and chemical properties of elements. Some of the oxidation states produce volatile compounds ($CO_2$, $CH_4$, $N_2O$, $N_2$, $NH_3$, $SO_2$, $H_2O$, etc.), so that C, N, and S cycle between soils and the atmosphere. Similar exchanges occur between the atmosphere and the oceans, so the behavior of C, N, and S is more complex than that of many other elements. Their exchanges are also more sensitive to environmental changes. Table 1.3 shows current estimates of the worldwide distribution of C, N, S, and oxygen in the atmosphere, the surface hydrosphere (freshwaters and oceanwaters < 50 m depth), the biosphere (living organisms), and the surface meter of soil. The amount of C as soil carbonate is a crude guess; the other estimates are slightly better. The amounts of C and S in soil organic matter are almost as large as the sums of C and S in all other reservoirs combined. The largest fraction of N, on the other hand, is in the atmosphere.

The annual exchanges of C, N, and S between the active parts of the environment (Table 1.4) are probably more meaningful than the total amounts in Table 1.3 but are also more difficult to estimate. The only rates known with any certainty are the rates of fossil fuel combustion and fertilizer denitrification. The other rates in Table 1.4 may be in error by as much as an order of magnitude. The environmental effects of human alterations of the natural carbon, nitrogen, and sulfur cycles are little known. Climate changes due to soil emissions of $CO_2$, $N_2O$, and $CH_4$ to the atmosphere are being hotly debated. Since water in the atmosphere has a much greater effect on atmospheric temperature than the other gases, the soil's effect on plant growth and the plant's great effect on water balance may be as important as the gas emissions. Soils buffer the composition of the C, N, and S gases in the atmosphere. Soil organic matter is the largest carbon reservoir in the environment and its amount changes inversely with the $CO_2$ concentration in the air, increases with rainfall, and decreases with increasing temperature. The accumulation of carbon as peat and soil organic matter in northern Canada has come at the expense of atmospheric $CO_2$ since the retreat of the glaciers in that area 12 000 years ago. The full extent of soil effectiveness in buffering $CO_2$ is poorly understood.

The oxidation–reduction reactions of C, N, S, and O in soils are, virtually without exception, catalyzed by microbial enzymes. These reaction rates without such catalysis are very slow (irreversible). Even with catalysis, the reactions are quite irreversible. The behavior of carbon and nitrogen in particular is dominated by nonequilibrium.

### 2.8.1   Nitrogen

In economic and agricultural terms, nitrogen is the most important element in fertilization and in optimizing plant yield per unit area. Water availability is more important to plant growth but is outside the scope of this book. Nitrogen in soil has been studied for centuries and is still the most-studied element in soil chemistry, microbiology, and fertility. It is the soil element that most commonly limits plant growth. Nitrogen reactions are treated in detail in texts on soil fertility and soil microbiology and in many review articles.

Nitrogen chemistry in soils is the changes of nitrogen during organic reactions, oxidation of organic nitrogen to $N_2$ and $N_2O$ (*denitrification*) or to $NO_3^-$ (*mineralization*), and the reduction of $N_2$ to amino acids (*nitrogen fixation*). All of these steps require microbial catalysis, but are nonetheless irreversible, and N fixation requires energy considerably in excess of the amount needed for a thermodynamically reversible process, that is, microbial N fixation is energy-inefficient. The thermodynamically stable states of nitrogen under soil conditions are shown in Chapter 4. Nitrate is stable only under strongly oxidizing conditions, and amino N, under strongly reducing conditions. $N_2$ is the dominant species in the environment, but the rates and mechanisms of nitrogen reactions inspire much more interest than the stable states.

Although many microorganisms are apparently capable of denitrification, only a few specialized species, including the free-living *Azotobacter*, blue-green algae, some anaerobic bacteria, and *Rhizobium* bacteria in root nodules of legumes, are capable of nitrogen fixation. Energy for nitrogen fixation is provided by photosynthesis. A goal of plant research is to enlarge the number of plants that can support rhizobial nitrogen fixation, so N fertilization would no longer be necessary. If attainable, this N fertilization would not be free. Crop yields would decrease by about 25% compared to N-fertilized plants, because energy would be diverted away from growth to reduce $N_2$.

The dependence of nitrogen transformations on photosynthetic energy gives rise to a C/N mole ratio of typically about 12, or mass ratio of 10, in soils. When the natural ratio is changed by nitrogen fertilization, plant uptake of nitrogen, or addition of low-nitrogen organic matter such as straw and wood, soil microorganisms restore the balance by carbon oxidation, nitrogen fixation, or denitrification over the course of several weeks to months. Soil microbes resist any long-term changes of the soil's natural N status.

Soil nitrogen contents fluctuate with soil and fertilizer management, including growth of leguminous versus nonleguminous crops. The production and use of nitrogen fertilizers is a substantial portion of the nitrogen cycle in agricultural areas. The 200-Tg $yr^{-1}$ rate of terrestrial nitrogen fixation corresponds to about 20 kg $ha^{-1}$ $yr^{-1}$ over the earth's land area of $1.2 \times 10^8$ $km^2$. The 30-Tg $yr^{-1}$ production of nitrogen fertilizers is applied to perhaps one-third of the 11% of the land area that is cultivated. The average nitrogen input to the fertilized soils is therefore about 75 kg $ha^{-1}$ N. Roughly half of the fertilizer N is recovered by plants. The remainder is denitrified or leached from the soil. Fertilization tends to increase the $N_2O/N_2$ ratio of the gases produced during denitrification.

Nitrogen is leached from soils as nitrate; $NH_4^+$ is retained by cation exchange but is oxidized eventually to nitrate if not absorbed by plants and microbes. Nitrate leaching is a potential pollution hazard to surface waters and groundwaters. Denitrification and avid plant and microbial uptake of nitrogen tend to minimize the $NO_3^-$ concentrations of soil solutions. Soil solution extracts from fertilized soils contain as much as 20 to 40 mg $L^{-1}$ of $NO_3$—N (nitrogen in the form of $NO_3^-$). The nitrate concentration is often much less below the root zone. The current $NO_3^-$ limit for drinking water in the United States is 45 mg $L^{-1}$ $NO_3^-$, or 10 mg $L^{-1}$ $NO_3$—N. Higher concentrations can lead to nitrite poisoning (methemoglobinemia or "blue baby" syndrome) in infants.

High $NO_3^-$ concentrations in groundwaters generally result from the following:

1. *Overfertilization with* $NH_4^+$, $NO_3^-$, *and urea fertilizers.* In well-aerated soils, with adequate moisture and moderate temperature, $NH_4^+$ and urea are converted to $NO_3^-$ in a matter of weeks. Groundwater pollution from overfertilization is most pronounced when early-season or over-winter leaching moves fertilizer nitrogen below the root zone of young or seasonally inactive plants. Some U. S. states have enacted "Best Management Practice" laws for agriculture, which include provisions to minimize overfertilization.

2. *High organic matter inputs to soils, including waste disposal operations.* Sewage sludge, manure disposal, and irrigation with wastewaters can load

the soil with more organic nitrogen than can be utilized by plants. When oxidized, these materials release $CO_2$, $N_2$, and $N_2O$ to the atmosphere and nitrate to the soil solution. At moderate loading rates, the $NO_3^-$ can be utilized by plants and soil microorganisms, which reduces its concentration to acceptable limits before leaching occurs. At higher loading rates, however, careful water management is necessary to avoid $NO_3^-$ pollution of water.

3. *Irrigating arid soils with waters containing natural nitrates.* Nitrate accumulation normally is relatively small compared to the accumulation of other salts, but may still be sufficient to raise the nitrate concentrations of groundwaters and drainage waters to unacceptable levels. If the downward migration of nitrate is slow, the nitrate needs many years before it reaches the groundwater table in these regions. That may be sufficient time for some of the nitrate to be removed by microbes in the *vadose* zone (the soil zone between the root zone and the groundwater table).

The complete mechanism of soil nitrogen redox changes is unknown despite much study. The overall oxidative reactions are

$$\text{Organic-N and } NH_3 = N_2 \text{ and } N_2O \tag{2.5}$$

or

$$\text{Organic-N and } NH_3 = NO_3^- \tag{2.6}$$

The amount of $N_2$ and $N_2O$ lost during oxidation under natural conditions is usually small. These gases are lost in greater amounts when nitrogen fertilizers are added to soils, and the fraction lost as $N_2O$ increases greatly.

The reductive pathways for nitrogen include $NO_3^-$ and $N_2$ reduction to amino acids in plants and in microbes. Soil microorganisms reduce nitrogen stepwise:

$$NO_3^- = NO_2^- = N_2 \text{ or } N_2O \tag{2.7}$$

or

$$NO_3^- = NO_2^- = \text{organic-N or } NH_3 \tag{2.8}$$

Unknown intermediate steps exist between the nitrite ($NO_2^-$), $N_2$, and $NH_3$ forms. The fraction denitrified and the $N_2O/N_2$ ratio increase with increasing amounts of initial $NO_3^-$. Nitrate reduction is the main pathway of denitrification, $N_2$ and $N_2O$ production, in soils.

The amounts of $NH_4^+$ and $NO_3^-$ in soils are small compared to the amounts of organic N. Because soil nitrogen contents tend toward steady states, the concentrations of $NO_3^-$ and $NH_4^+$ in the soil solution are rough indicators of nitrogen availability to plants.

Another portion of the nitrogen cycle involves the $NH_3$ that is liberated from soils and the oceans and is then reabsorbed. Current estimates are that 100 to 250 Tg yr$^{-1}$ of nitrogen are lost from the world's soils as $NH_3$, that 30 to 50 Tg yr$^{-1}$ of $NH_3$ enter soils in rainwater, and 600 to 1300 Tg yr$^{-1}$ of $NH_3$ enter the soil–plant system by direct absorption of the gas from the atmosphere. Considerable ammonia is released from animal feedlots and manure.

Ammonia concentrations in well-aerated soil solutions are normally $<10^{-6}$ M. The $NH_4^+$ ion is almost identical in size to $K^+$, and can be held between 2:1 layer silicate lattices ("fixed") in the same way as $K^+$. In recently fertilized soils, anaerobic soils and sediments, soils used for municipal and agricultural waste disposal, and forest soils of low pH, exchangeable $NH_4^+$ levels can be a significant portion of the soil's inorganic nitrogen. Relative to other exchangeable cations, however, the amount of exchangeable $NH_4^+$ is small. Ammonium ions are less mobile than $NO_3^-$ and are less likely to be lost through denitrification, although ammonia readily volatilizes from the surface of alkaline soils. Nitrification ($NH_4^+$ to $NO_3^-$) inhibitors are commercially available. They are being investigated as a means of maintaining fertilizer nitrogen in the $NH_4^+$ form, thus retarding denitrification and leaching losses.

## 2.8.2  Sulfur

Although sulfur is less volatile than carbon or nitrogen, organic decay in swampy soils may release a little sulfur as methyl mercaptan ($CH_3SH$), dimethyl sulfide ($CH_3$—S—$CH_3$), dimethyl disulfide ($CH_3$—S—S—$CH_3$), and hydrogen sulfide ($H_2S$), but the release is rare in aerobic soils. The organic sulfur gases are the distinctive odor from paper mills. Within the soil, these gases oxidize, or the sulfide reacts strongly with transition metal ions in soils and precipitates as $FeS_2$, MnS, and other transition metal sulfides rather than escaping to the atmosphere. If soil emissions of $H_2S$ and the organic sulfide gases were substantial, their strong odors would be noticeable. Sulfur emission worldwide from anaerobic soils and swamps has been estimated as 30 Tg yr$^{-1}$.

Sulfate is the stable sulfur oxidation state in aerobic soils, and sulfide is stable in anaerobic soils. Sulfur changes its oxidation state by microbial catalysis and the changes seem to be much more reversible than nitrogen and carbon reactions. Elemental sulfur is rare naturally in soils but is sometimes added to soils as an amendment, and sulfides are common in many mining wastes. When elemental sulfur and sulfides are exposed to oxygen, they oxidize to $H_2SO_4$. Soil acidities as high as pH 2 may persist until the sulfide or sulfur has all been oxidized and leached away.

Major sulfur inputs to soils include atmospheric $SO_2$ and its various oxidation products from coal combustion (100 Tg yr$^{-1}$ sulfur), petroleum processing (30 Tg yr$^{-1}$), ore smelting (15 Tg yr$^{-1}$), and sulfate from sea spray (20 Tg yr$^{-1}$). Most of the atmosphere's sulfur falls near the areas where it is produced. The sulfur fallout over West Germany, for example, is as high as 50 kg ha$^{-1}$ yr$^{-1}$ in nonforested areas and 80 kg ha$^{-1}$ yr$^{-1}$ in forests. The fallout occurs both as acid rain and as direct plant absorption; direct soil absorption of atmospheric sulfur is minor in humid and temperate regions. In arid regions $SO_2$ and its oxidation products, $H_2SO_3$ and $H_2SO_4$, are absorbed directly and rapidly by the basic soils and their dust. The acidity from the large amounts of S formerly emitted by copper smelters in arid Southern Arizona was neutralized in the atmosphere within 50–100 km of the sources. The largest emitter (2000 tons $H_2SO_4$/day for 70 years) acidified nearby surface soils to a depth of 10 mm.

Atmospheric sulfur from smokestacks can be carried hundreds of kilometers over flat terrain. In regions of sulfur-deficient soils, atmospheric sulfur at low concentrations can benefit plants. Benefits from the low concentration, however, must be weighed against the associated acidification of freshwater, phytotoxicity, health hazards, smog, and building deterioration at the higher concentrations near the sources.

Sulfate anions are retained only weakly by soils, but the retention increases with soil acidity. Sulfate anions are absorbed readily by plants and incorporated into biomass. Hence, biomass and SOM constitute large sulfur reservoirs at the earth's surface. The C/S mass ratio in soil organic matter is typically about 100/1. The sulfate content of soils increases with aridity and with salt accumulation.

Widespread sulfur deficiencies in many agricultural soils became more obvious in recent decades as "treble superphosphate" began to supplant "superphosphate" as a phosphate fertilizer. Superphosphate is made with $H_2SO_4$ and contains about 50 mol % sulfate. This inadvertent sulfur fertilization ended when higher purity treble superphosphate, made with $H_3PO_4$, was substituted. The "improved" treble superphosphate sometimes produced lower crop yields than did superphosphate, until the sulfur deficiency was recognized and corrected.

## BIBLIOGRAPHY

Black, C. A. 1968. *Soil–Plant Relationships*, 2d ed. Wiley, New York.

Mortvedt, J. M., P. Giordano, and W. L. Lindsay (eds.). 1972. *Micronutrients in Agriculture*. American Society of Agronomy, Madison, WI.

Russell, E. W. 1973. *Soil Conditions and Plant Growth*, 10th ed. Longmans, London.

## QUESTIONS AND PROBLEMS

1.  Why are total soil contents poor indicators of the amounts of ions that may enter the food chain? By what mechanisms are ions held by soils?

2.  What are the major exchangeable cations in soils? How do the relative proportions change between acid and basic soils? What are the rates of exchange for the various cations? What soil conditions are typical deficiency, sufficiency, and toxicity (if any) for these ions?

3.  Discuss the soil's role in the cycling of a given element in the food chain.

4.  How are Fe(III) and Al(III) reactions both different and similar in soils? Which transition metals are essential to

    (a) Plants?

    (b) Animals?

Which are only toxic, so far as is presently known? Which can be both essential and toxic?

**5.** What are the forms of the micronutrients in soil solutions? How does their availability change with pH?

**6.** For a given chemical element, trace the chemical states through which it might pass from an igneous mineral through weathering and back to igneous mineral. Discuss both solution and solid states and mention the locations (rock, various parts of the soil, marine water vs. sediment, etc.) in which these states exist.

**7.** Explain why $CaCO_3$ tends to accumulate at some depth in semiarid and arid region soils, while more soluble salts tend to be in deeper horizons or absent. Why is the $CaCO_3$ layer thought to represent the average depth of wetting?

**8.** If $K^+$ is released both from "fixed" forms and by weathering during the growing season, why is exchangeable $K^+$, as determined by a single extraction, well correlated with the soil's K-supplying power?

**9.** Why is $F^-$ a common constituent of igneous rocks?

**10.** Explain in your own words why the "lithosphere" and "soil" values of Table 2.2 often differ so markedly (different elements may require different answers).

**11.** Explain how and why plants tend to be much more tolerant of high trace-metal concentrations than are animals.

**12.** Why are crop sulfur deficiencies more common now than during the 1950s?

# 3

# WATER AND SOLUTIONS

Soil chemistry centers around the soil solution, the aqueous solution that reacts with, and is between, the solid phase and roots and soil microbes. A *solution* is de"ned as a mixture and usually means an *aqueous solution*, a mixture of solutes and water. Soil and rock minerals can be considered *solid solutions*, because they contain foreign ions that are mixed into the mineral's structure. Soil chemistry involves both aqueous and solid solutions. This chapter presents the chemistry of ions reacting with $H^+$, $OH^-$, $O^{2-}$, and $H_2O$ in the soil solution and soil minerals.

The state of ions in aqueous solution is similar to their states in soil minerals and solid solutions because both water and soil minerals are dominated by the $O^{2-}$ ion. Water is an oxide whose charge is countered by $H^+$. In minerals the $O^{2-}$ charge is countered by many other cations. Water reacts somewhat like $O^{2-}$ with cations, but much more weakly. The solute composition of the soil solution is the result of competition of the oxide groups in soil solids with $H^+$, $OH^-$, and $H_2O$ in the soil solution.

Water molecules also interact with each other. The interaction is indicated by water's relatively high boiling point and speci"c heat (the amount of energy needed to raise its temperature). The $H_2O$ molecule as written looks similar to $H_2S$, and water should boil at a lower temperature because oxygen has a lower atomic weight than sulfur. However, $H_2S$ boils at a much lower temperature, $-61°$ C, than water at $0°$ C, because the $H_2S$ molecules are much more independent and interact very weakly. The $H^+$ ions in $H_2S$ are $180°$ apart, creating a linear, *nonpolar* molecule. The $H^+$ ions in $H_2O$ are $105°$ apart, creating a nonlinear, *polar* molecule. This is a *dipole* with positive (the hydrogen side) end and a negative (the side opposite the H ions) end. Technically, water is a quadropole but that sophistication is unnecessary here. The positive end of one water molecule attracts the negative end of another water molecule. This electrostatic attraction is absent in the nonpolar $H_2S$ molecule, which has no positive and negative ends. Water molecules cluster together in groups averaging about six in number at room temperature. Because of thermal motion, the groups continually break apart and reform. In ice the structure is a rigid, open hexagonal packing of water molecules. In water, the small groups are thought to have a similar structure, but each group is not oriented to the next group. The groups can slide closer together so water is denser than ice. Water is like an ice slurry on a molecular scale.

Ions and charged surfaces can break down the "ice slurry" structure of water. The electric charges are stronger than dipole forces and tend to pull water molecules away from their groups by attracting the positive or negative ends of the water dipoles. Solutes and water molecules are constantly in motion, but they remain in the vicinity of each other for some period of time. If water molecules remain near an ion longer than the time required for the water molecules to dissociate from the water structure, the ion will have a sphere of water molecules (a *solvation sphere* or *sheath*) around itself. The number of water molecules in the closest solvation sphere is called the *primary hydration number*.

The value of the primary hydration number is in dispute. Different methods of measurement yield quite different values because they respond to different strengths and times of ion–water interaction. Careful measurements of the hydration number of $Na^+$, for example, yielded values of 1, 2, 2.5, 4.5, 6 to 7, 16.9, 44.5, and 71, depending on the method used. Table 3.1 shows hydration numbers for common ions determined by several methods that tend to agree. The last column shows Bockris and Reddy's estimates of primary hydration numbers for the univalent ions.

Outside of the primary solvation sphere is a second sphere of water molecules also affected by the ion's charge. These water molecules have also been torn away from their water structure to some extent, but are not so closely associated with the ion. The orientation of water molecules in the primary solvation sphere and the more random orientation of water molecules in the secondary sphere essentially dissipate

**Table 3.1. Primary hydration numbers**

|         | From ion compressibility | From thermodynamic calculations | Most probable value |
|---------|--------------------------|---------------------------------|---------------------|
| $Li^+$  | 5–6                      | 5                               | $5 \pm 1$           |
| $Na^+$  | 6–7                      | 4                               | $4 \pm 1$           |
| $K^+$   | 6–7                      | 3                               | $3 \pm 2$           |
| $F^-$   | 2                        | 5                               | $4 \pm 1$           |
| $Cl^-$  | 0–1                      | 3                               | $2 \pm 1$           |
| $Br^-$  | 0                        | 2                               | $2 \pm 1$           |
| $I^-$   | 0                        | 1                               | $1 \pm 1$           |

**Table 3.2. Crystallographic radii and heats and entropies of ion hydration at 25° C**

| Ion | Crystallo-graphic Ion Radius (nm) | Heat of Hydration, $\Delta H$ (kJ mol$^{-1}$) | Entropy of Hydration, $\Delta S$ (J mol$^{-1}$ K$^{-1}$) |
|---|---|---|---|
| $H^+$ | — | $-1090$ | 109 |
| $Li^+$ | 0.060 | $-506$ | 117 |
| $Na^+$ | 0.095 | $-397$ | 87.4 |
| $K^+$ | 0.133 | $-314$ | 51.9 |
| $Rb^+$ | 0.148 | $-289$ | 40.2 |
| $Cs^+$ | 0.169 | $-255$ | 36.8 |
| $Be^{2+}$ | 0.031 | $-2470$ | — |
| $Mg^{2+}$ | 0.065 | $-1910$ | 268 |
| $Ca^{2+}$ | 0.099 | $-1580$ | 209 |
| $Ba^{2+}$ | 0.135 | $-1290$ | 159 |
| $Mn^{2+}$ | 0.080 | $-1830$ | 243 |
| $Fe^{2+}$ | 0.076 | $-1910$ | 272 |
| $Cd^{2+}$ | 0.097 | $-1790$ | 230 |
| $Hg^{2+}$ | 0.110 | $-1780$ | 180 |
| $Pb^{2+}$ | 0.120 | $-1460$ | 155 |
| $Al^{3+}$ | 0.050 | $-4640$ | 464 |
| $Fe^{3+}$ | 0.064 | $-4360$ | 460 |
| $La^{3+}$ | 0.115 | $-3260$ | 368 |
| $F^-$ | 0.136 | $-506$ | 151 |
| $Cl^-$ | 0.181 | $-377$ | 98.3 |
| $Br^-$ | 0.195 | $-343$ | 82.8 |
| $I^-$ | 0.216 | $-297$ | 59.8 |
| $S^{2-}$ | 0.184 | $-1380$ | 130 |

the ion's charge within 1 to 2 nm of the central ion. This charge dissipation allows the ions to interact less electrostatically with each other.

The degree of breakdown of the water structure, plus the degree of formation of new ion– or solute–water structures, is measured by the heat of hydration (Table 3.2). Heats of hydration are less ambiguous than structural concepts of hydration numbers and ion–, solute–, or clay–water interactions.

The heat of hydration increases negatively, indicating increased interaction, with decreasing ion size for a given valence group, from $-506$ kJ mol$^{-1}$ for $Li^+$, atomic number 7, to $-255$ kJ mol$^{-1}$ for $Cs^+$, atomic number 133. The heat of hydration also increases negatively with increasing ionic charge. The heat of hydration for monovalent cations is several hundred kJ mol$^{-1}$; for divalent cations, it is about $-1600$ kJ mol$^{-1}$; and for trivalent cations, it is about $-4000$ kJ mol$^{-1}$.

The highly negative heats of hydration for the trivalent ions support the idea that they exist as $Fe(H_2O)_6^{3+}$ and $Al(H_2O)_6^{3+}$ (hexaquoiron(III) and hexaquoaluminium) ions in water. The ion and water molecules in such complexes are tightly bound and together behave like a single large ion. Divalent transition metal ions and $Mg^{2+}$ form similar but weaker structures with water. The other alkaline earth cations are larger and associate less strongly with water.

Soil particle surfaces are also charged so they attract ions and water dipoles. The charge can be an atomic layer beneath the crystal surface so the interaction between soil particles and water is weaker than between ions and water. At the edges of crystals, the charge is weaker but at the surface.

## 3.1 ACIDS AND BASES

Substances that liberate $H^+$ in solution are *Bronsted acids*. Strong acids like HCl, $HNO_3$, $HClO_4$, and $H_2SO_4$ completely dissociate into $H^+$ (which probably exists as the hydrated or *hydronium* ions $H_3O^+$ or $H_7O_3^+$) and an anion. Weak acids deprotonate less readily. Examples are $Al^{3+}$ ($pK_1 = 5$), HF ($pK = 3.2$), and organic acids such as acetic acid, $CH_3COOH$ ($pK = 5$), where $K$ is the dissociation constant of the acids. Their more tightly bound protons dissociate at higher pH than strong acids. For the generalized weak acid HA,

$$HA = H^+ + A^- \tag{3.1}$$

and its dissociation, deprotonation, or acidity constant is

$$K_{HA} = \frac{(H^+)(A^-)}{HA} \tag{3.2}$$

Just as acids differ in strength, bases similarly differ in their ability to give up hydroxyl ions. The strong bases NaOH and KOH dissociate completely in solution. Bases such as $Ca(OH)_2$ and $Mg(OH)_2$ are only slightly less dissociated in solution, but are considerably less soluble than either NaOH or KOH. This lower solubility accounts for the lower pH of $Ca(OH)_2$ and $Mg(OH)_2$ solutions more accurately than does weak dissociation.

Weak bases are more precisely compounds like $NH_4OH$ ($pK_b = 5$), where $pK_b$ is the negative log of the equilibrium constant:

$$K_{NH_4OH} = \frac{(NH_4^+)(OH^-)}{NH_4OH} = 10^{-5} \tag{3.3}$$

The $NH_4OH$ molecule is the hydrate of $NH_3$, which is very soluble in water. Most of the solute remains in solution as $NH_3$, but a small portion forms $NH_4OH$, which dissociates into $NH_4^+$ and $OH^-$. The denominator of Eq. 3.3 is really the sum of the dissolved $NH_3$ plus $NH_4OH$. Dissolved $CO_2$ and $H_2CO_3$ are identical with ammonia in this respect.

Because the pH change during titration of acid soils with $OH^-$ resembles the titration curves of weak acids with $OH^-$, soil clays were called weak acids for many years. For two main reasons, however, weak acid inadequately describes soil acidity.

Firstly, soil acidity changes slowly with time after liming, for example, while true weak acids react instantly with $OH^-$. Freshly prepared acid clays are strongly acidic but begin to decompose to weak acids within a few hours. The clay decomposition liberates $Al^{3+}$. The base added to the decomposed acid clay actually titrates $Al^{3+}$, which has a titration curve resembling that of a weak acid. Aluminium hydrolysis and clay decomposition are slow, however, so the suspension pH changes slowly with time. Secondly, the amount of soil acidity released depends on the salt concentration of the bulk solution, but salt concentration has little effect on true weak acids. The term weak acid is better restricted to those compounds whose acidity is independent of time and salt concentration, such as acetic acid and phosphoric acid. Soil organic matter satisfies the definition of a weak acid much better than do inorganic soil clays.

### 3.1.1  Hydrolysis and Deprotonation

The attraction of cations for water molecules is so strong that the cation's charge tends to repel hydrogen ions, or protons, from the water *ligands* in the solvation sphere and makes them Bronsted acids. As the solution is made more alkaline, more $H^+$ tends to dissociate. An example is phosphoric acid molecule formed by $P^{5+}$ in solution, $PO(OH)_3$, written more familiarly as $H_3PO_4$. This molecule is stable in acid solutions. The +5 charge has already repelled 5 H ions from the four water molecules that originally surrounded $P^{5+}$. The first $H^+$ ion from $H_3PO_4$ is repelled by $P^{5+}$ at about pH 3 to form $H_2PO_4^-$, the second at pH 7, and the last $H^+$ at pH 10.

The $S^{6+}$ in sulfuric acid ($SO_2(OH)_2$) and the $N^{5+}$ in nitric acid ($NO_2OH$) are stronger acids, and the $H^+$ from these strong acids is repelled even in extremely acid solutions. The remnants of the water molecules in $PO_4^{3-}$, $SO_4^{2-}$, and $NO_3^-$ are oxide ($O^{2-}$) ligands that are almost impossible to strip away from the central ion, unless the central ion is first reduced to a lower oxidation state.

A similar repulsion of $H^+$ occurs from the solvation sphere of hexaquoiron(III) ions,

$$Fe(H_2O_6)^{3+} = Fe(H_2O)_5^{2+} + H^+ \tag{3.4}$$

This reaction is called *hydrolysis*, because it splits a water molecule, or *deprotonation*. The hydrolysis constant, ignoring the water ligands, is

$$K_H = \frac{(H^+)(FeOH^{2+})}{Fe^{3+}} = 10^{-2.2} \tag{3.5}$$

The $Fe(H_2O)_6^{3+}$ ion is a weaker acid than sulfuric or nitric acid, but is a stronger acid than phosphoric acid. Table 3.3 gives the hydrolysis constants of several common cations. The smaller the hydrolysis constant (the more negative its exponent), the weaker the acid.

The hydrolysis of iron(III) continues progressively at still higher pH, giving

$$FeOH(H_2O)_5^{2+} = Fe(OH)_2(H_2O)_4^+ + H^+ \tag{3.6}$$

**Table 3.3. Solubility products and hydrolysis constants of metal ions[a]**

| Ion | $\log K_{sp}$ | $\log K_1{}^{b}$ | $\log K_2$ |
|---|---|---|---|
| $Be^{2+}$ | −21 | −6.5 | |
| $Mg^{2+}$ | −10.8 | −12 | |
| $Ca^{2+}$ | −5.0 | −12.5 | |
| $Mn^{2+}$ | −12.5 | −10.5 | |
| $Fe^{2+}$ | −14.8 | −7 | |
| $Ni^{2+}$ | −15 | −8 | |
| $Cu^{2+}$ | −19.5 | −7.5 | |
| $Zn^{2+}$ | −17 | −9.1 | |
| $Cd^{2+}$ | −14 | −10 | |
| $Hg^{2+}$ | −25.5 | −3.5 | |
| $Pb^{2+}$ | −18 | −8 | |
| $Al^{3+}$ | −33.5 | −5 | −5.5 |
| $Fe^{3+}$ | −39 | −2.9 | −3.3 |
| $La^{3+}$ | −20 | −9 | |
| $Ti^{4+}((TiO_2)(OH)^2)$ | −29 | $> -1$ | |
| $Th^{4+}$ | −44 | −4.1 | |

[a] From L. G. Sillen and A. E. Martell. 1974. Stability constants. *The Chemical Society Spec. Publ.* **25**, London.
[b] $K_1 = \dfrac{(MOH^{(n-1)+})(H^+)}{(M^{n+})}$

and

$$Fe(OH)_2(H_2O)_4^- = Fe(OH)_3(H_2O)_3 + H^+ \tag{3.7}$$

The uncharged $Fe(OH)_3(H_2O)_3$ species does not repel other iron(III)–water ions and is therefore less water soluble than the charged ions. It can lose four water molecules and precipitate as FeOOH. FeOOH is more stable than $Fe(OH)_3$. These hydrolysis reactions of iron and especially of aluminium are primary factors in the production and control of soil acidity.

The hydrolysis in Eqs. 3.4, 3.6, and 3.7 oversimplifies the actual mechanism of hydrolysis. The hydrolyzed ions begin to associate and form polynuclear ions, which enlarge with further hydrolysis. These polynuclear ions form in solutions and may exist in natural waters for a long time before they precipitate as a solid. Whether they exist in soil solutions, with the soil's large and adsorptive surface area and at the low concentrations in soil solutions, is less certain. When ions dissolve from solids, on the other hand, the ions formed most likely remain as mononuclear species because their concentrations are so low.

The hydroxyoxides are capable of still further hydrolysis if the pH becomes more alkaline. Iron hydroxide, for example, can dissolve according to the reaction

$$Fe(OH)_3(H_2O)_3 = Fe(OH)_4(H_2O)^- + H^+ \tag{3.8}$$

or

$$Fe(OH)_3(H_2O)_3 + OH^- = Fe(OH)_4(H_2O)^- \tag{3.9}$$

Equations 3.8 and 3.9 are equivalent, because producing 1 mol of $H^+$ is the same as consuming 1 mol of $OH^-$. The equilibrium constant is

$$K_h = \frac{Fe(OH)_4^-}{OH^-} = 10^{-5.5} \tag{3.10}$$

The $Fe(OH)_4^-$ ion, because it is charged, is more soluble than FeOOH. Hence, the solubility of Fe(III) increases at alkalinities above pH 8.5. Figure 3.1 shows the effect of pH on the distribution of aquohydroxyiron(III) species in solution.

The hydrolysis reactions of $Al^{3+}$ are analogous to those of $Fe^{3+}$, but $Al^{3+}$ is less acidic. The loss of the first $H^+$ from $Al(H_2O)_6^{3+}$ occurs at pH 5. The hydrolysis of the second and third protons occurs at slightly higher pH and is complicated by polymerization of the hydrolysis products and slow precipitation of $Al(OH)_3$. Figure 3.1

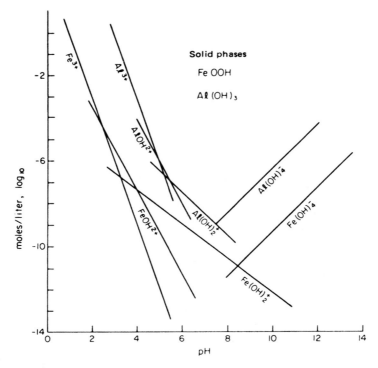

**FIGURE 3.1.** Equilibrium solubility of monomeric Fe(III) and Al(III) ions from FeOOH and $Al(OH)_3$ as a function of pH.

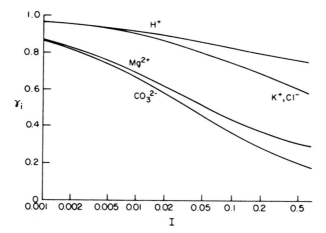

**FIGURE 3.2.** Ion activity coefficients versus ionic strength *I*. Calculated from the extended Debye–Hueckel equation.

disregards these complications. The $Al(OH)_4^-$ begins to form above pH 8, but the exact value is uncertain.

High solubility in acidic and basic solutions, and low solubility at neutral pH, as shown in Figures 3.1 and 3.2, is called *amphoterism.* Amphoteric ions can form positive or negative hydroxy complexes depending on solution pH. This allows hydroxyoxides to be both acidic and basic. $Al(OH)_3$ consumes $H^+$ when dissolving in acid solutions, thereby acting like a base. By consuming $OH^-$ and forming $Al(OH)_4^-$, aluminium hydroxide acts like an acid. This amphoteric behavior is important to charge development by some soil clays.

The hydrolysis illustrated for $Fe^{3+}$ and $Al^{3+}$ can happen to all cations, but the extent of hydrolysis varies widely. Hydrated alkali and alkaline earth cations should deprotonate at extremely high pH, but sufficiently high pH values do not occur in nature. In typical soil solutions of pH 5 to 9, P(V) exists as $H_2PO_4^-$ and $HPO_4^-$ in aqueous solutions, Fe(III) exists as $Fe(OH)_2^+$, and Ti(IV) probably exists as $TiO(OH)_2^{2-}$. Si(IV) in aqueous solution is $Si(OH)_4$, also written $H_4SiO_4$; it loses its first $H^+$ only above pH 9. The pH values at which $H^+$ ions are lost from the solvation sheath are indicated by the negative logarithms of the hydrolysis constant, p$K$, and are shown in Table 3.3; p$K_h < 7$ indicates that the cations are acidic.

### 3.1.2  Solubility Products

Table 3.3 gives the solubility products of the metal hydroxyoxides that are likely to exist in soils. The general form of the solubility products is

$$K_{sp} = (M^{n+})(OH^-)^n \tag{3.11}$$

A solid also becomes less water soluble (the solubility product decreases) when its crystals are purer, their structure is more ordered, their size increases, and as the crystals contain less water (less hydrated), but these effects are secondary to the solubility product. The solubility product decreases as crystals grow in size and lose waters of hydration and occluded or coprecipitated ions. The slow growth and recrystallization is much more pronounced in the mixture of ions in soil solutions than in the pure aqueous solutions of the chemistry laboratory. The solid-phase reactions are often exceedingly slow in soils compared to the formation rates of new, poorly crystalline material. Hence, soil-formed crystals tend to be small and amorphous and to contain many impurities.

Solubility products measured in pure systems may not represent soil conditions very well. The impurities in solids affect their aqueous solubility; soil minerals are characteristically impure. Nonetheless, predictions of soil solution concentrations usually assume the solubility products of pure minerals apply. The water molecule is ignored in stability constant and solubility product equations. The concentration of water is assumed to be unity because water is present in great excess and does not change significantly during the reaction. This assumption is good in all but the most concentrated aqueous solutions and in dry soils.

## 3.2 CHEMICAL ACTIVITY

Ions in water are not free and unattached. They interact with water and with each other. Close-range ($<0.5$ nm) ion–ion interactions are termed *complex ions* or *ion pairs* and are governed by specific interactions between ions. These close-range interactions are discussed later. Longer-range ($>0.5$ nm) interactions are treated by the concept of chemical activity.

The interactions of ions with water molecules and other ions affect the concentration-dependent (*colligative*) properties of solutions. Colligative properties include osmotic pressure, boiling point elevation, freezing point depression, and the *chemical potential*, or *activity*, of the water and the ions. The activity is the driving force of reactions. Colligative properties and activities of solutions vary nonlinearly with concentration in the real world of nonideal solutions.

Solutes can be thought of as ideal by considering their activities rather than their concentrations. The activity is defined as

$$\alpha = \gamma M \tag{3.12}$$

where $\alpha$ is the activity, $\gamma$ is the *activity coefficient*, and $M$ is molarity. The activity can be regarded as an ideal concentration, but is more correctly defined as a ratio related to concentration and is unitless. The units of $\gamma$ are then $L \, mol^{-1}$.

The colligative properties of an *ideal solution* are equal to the concentrations of the components, and their activity coefficients equal one. The deviation of the activity coefficient from one expresses the degree of nonideality. Figure 3.2 shows the change in the aqueous activity coefficients of several ions over the concentrations found in soil solutions and groundwater.

The activity coefficient is defined so that

$$\lim_{M \to 0} \gamma = 1 \qquad (3.13)$$

As the concentration approaches zero, $\gamma$ approaches one. As the solute is diluted by the solvent and the solute ions or molecules are farther apart, they interact less with each other and behave more ideally.

Water and ions are affected by the amounts and charges of all of the ions in the solution; the *ionic strength* $I$ combines the effects of concentration and ion charge

$$I = \frac{1}{2} \sum M_i Z_i^2 \qquad (3.14)$$

where $M_i$ is the molarity, and $Z_i$ is the charge, of each ion i. The ionic strength estimates the effective ion concentration by taking into account the large effect of ion charge on solution properties. A solution has only one ionic strength, but each ion may have a different activity coefficient (Fig. 3.2).

The properties of individual ions cannot be measured; we can only accurately measure the properties of salts and from them estimate the properties of ions. From a fundamental or thermodynamic standpoint, ions, therefore, do not exist, but we accept that they exist. We estimate the activities of ions by arbitrarily dividing up the activities of their salts. The activity of $CaCl_2$ in a $CaCl_2$ solution, for example, is

$$a_{CaCl_2} = \gamma_{Ca} \cdot M_{Ca^{2+}} (\gamma_{Cl} \cdot M_{Cl^-})^2 \qquad (3.15)$$

remembering that the molarity of $Cl^-$ is twice that of $Ca^{2+}$. The interaction between ions increases with concentration and with the square of the ion charge.

The activity of a salt can then be divided between cation and anion. By measuring the activities of NaCl, NaBr, $NaNO_3$, and $NaClO_4$, for example, the contribution of $Na^+$ to the total activity can be sorted out. This process is very laborious; calculating the ion activity directly would be much quicker.

Debye and Hueckel in 1924 proposed Eq. 3.16 using the ionic strength to account for the effect of all the ions in solution on the activity coefficient of ion i:

$$\log \gamma = -A Z_i^2 I^{1/2} \qquad (3.16)$$

where $\gamma$ is the ion's activity coefficient, $A$ is a constant ($= 0.511$ for aqueous solutions at 25° C and is relatively insensitive to temperature), $Z_i$ is the ion's charge, and $I$ is the ionic strength. Equation 3.16 is called the Debye–Hueckel limiting law, because it predicts ion activity coefficients only in dilute solutions (solutions that have been cynically called "slightly contaminated distilled water"). Equation 3.16 was nonetheless a great breakthrough in the understanding of ion behavior in solutions.

Debye and Hueckel assumed that ions interact electrostatically like charged particles of zero size, but ions and their associated water molecules have significant physical size. These and other problems make Eq. 3.16 unreliable at ionic strengths greater than about 0.01. The equation has been modified to be useful up to $I = 0.1$:

$$\log \gamma = -AZ_i^2 \left( \frac{I^{1/2}}{1 + Ba_i I^{1/2}} \right) \tag{3.17}$$

where $B = 0.33$ for aqueous solutions at 25° C and $a_i$ is an individual ion parameter determined experimentally. Table 3.4 gives $a_i$ values for some common ions. Because values of $a_i$ somewhat resemble the diameters of ions plus their associated water molecules, $a_i$ is thought by some to have physical significance and has been termed the "distance of closest approach."

The empirical Davies equation

$$\log \gamma = AZ_i^2 \left( \frac{I^{1/2}}{1 + I^{1/2}} - 0.3I \right) \tag{3.18}$$

also yields satisfactory values for individual ion activity coefficients over the range of concentrations normally encountered in soil solutions and freshwaters.

Debye and Hueckel's work is historically important to soil chemistry because their derivation was similar to that of Guoy and Chapman, published independently about 1910, who tried to predict the ion distribution in the aqueous solution around a charged surface such as a soil particle. Although the Guoy–Chapman theory and its

**Table 3.4. Values of $a_i$, the Debye–Hueckel "distance of closest approach"[a]**

| Inorganic Ions | $a_i$ $(10^{-9}$ m) |
|---|---|
| $NH_4^+$ | 0.25 |
| $Cl^-, NO_3^-, K^+$ | 0.3 |
| $F^-, HS^-, OH^-$ | 0.35 |
| $HCO_3^-, H_2PO_4^-, Na^+$ | 0.4–0.45 |
| $HPO_4^{2-}, PO_4^{3-}, SO_4^{2-}$ | 0.4 |
| $CO_3^{2-}$ | 0.45 |
| $Cd^{2+}, Hg^{2+}, S^{2-}$ | 0.5 |
| $Li^+, Ca^{2+}, Cu^{2+}, Fe^{2+}, Mn^{2+}, Zn^{2+}$ | 0.6 |
| $Be^{2+}, Mg^{2+}$ | 0.8 |
| $H^+, Al^{3+}, Fe^{3+}, La^{3+}$ | 0.9 |
| $Th^{4+}, Zr^{4+}$ | 1.1 |
| Organic Ions | |
| $HCOO^-$ | 0.35 |
| $CH_3COO^-, (COO)_2^{2-}$ | 0.45 |
| $Citrate^{3-}$ | 0.5 |
| $C_6H_5COO^-$ | 0.6 |

[a] From J. Kielland. 1937. *J. Am. Chem. Soc.* **59**:1675–1678.

later modification have shortcomings for soil–water systems, Debye–Hueckel successfully describes the less complicated conditions of dilute aqueous solutions.

Nonelectrolytes—dissolved gases, organic molecules, neutral ion pairs, and undissociated weak acids and bases—are also nonideal solutes in water and are common constituents of soil solutions. Their activities also vary nonlinearly with concentration, particularly at high concentrations. The activity coefficients of nonelectrolytes at low concentrations are approximated by

$$log\gamma = -k_\mathrm{m}I \tag{3.19}$$

where $k_\mathrm{m}$ is called the *salting coefficient* and $I$ is ionic strength. Measured values of $k_\mathrm{m}$ range from 0.01 to 0.2 for common nonelectrolytes. The name salting coefficient comes from the tendency of nonelectrolytes to be less water soluble at increasing salt concentration, so that nonelectrolytes can be "salted out" of solution.

Like ions in aqueous solutions, the chemical activities of ions solids, and of water, also vary with concentration. Over the range of solute concentrations in soil solutions, however, the activity of water changes only negligibly from that of pure water. The chemical potentials of pure solids are defined as one, because any amount of the solid fixes the equilibrium activity of that substance in the aqueous solution. The activity of the aqueous solution is therefore independent of the amount of solid present.

The chemical potential of solid solutions, including impure minerals such as those in soils and rocks, is more difficult to define. Isomorphously substituted ions in a mineral change its activity and aqueous solubility from that of the pure mineral. Progress in defining solid activities has been slow. Soil minerals have often been assumed, by necessity, to be pure minerals and assumed to have activity $= 1$. This assumption is weak and is discussed later in this chapter.

## 3.3  COMPLEX IONS AND ION PAIRS

When ions and molecules interact closely, they lose their separate identities and are better thought of as complex ions or ion pairs. The $Al(H_2O)_6^{3+}$ and $Fe(H_2O)_6^{3+}$ ions are complex ions: The water molecules are closely attached to the central ion, with the group acting as one entity.

Complex ions are the combination of a central cation with one or more ligands. A *ligand* is any ion or molecule in the coordination sphere around the central ion, $H_2O$ in the case of $Fe(H_2O)_6^{3+}$. Water is usually taken for granted in complex ions and often disregarded. Ligands replace one or more of the water molecules in the primary hydration sphere. Ion pairs, on the other hand, are thought to form by ligand attachment in the second solvation sphere (Figure 3.3) and the bonding is weaker than in complex ions. Complex ions and ion pairs are synonymous with *inner-* and *outer-sphere complexes.* Many alkaline earth and transition metal cations are present in soil solutions as complex ions or ion pairs.

FIGURE 3.3. Diagram of an ion pair and a complex ion.

To associate with a central ion, ligands must compete with the water molecules in the central ion's solvation sphere and must lose some of the water molecules in their own solvation sphere. In addition, since many ligands are the anions of weak acids, $H^+$ competes with the central cation for the ligand. Forming a complex ion or ion pair involves competition between the cation and $H^+$ for the ligands, and between the water, $OH^-$, and ligands for the central cation.

The strength of association between the ions in solution is expressed by various equilibrium constants. *Stability (formation) constants* refer to complex ions and ion pairs; *hydrolysis (deprotonation) constants* refer to the loss of $H^+$ from the water ligands surrounding central cations. *Solubility products* refer to the aqueous ion activities in equilibrium with solid phases. Some "constants" are reported in the literature in terms of concentrations rather than activities. Such constants are misnamed, since they depend both on the concentration and on the nature of other ions in solution. Converting concentrations to activities gives a much more useful value.

The formation of complex ions is the result of cation–anion attractive forces winning out in the competition between cations and $H^+$ for the various ligands, including water. An example is the formation of the monofluoroaluminium complex ion

$$Al(H_2O)_6^{3+} + F^- = AlF(H_2O)_5^{2+} + H_2O \tag{3.20}$$

This reaction is exploited to extract reactive $Al^{3+}$ from soils. Forming $AlF^{2+}$ lowers the activity of $Al(H_2O)_6^{3+}$ in the water so the fluoride dissolves some $Al^{3+}$ from the solid phase. The stability constant of this complex ion is

$$K_{AlF^{2+}} = \frac{AlF^{2+}}{(Al^{3+})(F^-)} \tag{3.21}$$

The waters of the hydration sphere are usually ignored in the equilibrium constant, because excess water is present in aqueous solutions, and the energy of the Al–F bond must be greater than that of the ion–water bonds for the complex to form. The concentration of the $AlF^{2+}$ complex ion increases with increasing concentrations of $Al^{3+}$ and $F^-$.

At the same time, $H^+$ competes for the fluoride ion. HF is a weak acid and its dissociation or acidity constant is

$$K_{HF} = \frac{(H^+)(F^-)}{HF} \tag{3.22}$$

Substituting Eq. 3.22 into Eq. 3.21 and rearranging yields

$$(AlF^{2+}) = K_{AlF^{2+}} K_{HF} \frac{(Al^{3+})(HF)}{H^+} \qquad (3.23)$$

The concentration of the $AlF^{2+}$ complex ion increases with increasing HF and $Al^{3+}$ concentrations and decreases with increasing acidity.

Increasing $F^-$ concentrations encourage more $F^-$ ligands to replace water ligands around $Al^{3+}$, to a limit of $AlF_6^{3-}$. The hexafluoroaluminium ion is, in fact, the complex ion removed during fluoride extraction of aluminium from soils, because of the high fluoride concentrations employed.

Ligands such as $H_2O$, $OH^-$, $F^-$, and $CN^-$ occupy only one position around a central cation (Fig. 3.4) and are called *unidentate ligands*. Four of the six $F^-$ ligands are in the plane of the $Al^{3+}$ cation; the other two ligands are above and below the plane. *Bridging ligands* such as $O^{2-}$, $CO_3^{2-}$, and $PO_4^{3-}$ can occupy one position in the coordination spheres of two different cations. This produces a *polynuclear complex* (nuclear referring to the central ion). The solubility of polynuclear complexes is usually less than mononuclear complex ions. Polynuclear complexes tend to polymerize further and precipitate from solution.

A *polydentate ligand* can occupy two or more positions around a cation and can surround the cation. Such ligands are usually large organic molecules called *chelates*, from the Greek word for claw. Some enzymes, for example, are polydentate ligands occupying several positions around a central cation while also bonding to substrate molecules. This results in a configuration that catalyzes chemical changes in the substrate. Soil organic matter strongly adsorbs $Cu^{2+}$, $Zn^{2+}$, $Fe^{2+}$, and other transition metal ions, probably by acting as a chelate.

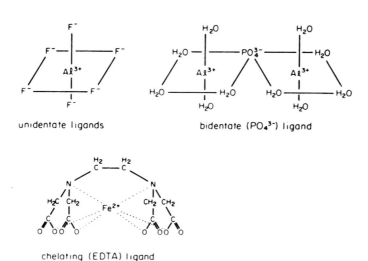

**FIGURE 3.4.** Schematic structure of the $AlF_6^{3-}$, binuclear Al–PO$_4$, and EDTA–iron(II) complex ions.

An example of a chelating ligand is ethylenediaminetetraacetic acid ($H_4EDTA$) and its many relatives (Fig. 3.4). The six positions around $Fe^{2+}$ are occupied by the two amine and four acetate groups. Chlorophyll and hemoglobin are also chelates. Chelates are quite soluble and tend to keep Fe, Zn, and Cu in solution for plant absorption. Chelates are also used to extract microelement and heavy metal ions from soils. The stability constant of the Fe(III)–EDTA complex is

$$K_{FeEDTA} = \frac{FeEDTA^-}{(Fe^{3+})(EDTA^{4-})} \tag{3.24}$$

The competing reaction for Fe(III) in soils is its dissolution/precipitation as FeOOH:

$$Fe^{3+} + 3H_2O = Fe(OH)_3 + 3H^+ \tag{3.25}$$

The solubility product of $Fe(OH)_3$ is

$$K_{sp} = (Fe^{3+})(OH^-)^3 = (Fe^{3+})\left(\frac{K_w}{H^+}\right)^3 \tag{3.26}$$

where $K_{sp}$ is the solubility product of $Fe(OH)_3$ and $K_w$ is the dissociation constant of water, $K_w = (H^+)(OH^-) = 10^{-14}$.

$H^+$ competes by reacting/dissociating as $H_4EDTA$:

$$H_4EDTA = 4H^+ + EDTA^{4-} \tag{3.27}$$

where the dissociation constant is

$$K_{H_4EDTA} = \frac{(H^+)^4(EDTA^{4-})}{H_4EDTA} \tag{3.28}$$

Substituting Eqs. 3.26 and 3.28 into 3.24 and rearranging gives the solubility of the Fe(III)–EDTA complex ion in equilibrium with $Fe(OH)_3$:

$$(FeEDTA) = K_{FeEDTA} K_{sp} K_{H_4EDTA} \frac{H_4EDTA}{K_w^3} \cdot H^+ \tag{3.29}$$

where $K_w$ is the dissociation constant of water. Equation 3.29 can be further extended to include the reduction of Fe(III) to Fe(II). This reduction changes the solubility of hydrated iron and hence the stability of the EDTA complex.

The relations between EDTA and other chelates with cations in pure solution can often be applied to soil solutions by including suitable solid-phase controls and competition from other cations and hence related to ion uptake by plants. Lindsay and co-workers (1979) have carried out such calculations in detail.

At equal cation concentrations, transition metal cations compete more effectively for ligands than can alkali and alkaline earth cations. Transition metal ions have the advantage of being able to shift some electrons to better accommodate ligand configurations. The ability of unidentate ligands to shift electron orbitals and thus form stronger complex ions generally increases in the following order: $I^- < Br^- <$

**Table 3.5. Stability constants of EDTA (ethylenediamine tetraacetic acid)**[a]

| Metal Ion | $\log K$ |
|---|---|
| $Li^+$ | 2.8 |
| $Na^+$ | 1.7 |
| $Mg^{2+}$ | 9.0 |
| $Ca^{2+}$ | 10.7 |
| $Ba^{2+}$ | 7.8 |
| $Mn^{2+}$ | 13.8 |
| $Fe^{2+}$ | 14.0 |
| $Co^{2+}$ | 16.0 |
| $Cu^{2+}$ | 18.5 |
| $Zn^{2+}$ | 16.3 |
| $Al^{3+}$ | 16.1 |
| $Fe^{3+}$ | 25.00 |
| $La^{3+}$ | 15.4 |
| $Th^{4+}$ | 23.2 |
| $(pK_1 = 2.0, pK_2 = 2.7, pK_3 = 6.2, pK_4 = 10.3)$ | |

[a]From L. G. Sillen and A. E. Martell. 1974. *The Chemical Society Spec. Publ.* **25**, London.

$Cl^- < F^- < C_2H_5OH < H_2O < NH_3 <$ ethylenediamine $< CN^-$. The strength of nitrogen-containing ligands is noteworthy in this list.

The relative ability of the transition metal ions to form complex ions is $Mn^{2+} < Fe^{2+} < Co^{2+} < Ni^{2+} < Cu^{2+} > Zn^{2+}$ for the divalent cations and $Cr^{3+} = Mn^{3+} > Fe^{3+} < Co^{3+}$ for the trivalent cations. The strongest complexing divalent cation is Cu(II). Fe(III) is the weakest complexing trivalent transition metal ion, but is stronger than other trivalent cations such as $Al^{3+}$ and the lanthanides. The heats of hydration (Table 3.2), strengths of EDTA complexes (Table 3.5), and solubility products of metal hydroxyoxides (Table 3.3) also follow this general order, with water, EDTA, and $OH^-$ as the respective ligands. Stability constants less than $10^9$ indicate the weaker ion–ion interaction of ion pairs.

Figure 3.5 shows how the distribution of EDTA complex ions changes with pH under representative soil solution conditions. The changes are due to competition between the cations for all of the ligands. The Fe–EDTA complex predominates in acid solutions, because of the great stability of Fe–EDTA complexes and weak competition from the low $OH^-$ concentrations for Fe(III). The EDTA ligand prefers Fe(II and III) despite the high $Ca^{2+}$ and $Mg^{2+}$ concentrations in soil solution. Other transition metal ions are generally in low concentrations in soil solutions. In alkaline soils, however, the higher $Ca^{2+}$ and $Mg^{2+}$ concentrations and the very low solubility of Fe(III) hydroxide favor the formation of CaEDTA and Mg–EDTA complexes

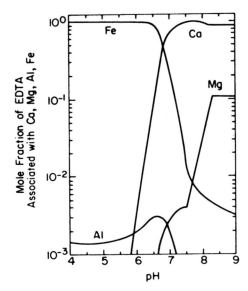

**FIGURE 3.5.** Mole fraction of EDTA in various complexes versus pH of hypothetical soil solutions. (From W. A. Norvell. 1974. In *Micronutrients in Agricultures* J. J. Mortvedt, P. M. Giordano, and W. A. Lindsay, (Eds.) Soil Science Society of America, Madison, WI.)

instead. This general picture holds true for many complexing ligands but shifts according to the values of the specific stability constants.

Another illustration of competition is during laboratory measurements of $Ca^{2+}$ and $Mg^{2+}$ in soil extracts. To ensure that all $Ca^{2+}$, $Mg^{2+}$, and EDTA, but only these ions, are present as complex ions, the solution is made alkaline to about pH 10 to precipitate transition metal ions. Then $CN^-$ (which complexes strongly with the transition metals but weakly with $Ca^{2+}$ or $Mg^{2+}$) is added to complex any remaining transition metal cations.

Ion pairs, or outer-sphere complexes, are written as $CaSO_4^0$ and $CaCO_3^0$ to distinguish them from their respective solids. The $CaSO_4^0$ and $CaCO_3^0$ ion pairs have been found to be particularly important in accounting for apparent supersaturation of $CaCO_3$ in groundwaters and drainage waters of arid regions. Magnesium also forms sulfate and carbonate ion pairs in many natural waters. Ion pair formation increases with increasing ion charge and concentration. Alkali metal ions such as $Na^+$ and $K^+$ form ion pairs only in highly saline soils and brines. Taking ion pair and complex ion formation into account has greatly increased our knowledge of the solid phases that govern ion concentrations in soil solutions.

## 3.4 HARD AND SOFT LEWIS ACIDS AND BASES

The reactions of water, $H^+$, $OH^-$, and $O^{2-}$ describe the aqueous solution behavior of many cations—alkali, alkaline earth, Al, and others. This is *Bronsted acid–base*

**Table 3.6. Hard and soft Lewis acids and bases**

| Lewis Acids | Lewis Bases |
|---|---|
| HARD | |
| $H^+$, $Li^+$, $Na^+$, $K^+$ | $H_2O$, $OH^-$, $O^{2-}$, $CO_3^{2-}$, $PO_4^{3-}$, $SO_4^{2-}$ |
| $Mg^{2+}$, $Ca^{2+}$, $Sr^{2+}$ | $SiO_4^{4-}$, $F^-$, $NH_3$ |
| $Al^{3+}$, $Be^{3+}$, $Si^{4+}$, $Ti^{3-4+}$ | aluminosilicates |
| $Mn^{2+}$, $Fe^{3+}$, $Co^{3+}$, $Cr^{3+}$ | |
| INTERMEDIATE | |
| $Fe^{2+}$, $Co^{2+}$, $Ni^{2+}$, $Cu^{2+}$ | $NO_2^-$, $SO_3^-$, $Cl^-$, $Br^-$, pyridine |
| $Zn^{2+}$, $Pb^{2+}$ | soil organic matter |
| SOFT | |
| $Cd^{2+}$, $Hg^{1-2+}$, $Cu^+$, $Ag^+$ | $S^{2-}$, $CN^-$ |

behavior. These cations also react strongly with oxygen-dominated ligands such as $CO_3^{2-}$, $SO_4^{2-}$, $NO_3^-$, and silicates. These cations are the predominant exchangeable ions and the soluble and exchangeable anions in soil solutions.

The above cations and anions interact weakly, however, with another interesting ion group—$Cu^{2+}$, $Cd^{2+}$, $Hg^{2+}$, $S^{2-}$, $CN^-$, and other organic groups—which tends to react within itself in preference to the oxygen-dominated group. Pearson called the oxygen-dominated group hard Lewis acids (cations) and bases (anions) and the second group, soft Lewis acids and bases (Table 3.6). Pearson suggested the general rule: "Hard Lewis acids tend to associate with hard Lewis bases; soft Lewis acids tend to associate with soft Lewis bases."

Hard Lewis acids and bases have inflexible electron orbitals that form ionic bonds. The electron orbitals of soft Lewis acids and bases are more polarizable and more likely to form covalent bonds. Soft Lewis acids and bases are also called covalent-bonding ions and are "siderophile" (sulfur-loving) ions in the geology literature. Organic ligands and soil organic matter range from hard to soft Lewis bases.

This classification explains why, for example, $Fe^{3+}$ reacts differently than $Fe^{2+}$ in soils. Reduced oxidation states tend to be softer Lewis acids and bases. Hard and soft also explains why $Cd^{2+}$ reacts quite differently than other cations of similar charge and size such as Ca, and why soil organic matter reacts with soft Lewis acids and also contributes greatly to the exchange capacity of hard Lewis acids.

## 3.5   SOIL REACTION COEFFICIENTS

Many of the hydroxyoxides listed in Table 3.4 exist in soils. Their ion activity products, as well as those of phosphates, carbonates, sulfides, and silicates, have been measured in soil solutions. Unfortunately, the ion activity products often differ widely from accepted solubility products in pure solutions, and also from soil to soil. The differences between ion activity products and solubility products is due to nonequilibrium and the formation of solid solutions whose aqueous solubilities differ from the solubilities of pure compounds. The lack of equilibrium is due to slow diffusion of ions in the weathered surfaces of soil particles. As the ion diffuses into the denser, more crystalline and less weathered interior, its diffusion rate slows dramatically. Diffusion rates in the truly solid phase are essentially zero at room temperature. Diffusion is only significant where water enters into the solid and where weathering breaks up the crystal structure.

The ion activity products in soil solutions and tabulated solubility products should agree only if the solubility of a single phase dominates the system, if competing reactions are insignificant, if the single phase is reasonably pure, and if the system is close to equilibrium. This apparently holds true for gypsum, $CaSO_4 \cdot 2H_2O$, and FeOOH. Measurements of the $(Fe)(OH)^3$ ion product in soil suspensions agree fairly well with the solubility product of amorphous FeOOH, $K_{sp} = 10^{-39}$. Apparently Fe(III) reacts rapidly and only with $OH^-$ so that a single solid controls its solubility, including ion-exchange equilibria. Aluminium concentrations, on the other hand, appear to be controlled by slow processes, including weathering of aluminosilicates. The solubility product of gypsum probably holds because its solubility is great enough to swamp out any competing reaction.

Phosphate is associated with many phases of the soil, including organic matter. None of these phases predominates in all soils, and all have different dissociation strengths for phosphate. Hence, each should support a different phosphate concentration, and the strength of association decreases as the phosphate concentration increases for all of the phases. As a result, phosphate ions should distribute themselves among the various retention sites until, at equilibrium, all the ions have the same dissociation energy. The speed of these transformations may control soil phosphate concentrations rather than the equilibrium solubility of this distribution.

The rates of soil phosphate reactions also may differ from the rates of phosphate uptake by plants and of phosphate release by organic matter decay. This phosphate turnover would further upset soil phosphate equilibria. If a steady state (concentration is constant with time) existed between the soil and dissolved phosphate ions, it might be described by a reaction such as

$$\text{soil-OH} + H_2PO_4^- = \text{soil-}H_2PO_4 + OH^- \tag{3.30}$$

where all of the many phosphate interactions with soil are combined into a single generalized reaction, so that

$$K_r = \frac{(\text{soil-}H_2PO_4)(OH^-)}{(\text{soil-OH})(H_2PO_4^-)} \tag{3.31}$$

where $K_r$ is a reaction coefficient rather than an equilibrium constant. Equations 3.30 and 3.31 can be misleading. They have the form of equilibrium equations, but equilibrium does not exist. Also, the activities of soil-adsorbed ion such as soil-OH and soil-$H_2PO_4$ cannot be defined precisely. Equations like 3.30 describe an ongoing process rather than equilibrium.

## 3.6   MODELS OF THE SOIL SOLUTION

Strictly speaking, soils are always nonequilibrium systems. With care, however, a partial equilibrium or steady state can be attained by assuming that the soil solids do not change. This is the usual assumption in cation exchange and adsorption studies. Kittrick and co-workers were able to obtain near-equilibrium measurements of some soil minerals in studies requiring several years. From the resulting ion activities in solution, they were able to calculate some of the equilibrium constants used for the mineral stability diagrams shown later in this book.

Ion hydrolysis and solid dissolution reactions occur at the same time in the soil solution and many of these reactions are interdependent. One hydrolysis reaction that releases $H^+$, for example, affects the other hydrolysis reactions and solids containing $OH^-$ ligands. The advent of computers allowed rapid calculation of many simultaneous reactions, and this was soon applied to models that try to calculate the composition of the soil solution and natural waters.

The early models yielded approximate concentrations that reflected the understanding of the soil solution at the time. Later models have yielded better predictions of the soil solution's composition, but they are still only approximate. That reflects the complexity of the soil more than the inadequacy of modeling. The models predict ion interactions in the aqueous solution quite well. Reactions at the surface of colloidal particles are more complex, less understood, slower, and hence are more difficult to formulate. In addition, the models are forced to use the solubility products of pure, simple solids. Soil inorganic particles are far from pure compounds, are often poorly crystalline to amorphous, are not at internal equilibrium, and may not be in equilibrium with the aqueous phase. In addition, the reactions of soil organic matter are not known quantitatively; and soils are open systems, meaning that matter is continually being added and removed.

The models therefore reflect our level of knowledge of the soil solution and its interaction with soil solids. Since these models have the potential to predict the composition of natural waters (groundwater, lakes and streams, oceans as well as the soil solution), soil fertility, the effects of fertilizers and soil amendments, the effects of acid rain, and the attenuation and release of pollutants in soils, this important area of research should be actively pursued. The accuracy of the models, however, is still based on our understanding of the soil's chemistry and cannot be more accurate than that.

## APPENDIX 3.1   THERMODYNAMICS

Chemical activity, heat of hydration, and equilibrium constant are parts of the very useful discipline called thermodynamics. Thermodynamics is the relation of matter and energy that predicts the direction and final result of chemical and physical reactions, but does not predict the rate or the path of reactions. This section introduces thermodynamic terms commonly encountered in the soils literature. Thermodynamic relationships are derived in detail in many physical chemistry texts.

Three disarmingly simple "laws" of thermodynamics that describe the relations of matter and energy have evolved into an elaborate structure that can obscure their initial simplicity. If an exception is found, the laws will be modified to include the exception. The laws have some background assumptions and require careful definitions. Chemical thermodynamics is dominated by the concept of *equilibrium*. Equilibrium is a state or condition that remains unchanged as long as energy and matter are neither gained nor lost from the system. An equilibrium system will return to equilibrium after a slight perturbation, such as a change in temperature or pressure.

Equilibrium principles require that a system be completely described by easily measured variables, such as temperature, pressure, and composition. These properties of the system define its *state*. The state is relatively easy to define for simple systems in the chemical laboratory, but is much more difficult to define in the complex systems of nature and soils. For example, the complete mineralogical composition of soils is difficult, if not impossible, to determine. The amounts and types of minerals in the clay fraction are known at best only semiquantitatively. In addition, the composition of each mineral can vary over a wide range and the weathered surfaces are different than the unweathered portion. Soil behavior also depends on the size, matrix, and interactions of soil particles. These are at present inadequately measured or defined. Wetting and drying can irreversibly change the arrangement of soil particles. Heating and drying can destroy organic compounds, change soil minerals, and markedly modify the composition of the soil solution.

Soils are nonequilibrium systems, but sometimes can be considered close enough to equilibrium to let equilibrium principles apply. The requirements are that (1) the rates of soil change are negligible in the time scale under consideration, (2) the perturbation does not change the composition of the solid phase, (3) the soil is homogeneous, and (4) the soil is a *closed system*—it does not gain or lose matter—during the experiment. None of these assumptions can be completely true, but the errors can be small if the experiments are done carefully.

The "laws" of thermodynamics describe physical and chemical behavior to which we have not yet found exception. Claimed exceptions to the laws of thermodynamics have thus far proved to be unrecognized, and often very subtle, violations of the conventions of thermodynamics. The first law—matter and energy are conserved (neither created nor destroyed) during a process—is an idea that already appealed to medieval philosophers. During the late 1700s and early 1800s, careful measurements of water pumps and steam engines showed, however, that although mass was conserved, some energy was always lost. Furthermore, the faster the pumps and engines ran, the greater the loss of this energy; that is, the process was less efficient at greater speeds.

Rather than discard the first law despite this apparent flaw, people changed the definitions, and introduced the second law, so the first law would still be true. The first law now deals with total energy, which is conserved.

If a reaction proceeded infinitely slowly, the first law would still hold true. Real-world reactions, however, proceed at a finite pace. The second law takes this into account by stating that some energy, called *entropy S*, is irretrievably lost during any process. This brings thermodynamics much closer to the real world. The available energy (*Gibbs free energy* G) obtained from a process is always less than the energy input *H* (*enthalpy*) or first law energy

$$\Delta G = \Delta H - T \Delta S \qquad (3.32)$$

where $T$ is the absolute temperature and the $\Delta$ refers to a difference between two values. We do not know the absolute energies; we can only measure the difference in energy between two states. The terms in Eq. 3.32 are given positive or negative signs according to the change within the *system*. The system might be a volume of soil or a flask containing a soil suspension, considered as separate from its surroundings. If the system loses energy such as heat or work to the surroundings, the sign is negative because the system has less energy than before. If the system gains energy from the surroundings, the sign of the energy terms is positive.

Entropy is defined so that it is positive and increasing during spontaneous reactions, reactions that proceed without energy input. Entropy is the energy lost during the reaction. This energy ultimately is radiated into space and is made up for by solar energy plus some energy from the earth's interior. Without solar and earth energy to compensate for the continual loss of entropy, the earth would eventually reach equilibrium and life would end. Some people consider entropy to be the unavailable energy or a bookkeeping entry that corrects the first law. Others prefer to describe entropy in more physical terms, as the friction in all processes, or as the energy lost in rearranging ions and molecules during chemical reactions. Increasing entropy implies increasing randomness or disorder of matter; decreasing entropy implies structural ordering.

Soon a loophole was discovered in the second law. When the temperature is absolute zero, the second law reduces to the first law and becomes unnecessary. So the third law—that we cannot reach absolute zero—was invoked to keep the second law universally valid. Corollaries of the laws are that entropy strives toward a maximum, free energy strives toward a minimum, and when the free energy is at a minimum, the system is at equilibrium.

Systems far from equilibrium, such as living systems that are negative entropy because they are so ordered, may be better described by *irreversible thermodynamics*. Living systems are greatly ordered and generally have active energy flows; examples are ecological systems having organisms with elaborate organic compounds closely interacting with other compounds, active population growth, and photosynthesis. Irreversible thermodynamics attempts to describe this energy flow. Irreversible thermodynamics has found little application in soil chemistry. In many cases, soil minerals and solutions are close enough to equilibrium that equilibrium and reversible thermodynamics suffice. Often a soil process in the laboratory or field is viewed as reacting much more rapidly than the overall soil. Then the process can be treated essentially as if it were an equilibrium process.

The three laws of thermodynamics and Newton's laws of physics govern the behavior of matter and energy in the systems that we normally deal with. The 20th-century developments of quantum and statistical mechanics deal with atomic and subatomic behavior. One test of their validity is that they yield the laws of thermodynamics when expanded to larger systems. So far, soil behavior has not required quantum or statistical mechanics for explanation.

Soils are sensitive to changes brought about by drying, wetting, leaching with monovalent salt solutions, acidification and basification, changing ionic strength, changing oxidizing-reducing conditions, and changing the soil–solution ratio. The bulk of the soil is quite resistant to change, but soil surfaces are vulnerable. Some of these reactions revert back slowly if at all to the previous state. This irreversibility weakens assumptions of equilibrium. Experiments that hope to utilize equilibrium principles should not inadvertently change the chemistry of the soil's surface.

## A3.1.1  Gibbs Free Energy

The thermodynamic term of widest use in soil chemistry is the *free energy*, or more explicitly, the *Gibbs free energy*. This is the energy of a substance or a reaction that, at constant temperature and pressure, is available for subsequent use. Energy drives chemical reactions and $\Delta G$ is the most widely useful. It is directly related to (1) the activity or chemical potential, (2) the energy of formation of compounds, (3) the equilibrium constant of a reaction, and (4) the electrode potential. The first three are discussed here; the electrode potential is discussed in Chapter 4.

The $\Delta G$ can be defined as

$$\Delta G = V \, \Delta P + S \, \Delta T + f \text{(compositional, electrical, gravitational potentials)}$$
(3.33)

where $V$ is the volume, $\Delta P$ the pressure change, and $\Delta T$ the temperature change. At constant temperature and pressure

$$\Delta G_{T,P} = f \text{(compositional, electrical, gravitational potentials)} \qquad (3.34)$$

The Gibbs free energy is determined by changes in composition, gravity, and electrical potentials. Chemistry is usually concerned only with variations in composition potentials. The gravitational potential arises from differences in elevation, such as the "water head" or hydraulic potential of soil physics, but can usually be ignored by soil chemists. Electrical potential is an important consideration near charged surfaces such as soil particles. When dealing with an aqueous solution, however, the Gibbs free energy at constant temperature and pressure is determined solely by the composition and concentration of the solution.

The most useful concentration unit for solutions is the chemical activity $a$ (Section 3.2). The change in free energy with the amount of solute is

$$\Delta G = RT \ln a = 5.71 \log a \qquad (3.35)$$

where the units of $\Delta G$ are kJ mol$^{-1}$. The free energy decreases as the solute becomes more dilute. The change is calculated from an arbitrary standard state of solute, usually defined as an ideal 1 M solution, where $a = 1$. At solute activities up to unity, the activity of the solvent (water) is usually assumed to be unity, that is, to be unaffected by solute concentration.

The $\Delta G$ value of a certain state is the difference in free energy between that state and a *standard state*. For the free energy possessed by a chemical compound, the usual standard state is its free energy at 25° C (298.15 K) and $10^5$ Pa (1 atm) pressure. The elements are assigned free energies of zero. The energy of formation is absorbed or released by compounds when they form from their elements. For example, careful measurement of the reaction

$$H_2 + \tfrac{1}{2}O_2 = H_2O \qquad (3.36)$$

at standard conditions (298.15 K and $10^5$ Pa) yields 237 kJ of available energy per mole of water formed. Thus, $\Delta G° = -237$ kJ mol$^{-1}$ with the superscript denoting standard conditions. This reaction releases energy to its surroundings, so the sign of $\Delta G°$ is negative, indicating that the system has less energy than before.

Reactions that release energy and leave the system in a lower energy state, a more stable state, than before are *spontaneous*. The second law states that systems will strive to reach the lowest energy level. Thermodynamics says only that a reaction will proceed and not what the reaction rate will be. The hydrogen and oxygen in reaction 3.36 can coexist for centuries without reacting until a catalyst or spark is introduced.

The free energies of formation of many compounds and ions have been measured and compiled. Table 3.7 contains values for the $\Delta G°$ of formation of some compounds relevant to soil chemistry.

The change of free energy during a reaction is the difference between the free energies of the products and those of the reactants. An important energy reaction in nature is photosynthesis, the formation of glucose:

$$6CO_2 + 6H_2O = C_6H_{12}O_6 \text{ (glucose)} + 6O_2 \qquad \Delta G° = 2879 \text{ kJ mol}^{-1} \quad (3.37)$$

The sign of the free energy change of this reaction is positive, because solar energy has come from the surroundings and been trapped in the glucose molecule. The free energy of the reaction is the difference between the free energies of formation of the products (glucose and oxygen) and of the reactants (carbon dioxide and water):

$$\Delta G_{3.37} = \Delta G_{\text{glucose}} + 6(\Delta G_{O_2}) - 6(\Delta G_{CO_2} + \Delta G_{H_2O}) \qquad (3.38)$$

Each term is the value per mole so it is multiplied by the appropriate coefficient from Eq. 3.37. Rearrangement yields

$$\Delta G_{\text{glucose}} = \Delta G_{2.37} - 6(\Delta G_{O_2}) + 6(\Delta G_{CO_2} + \Delta G_{H_2O}) \qquad (3.39)$$

The $\Delta G$ of formation of oxygen is assigned a value of zero, because it is a pure element. The $\Delta G$ of formation of carbon dioxide ($-394.3$ kJ mol$^{-1}$) and other compounds is available from many handbooks. The $\Delta G$ of formation of glucose is thus

$$\Delta G_{\text{glucose}} = +2879 - 0 + 6(-394.3) + 6(-237.2) = -910.4 \text{ kJ mol}^{-1} \quad (3.40)$$

**Table 3.7. Standard Gibbs free energies of selected compounds at 25° C ($a$ is activity and $P$ is partial pressure or mole fraction of gas)**

| Formula | Name or State | $\Delta G°$ (kJ mol$^{-1}$) | Source |
|---|---|---|---|
| $Al(OH)_3$ | Gibbsite | $-1\,151$ | 3 |
| $Al_2Si_2O_5(OH)_4$ | Kaolinite | $-3\,783$ | 3 |
| $M_{0.56}(Al_{3.03}Mg_{0.58}Fe_{0.45}$ $(Si_{7.87}Al_{0.13})O_{20}(OH)_4$ | Montmorillonite | $-10\,330$ | 4 |
| $CaCO_3$ | Calcite | $-1\,129$ | 1 |
| $CaSO_4 \cdot 2H_2O$ | Gypsum | $-1\,797$ | 1 |
| $Fe_2O_3$ | Hematite | $-741$ | 2 |
| $Fe(OH)_3$ | (Crystalline?) | $-694.5$ | 2 |
| $FeCO_3$ | Siderite | $-674.0$ | 2 |
| $FeS_2$ | Pyrite | $-150.6$ | 2 |
| $MgCO_3$ | Magnesite | $-1\,029$ | 2 |
| $MnO_2$ | Pyrolusite | $-464.8$ | 2 |
| $MnCO_3$ | Rhodochrosite | $-817.6$ | 2 |
| $MnS$ | Alabandite | $-233$ | 2 |
| $HNO_3$ and $NO_3^-$ | $a = 1$ | $-110.6$ | 1 |
| $NH_3$ | $P = 1$ | $-16.64$ | 1 |
| $NH_4OH$ | $a = 1$ | $-263.8$ | 1 |
| $H^+$ | $a = 1$ | $0.0$ | 1 |
| $OH^-$ | $a = 1$ | $-157.1$ | 2 |
| $H_2O$ | $a = 1$ | $-237.2$ | 1 |
| $H_3PO_4$ | $a = 1$ | $-1\,143$ | 1 |
| $PO_4^{3-}$ | $a = 1$ | $-1\,019$ | 1 |
| $K^+$ | $a = 1$ | $-282.0$ | 2 |
| $KAlSi_3O_8$ | Feldspar | $-3\,581$ | 2 |
| $KAl_3Si_3O_{10}(OH)_2$ | Muscovite | $-5\,558$ | 3 |
| $SiO_2$ | Quartz | $-8\,567$ | 1 |
| $Si(OH)_4$ | Soluble silica, $a = 1$ | $-1\,317$ | 1 |
| $SO_2$ | $P = 1$ | $-300.2$ | 1 |
| $H_2S$ | $P = 1$ | $-33.6$ | 1 |
| $H_2SO_4$ and $SO_4^{2-}$ | $a = 1$ | $-744.6$ | 1 |
| $TiO_2$ | Rutile | $-888.2$ | 2 |
| $CO_2$ | $P = 1$ | $-394.3$ | 1 |
| $CH_4$ | $P = 1$ | $-50.75$ | 1 |
| $H_2CO_3$ | $a = 1$ | $-623.2$ | 1 |

1. From D. D. Wagman et al. 1968. Selected values of chemical thermodynamic properties. *U.S. Bur. Standards Tech. Notes 270-3 and -6*, Washington, DC.

2. From R. M. Garrels. 1960. *Mineral Equilibria*. Harper, New York.

3. From S. V. Mattigod. 1976. Ph.D. Dissertation, Washington State University, Pullman.

4. J. A. Kittrick. 1971. *Soil Sci. Soc. Am. Proc.* **35**:140.

When reaction 3.37 reverses during respiration, 1 mol of glucose liberates $-2879$ kJ of available energy to fuel the life processes of living organisms.

A third important facet of the Gibbs free energy is its relation to the equilibrium constant of a reaction. A reaction proceeds until the components are at their lowest energy level, the most stable state. This state is defined by the equilibrium constant $K$:

$$\Delta G = RT \ln K = 5.71 \log K \qquad (3.41)$$

Equation 3.41 requires that the standard states of the products and reactants be known, that the components can be defined quantitively and in a thermodynamic sense. In soils and much of nature these definitions are rarely possible. The states of ions or molecules in soil systems, and in probably all colloidal systems, are ill-defined thermodynamically. In rigorous thermodynamic terms even ions are undefined. Soil reactions, because of the nonequilibrium in soils and the lack of defined standard states, yield reaction coefficients, rather than reaction constants, and their values vary with soil conditions.


## APPENDIX 3.2   SOLID SOLUTIONS AND OPEN SYSTEMS

The thermodynamics of solid mixtures and solid solutions would seem to hold great promise for soil chemistry, as much as the thermodynamics of aqueous solutions has proved useful, but it has been largely neglected. The neglect of solid solutions is partly due to an incompatibility between classical thermodynamics and nature. Nature and soils are more complex and they are *open systems*—energy and matter flow in and out of the system being studied. Nonetheless, some of the principles of solid mixtures are tacitly applied to ion exchange equations and adsorption studies. Direct applications of the thermodynamics of mixtures to soils may have considerable merit. The necessary assumptions are similar to those that have already been accepted in adsorption and cation exchange studies.

Strictly speaking, thermodynamics applies only to total equilibrium and to *closed systems*. A closed system gains and loses no matter during the reaction. Soils steadily lose matter during weathering and gain matter by aerial deposition, rain, and tectonic movements. Soils also are not at total equilibrium; the world would be sterile if nature were at total equilibrium.

Soil chemistry cannot afford the luxury of rigorous thermodynamics and instead has to stretch and bend the rules into what is called extrathermodynamics, but the bending and stretching must be done wisely. Cation exchange, solubility, Donnan, and adsorption studies find use in soils and assume equilibrium and a closed system during the experiment. Applying the thermodynamics of solid solutions to the reaction between the soil solution and soil particles requires the same rule bending as the other studies.

When substances in liquids or solids mix completely on an atomic scale, that is, when they mix homogeneously and randomly, the mixing usually decreases the escaping tendency of the components. The escape can be in the form of molecules evaporating from a liquid mixture, or of Ca and Mg ions dissolving into water from dolomite $CaMg(CO_3)_2$. Other examples of escaping tendency are cations exchanging from the mixture of cations on soil surfaces, phosphate desorbing from soil solids, and trace ions dissolving from silicate minerals.

Ions in solids cannot mix as randomly as can the components of gases and liquids. Mixing in glasses is close to random; ions in crystals are mixed but confined to fixed positions. This degree of mixing in crystals is sufficient to allow the thermodynamics of mixing to apply, and the mixing greatly affects the aqueous solubility of the ions in these solids. The general name of these solid mixtures is *solid solutions*; Hildebrand named crystalline solid solutions with their more limited mixing as *regular solutions*; the mixing of ions in solids is regular and repeated.

One example of the effect of mixing on the free energy or escaping tendency is the evaporation of a mixture of two organic liquids. When hexane is mixed with heptane, hexane's escaping tendency decreases. The escaping tendency is measured by its partial pressure $P$, which depends on hexane's volatility in the pure state and its concentration in the mixture, its degree of mixing:

$$P_{\text{hexane}} = g P_0 C_{\text{hexane}} \tag{3.42}$$

where $P_0$ is the vapor pressure of pure hexane, $C$ is its mole fraction concentration in the mixture, and $g$ is the activity coefficient that accounts for deviation from ideal mixing. Mixing increases the randomness of the hexane molecules, increases hexane's entropy, and therefore decreases its free energy, escaping tendency, and vapor pressure. Alternatively, the probability of a hexane molecule evaporating from a mixture is less than if evaporating from pure hexane, because fewer hexane molecules are at the surface.

Similarly, the escaping tendency, or aqueous solubility, of an ion depends on its bonding strength to the other ions of the solid. For ions on the surface, the escaping tendency depends on the extent of its mixing on the surface. As an example of the effect of solid mixing, the aqueous solubility of $AlPO_4$ is expressed by $AlPO_4$'s solubility product when the mineral is pure:

$$K_{\text{sp}} = \frac{(Al)(PO_4)}{a_{AlPO_4}} \tag{3.43}$$

For pure $AlPO_4$, the activity $a = 1$. When the phosphate ions are instead adsorbed on the surface of $Al(OH)_3$, the phosphate ions behave as if they are $AlPO_4$ mixed with $Al(OH)_3$. The solubility of $AlPO_4$ expressed by its *ion activity product* $(Al)(PO_4)$ in the aqueous phase will be less than its solubility product because mixing increases the entropy of $AlPO_4$ in the solid phase and reduces its aqueous solubility. Its solubility in the mixture on the surface is

$$\text{IAP}_{AlPO_4} = g K_{\text{sp}} C_{AlPO_4} \tag{3.44}$$

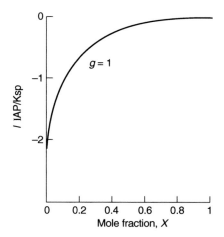

**FIGURE 3.6.** Change of relative equilibrium aqueous solubility, $IAP/K_{sp}$, of a component in a solid solution as a function of its composition in the solid. The solid solution is assumed to be ideal. (From H. L. Bohn. 1992. *Soil Sci.* 154:357.)

where $g$ is the activity coefficient, $K_{sp}$ is the solubility product of pure $AlPO_4$, and $C$ is the mole fraction of phosphate on the $Al(OH)_3$ surface.

Figure 3.6 shows the effect of this solid solution mixing on the aqueous solubility of a substance assuming ideal mixing, $g = 1$. Mixing has little effect on the $IAP/K_{sp}$ ratio of a substance until its mole fraction is <0.5. The reduced aqueous solubility due to this mixing should be pronounced for trace ions in the soil solution such as phosphate and the trace metals, but insignificant for Si, Al, and Fe.

The activity coefficient $g$ is an index of the deviation from ideal mixing. In ideal mixing, only the mixing entropy affects the escaping tendency. Mixing in nature is nonideal; molecular and ionic interactions change the amount of interaction and therefore affect the activity coefficient. The activity coefficients of solutes in water are generally <1 because ions interact with each other so their effective concentration is less than their actual concentration. Activity coefficients of solid components are usually >1 because substituted ions usually do not fit easily into the solid's structure. The structural stresses imposed by substitution tend to expel the ions into the aqueous solution. This somewhat counteracts the decreased aqueous solubility caused by the entropy of mixing and yields activity coefficients $> 1$. Each ion is different and fits differently into soil mineral structures. The usual result is that the ion's solubility is greater than if it mixed easily into the structure, hence $g > 1$, but not enough to overcome the effects of the mixing entropy. Early measurements indicate that the $g$ values for ions like Al that are common in soil minerals, and presumably substitute easily, are about $g = 3$ to 5. For ions like phosphate and Ca that do not substitute easily into aluminosilicates, $g \geq 20$.

Like activity coefficients in aqueous solutions, solid activity coefficients are concentration dependent. As more phosphate ions mix on the surface of $Al(OH)_3$, for example, the hydroxide structure becomes more and more strained. The strain is reflected in a positive deviation from ideal mixing. $PO_4$ ions mix less easily because other $PO_4$ ions are already on the surface, and the activity coefficient increases in value. The $Al(OH)_3$ tries to release the strain by trying to expel phosphate from the surface. As more phosphate is added, $AlPO_4$ eventually becomes more stable than the mixture and $AlPO_4$ precipitates as a separate phase.

Solid activities have been ignored largely because of a misunderstanding in our early chemistry training. The solubility product of Al hydroxide in chemistry texts, for example, is written as

$$K_{sp} = (Al)(OH)_3 \tag{3.45}$$

Equation 3.45 is accurate only in a pure system that contains only Al and OH ions and water. In that system the activity of Al hydroxide is one. For systems containing other components, such as aluminosilicates and Al substituted into FeOOH, a more complete expression of Al solubility is

$$K_{sp} = \frac{(Al)(OH)_3}{Al(OH)_{3,\,solid}} \tag{3.46}$$

and

$$IAP_{Al(OH)_3} = g K_{sp} C_{Al(OH)_3} \tag{3.47}$$

The aqueous solubility of Al can be related to the solubility of $Al(OH)_3$ even though that mineral may not be present. The solid activity of $Al(OH)_3$ is $gC$. The aqueous solubility of Al depends on the equivalent concentration of $Al(OH)_3$ in the solid phase. The activity of the solid is defined by the activity of the ions in the aqueous solution. The IAP is a measurable property of the system, so the solid activity is a thermodynamic property.

The concentration of Al hydroxide in soils is usually high enough that Eq. 3.45 is adequate and Eq. 3.46 is unnecessary. That is probably not true for ions in trace concentrations in soils. The mole fraction concentrations of phosphate and transition metal ions in soils are $\ll 0.01$. So soil phosphate solubility at "equilibrium" can be orders of magnitude less than the solubility of pure Al phosphate, as many soil measurements show. We can call it equilibrium because the solubility changes slowly to imperceptibly with time. If anything, the aqueous phosphate concentration decreases with time as the surface Al ions slowly diffuse into the weathered surface, mix further with other ions, and further increase their entropy. Simple $AlPO_4$ solubility product equivalent to Eq. 3.45, on the other hand, predicts that the phosphate concentration will increase with time.

Mixing on soil surfaces explains some soil phenomena very well. Mixing explains why soils can adsorb virtually every ion from soil solutions and can retain those ions much more tightly than can the ion's own hydroxyoxides, or retain phosphate more tightly than Al phosphate. This happens even though soil minerals are continuously losing their own components by weathering. Soil "adsorption sites" are areas where ions from the soil solution can mix with ions on soil surfaces. The mixing ability varies with the kind of ions on the soil's surfaces and with the ability of the aqueous ion to bond strongly with those surface ions.

Soil is an open system. Weathering carries away substances from the soil; wind and rain add others. The surfaces of soil particles reflect these processes as well as the composition of the internal part of the particles. Because the particle and its surface do not have the same composition, the particle is not at equilibrium with itself. Ion diffusion within crystals is slow enough that this disequilibrium can be ignored.

In the weathered surfaces of soil particles where some semiliquid water may be present, ion diffusion may be fast enough to affect laboratory and field experiments. The slow removal of phosphate ions by soils from the aqueous phase may be due to the slow diffusion of $PO_4$ to Al and Fe ions within the weathered surface. The increasing strength of trace metal retention by soils with time may similarly be due to such diffusion into the semisolid, weathered surfaces of soil particles.

Ion activities in the soil solution can also be treated as being governed by the *saturation index* of a mineral:

$$\text{Saturation Index} = \frac{\text{IAP}}{K_{sp}} \tag{3.48}$$

From Eqs. 3.47 and 3.48, the saturation index equals the mineral's solid activity coefficient times its mole fraction in the solid phase:

$$\text{Saturation Index}_i = g_i C_i \tag{3.49}$$

## APPENDIX 3.3   KINETICS

Thermodynamics predicts that substances will react until they reach their most stable states, but does not say how the most stable state will be achieved or how long it will take. Not all reactions lead immediately to the most stable states, and some reactions are exceedingly slow. Kinetics is the study of these reaction *mechanisms*: the rates, paths, and intermediate products of chemical reactions.

A substance put in conditions in which it is unstable will sometimes not react at all. A mechanical example of such "metastable equilibrium" is a rectangular block standing on end. It will not reach the more stable state of lying on its side until it is pushed and lifted so that its center of gravity is beyond its edge.

In chemical reactions, pushing and lifting the block correspond to activation energy. A mixture of $H_2$ and $O_2$ gases will not react until a spark or high temperature provides sufficient activation energy to greatly perturb the metastable equilibrium and allow the gases to react. Figure 3.7 shows the change in energy of a mixture of substances A and B that requires some activation energy to create the activated state $AB^*$ before it degrades to C and D. Photosynthesis is an example of this process. Sunlight provides the activation energy that creates an activated state, glucose, which is metastable. The activated state returns to the stable initial states, $CO_2$ and $H_2O$, through a path of metabolism and decay that is as intricate as photosynthesis.

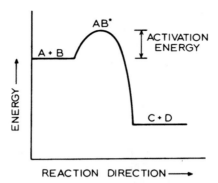

**FIGURE 3.7.** A representation of the extra energy (activation energy) needed to carry out the reaction $A + B \rightarrow C + D$. $AB^*$ is the intermediate, activated complex.

Nitrogen fixation, denitrification, soil weathering, phosphate fixation, clay mineral degradation, and potassium and transition metal fixation are problems for which the reaction rates are usually as, or more, important than equilibrium. Most soil chemical applications of kinetics have been in soil microbiology and soil biochemistry, where the lack of equilibrium is more obvious. The use of kinetics in inorganic soil chemistry will undoubtedly broaden in the future. It can even be argued that kinetics is basic to thermodynamics, because equilibrium is the condition where opposing reaction rates are equal.

Small amounts of some substances can increase reaction rates enormously. These substances, when left unchanged by the reaction, are called *catalysts*. Perhaps the simplest type of catalytic action occurs when a surface adsorbs the reactants so that they remain in close proximity for relatively long periods of time. The probability of forming a new compound from the reactants then is much greater than if they merely collide and rebound in a gaseous or liquid phase. Soil surfaces may act as catalysts in this way.

Catalysts lower the activation energy barriers that hinder reactions. The activation energy requirement can arise from many chemical and physical factors, and the mechanisms by which catalysts lower the activation energy are probably just as numerous. Iron, manganese, and other transition-metal ions catalyze electron transfers during oxidation–reduction reactions. Enzymes are the organic catalysts indispensable for most reactions of living organisms.

Reaction inhibitors slow reaction rates. Nitrogen mineralization and nitrification (conversion of organic nitrogen and ammonium to nitrate) rates in soils, for example, can be slowed temporarily by chemicals that specifically slow or stop the microorganisms involved. Toxic metals can also operate as enzyme inhibitors, by replacing the metal coenzyme portion of an enzyme and thereby inactivating it.

Kinetics is being employed increasingly to study the soil chemistry of carbon, nitrogen, potassium, phosphate, and trace metals. The soil reactions of these elements are often slow enough to be experimentally measurable. Because carbon and nitrogen cycle back and forth between soil, water, plants, animals, and atmosphere faster than the rates at which they reach their most stable thermodynamic states, kinetics is an appropriate tool with which to investigate their behavior.

For inorganic ions, the reactions themselves can be very fast, but the ions may have to diffuse through soil pores before they reach a reaction site. The ions may also have to diffuse through the weathered surface. Diffusion processes lend themselves to kinetic treatment. With multiple diffusion and reaction processes going on simultaneously, the kinetic treatment can become very complex.

## A3.3.1   Reaction Order and Rate Constants

One approach of kinetics is to describe the dependence of reaction rates on reactant concentrations. For instance, the rate of phosphate fixation depends at least partly on the amount of fertilizer added, and the rate of denitrification (the conversion of soil nitrogen, usually nitrate to $N_2$ and $N_2O$) depends on soil solution nitrate concentrations. Kinetics relates reaction rates and reactant concentrations by means of the reaction order and the reaction rate constant. The denitrification rate $(-\Delta NO_3^- / \Delta t)$ is presumably related to soil nitrate concentration by

$$-\frac{\Delta NO_3^-}{\Delta t} = k\,\Delta\{N_2O\}^n = k'\{\text{soil } NO_3^-\}^n \qquad (3.50)$$

where $k$ and $k'$ are the reaction rate constants or coefficients, $n$ is the reaction order, and braces denote concentrations. The negative sign indicates that nitrate is disappearing, and the positive sign indicates the production of $N_2$ and $N_2O$ gases.

In pure systems under carefully controlled conditions, the reaction order is often 0, 1, or 2. These reaction rates can be plotted linearly with respect to time by choosing the appropriate concentration axis (Fig. 3.8). The reaction order can be obtained by fitting data to such plots. Zero-order reaction rates are independent of the amount or concentration of the reactant studied:

$$-\frac{\Delta C}{\Delta t} = k \qquad (3.51)$$

where $C$ is the concentration of some substance that is disappearing at a rate that is constant with time.

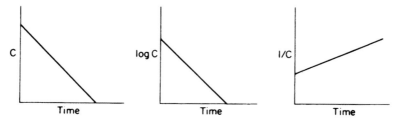

**FIGURE 3.8.** Zero-, first-, and second-order reactions plotted linearly with time. Note the varying units of the vertical axis.

The most common order found is first order, in which the reaction rate depends on the concentration of one reactant A:

$$-\frac{\Delta[A]}{\Delta t} = k\,[A][B] \tag{3.52}$$

This order can be obtained by "swamping" the system with the other components so that reactant A is rate limiting.

Second-order reactions depend on the concentrations of two reactants, or on the concentration of one reactant squared. The reaction of A and B to form D, for example, might follow the equation

$$\frac{\Delta D}{\Delta t} = k\,[A][B] \tag{3.53}$$

The sign of the left side is positive because $D$ is increasing. If the concentration of B were much higher than that of A, the reaction rate would appear to depend solely on A. The reaction rate would be pseudo-first-order with respect to A and almost pseudo-zero-order with respect to B. The order of a reaction therefore depends on the conditions of the experiment.

The rates of soil adsorption reactions may also depend on the exponential of the amount already adsorbed. Phosphate adsorption by soils, for example, sometimes follows the Elovich equation:

$$\frac{\Delta A_{ads}}{\Delta t} = \alpha e^{-\beta} A_{ads} \tag{3.54}$$

where $A_{ads}$ is the amount already adsorbed and $\alpha$ and $\beta$ are empirical constants. Because of its two constants, this equation is more easily fitted to experimental data. The equation is plotted in Fig. 3.9, where $\alpha$ is the slope of the line. Reaction mecha-

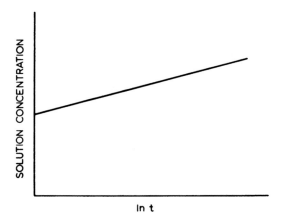

**FIGURE 3.9.** Plot of the Elovich adsorption equation 3.54.

nisms sometimes can be inferred from measured reaction orders. After ensuring that no other reactions are rate limiting, the reaction mechanisms are inferred by splitting the reaction into hypothetical step reactions until the sum of the orders equals the measured order. Insuring that the reaction in question limits the reaction rate is difficult for soils. The number of steps between reactant and product can be very large.

Reaction rates in soils are more complex than those in pure systems. Secondary and side reactions are difficult, if not impossible, to control. The measured reaction order is therefore usually fractional rather than a whole number because other reactions are usually going on at rates different from the one in question. If only the total change of a component is measurable, the overall reaction rate and order are weighted according to the relative contribution of each reaction.

In the denitrification example (Eq. 3.50), the rate of nitrate loss depends also on microbial activity and therefore on the presence of an energy source:

$$-\frac{\Delta[NO_3^-]}{\Delta t} = k[\text{soil } NO_3^-]^n[\text{available C}]^m \tag{3.55}$$

The reaction order is the sum of $n$ plus $m$. Competing reactions of different rates, and reverse reactions, are the rule in nature rather than the exception. Reaction orders of such complex systems are usually nonintegral. Laboratory or field measurements are often possible only if all variables except the one of interest are held constant. Because the effects of only one or a few variables are measured, the reaction rate is incompletely described and the order is actually a pseudo-order. The term "pseudo" unfortunately has a disparaging connotation. Here it implies only that the system is too complicated to measure completely.

Determining the rate constant and order of a reaction is tedious and time-consuming. For many studies, this detail is unwarranted and the *half-life* is measured instead. The half-life is the time required for half of the original concentration of reactant to disappear. For the particular case of a first-order reaction, the half-life $(t_{1/2})$ is directly related to the reaction rate constant $k$ by

$$t_{1/2} = \frac{\ln C_0/2}{k} = \frac{0.693}{k} \tag{3.56}$$

where $C_0$ is the original reactant concentration. Because many soil studies are carried out under conditions involving only one experimental variable, the pseudo-orders of the reactions may be close to unity. In these cases the half-life is a useful semiquantitative indicator of reaction rate.

## A3.3.2 Temperature Effects

Higher temperatures increase the energy and probability of particle collisions. These, in turn, generally increase reaction rates. Measuring reaction rates at various temperatures can provide useful clues about reaction mechanisms. Figure 3.10 shows the effect of temperature on the rates of three types of reactions. The exponential rise

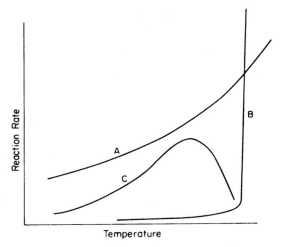

**FIGURE 3.10.** Typical effects of temperature on reaction rates of (A) inorganic reactions, (B) explosive reactions, and (C) biochemical reactions.

of curve A is characteristic of inorganic reactions, the rate increases two- or three-fold with each 10° C value, or the increase in reaction rate with a 10° C increase in temperature, $Q_{10} = 2$ to 3. The plot of $\log k$ versus $1/T$ is often linear for such reactions. Such data can yield the activation energy $E_a$ from the Arrhenius equation:

$$\frac{\Delta \log k}{\Delta T} = \frac{E_a}{2.303RT^2} \tag{3.57}$$

Another response of rate to temperature is an explosion, curve B of Fig. 3.10. The reaction rate increases only slowly until a critical temperature is reached, whereupon the rate approaches infinity.

### A3.3.3 Microbially Catalyzed Reactions

Many chemical reactions in soils occur at measurable rates only because of enzymatic or microbial catalysis. Curve C of Fig. 3.10 is characteristic of biological reactions. Biological reactions generally increase about threefold per 10° C rise in temperature ($Q_{10} = 3$) up to an optimum temperature and then decrease rapidly at higher temperatures. Most soil organisms are *mesophiles*, whose optimal temperatures are 30 to 37° C; 37° C is particularly common. Soils also contain *thermophiles*, whose optimum temperatures are 55 to 60° C, and *psychrophiles*, whose optimal range is 5 to 15° C. Psychrophiles were once thought to be rare in soils, but recent work suggests that their incidence and importance has been underestimated. The mean annual temperature of most soils is in the 5 to 15° C range.

When biochemical reactions are studied with time, two situations commonly occur. One is when enzyme concentrations remain constant (the biological population in effect remains constant), and Michaelis–Menten kinetics often apply. This treatment assumes that the enzyme (E) and the reactant or substrate (S) form a complex (ES) that dissociates either to the original substrate and the enzyme, or to a new product (P) and the enzyme. Thus,

$$S + E \underset{k_2}{\overset{k_1}{\rightleftharpoons}} ES \overset{k_3}{\rightarrow} E + P \qquad (3.58)$$

Michaelis and Menten found that the rate of substrate disappearance with time could be expressed as

$$-\frac{\Delta[S]}{\Delta t} = \frac{k_3[E][S]}{K_m + [S]} \qquad (3.59)$$

where the brackets indicate concentrations and where $K_m$ is the Michaelis constant. Equation 3.59 describes the curve of Fig. 3.11a. Equation 3.59 assumes that all nonspecified rate factors are at steady state during the period of measurement.

When the substrate concentration is much smaller than $K_m$, the denominator of Eq. 3.59 reduces to $K_m$ and the reaction rate, at constant enzyme concentration, is first order with respect to the substrate concentration [S]:

$$-\frac{\Delta[S]}{\Delta t} = \frac{k_3}{K_m[E][S]} = k'[S] \qquad (3.60)$$

This equation describes the steep portion of the curve near the origin in Fig. 3.11a. When [S] is much larger than $K_m$, Eq. 3.60 reduces to

$$-\frac{\Delta[S]}{\Delta t} = k_3[E] \qquad (3.61)$$

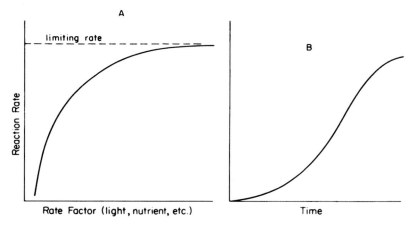

**FIGURE 3.11.** Change of reaction rate (A) with a rate factor according to the Michaelis–Menten kinetics and (B) with time and population response.

which means that the rate is independent of, or zero order with respect to, S. The rate then depends only on the steady-state concentration of the enzyme, that is, on the plant or microbial population.

The rate of an enzymatic reaction depends on many factors. Increasing one factor generally increases the rate less than the increased amount of the factor, because the rate is now hindered by other growth factors. For instance, nitrogen fertilization may increase crop growth rates, but plants must still contend with limited water and nutrient supplies, disease, and climate. Each successive increment of added nitrogen will have less effect on growth, because other factors are now rate limiting. Only by changing other rate-limiting conditions can an organism take better advantage of an added growth factor. Michaelis–Menten kinetics apply both to short-term, single-plant experiments and to the worldwide plant response to changing $P_{CO_2}$ in the atmosphere. In both cases the population and enzyme concentrations remain constant.

The second situation is more common in experimental work; the biological population increases during the study period in response to some growth factor. This requires time and causes the characteristic lag time before biological reactions begin. This lag distinguishes reactions that are primarily biochemical. Inorganic reactions, in contrast, are usually most rapid at the beginning of the experiment.

During the lag period, the growth of the organism and the corresponding reaction rate are slow but increase exponentially with time. Later the rate slows to zero when a new set of limitations is reached. This results in the familiar S-shaped or sigmoid curve (Fig. 3.11b). The curve is described by equations such as

$$\frac{\Delta N}{\Delta t} = \frac{\Delta [S]}{\Delta t} = \frac{rN(N_{lim} - N)}{N_{lim}} \tag{3.62}$$

where $N$ is the number of organisms, $N_{lim}$ is the maximum population of organisms, and $r$ is the difference between the organism's coefficients of birth and death rates (or forward and backward reaction rates). Equation 3.62 applies to such situations as the soil's evolution of $CO_2$ from freshly introduced organic matter, and the appearance of nitrate in soils recently fertilized with ammonium salts.

The application of kinetics and thermodynamics requires a deeper understanding than the brief introduction given here. Although best suited to simple systems, thermodynamics and kinetics are also unexcelled as tools for the understanding of chemical phenomena in nature.

## BIBLIOGRAPHY

Bockris, J. O., and A. K. N. Reddy. 1970. *Modern Electrochemistry*, Vol. 1. Plenum/Rosetta, New York.

Bohn, H. L. 1992. Chemical activity and aqueous solubility of soil and solid solutions. *Soil Sci.* 154:357–365.

Lindsay, W. L. 1979. *Chemical Equilibria in Soils*. Wiley–Interscience, New York.

Sposito, G. 1981. *The Thermodynamics of Soil Solutions*. Clarendon Press, Oxford, UK.

Stumm, W., and J. J. Morgan. 1995. *Aquatic Chemistry*, 3d ed. Wiley, New York.

## QUESTIONS AND PROBLEMS

1. What is the pH of 0.1 M HCl, 0.02 M $H_2SO_4$, 0.1 M acetic acid ($K = (H^+)(Ac^-)/(HAc) = 10^{-5}$), and 0.1 M $NH_4OH$ ($K = (NH_4^+)(OH^-)/(NH_4OH) = 1.8 \times 10^{-5}$)?

2. What is the pH of a solution containing 0.1 M acetic acid and 0.01 M, 0.1 M, or 0.5 M sodium acetate?

3. A solution initially contains 0.1 M $NH_4OH$ and 0.1 M $NH_4Cl$. Plot the pH change when acid and base are added to this solution. Compare this curve to the pH curve of a $NH_4OH$ solution titrated with HCl. Why are they different?

4. What is the ionic strength of a solution containing 0.015 M $Ca^{2+}$, 0.05 M $Na^+$, 0.030 M $Cl^-$, 0.02 M $SO_4^{2-}$, and 0.01 M $HCO_3^-$? What are the activity coefficients of the ions in this mixture?

5. What is the aqueous concentration of $H_2CO_3$ plus $CO_{2aq}$ in equilibrium with air? In equilibrium with a soil atmosphere of $P_{CO_2} = 0.1$? What are the concentrations of $HCO_3^-$ and $CO_3^{2-}$ in equilibrium with $P_{CO_2} = 0.1$ at pH 6, 7, and 8?

6. Convert the following values to SI units:
   (a) 10 meq/100 g
   (b) 2 mmho/cm
   (c) 100 kg/ha-yr
   (d) 150 lb/acre
   (e) 9 millimicrons
   (f) 7.4 A
   (g) 10 me/100 g

7. Calculate the $\Delta G$ change during the change from feldspar to gibbsite, soluble silica, and $K^+$ (all components in their standard states). What is the further $\Delta G$ change if all the activities of soluble silica and $K^+$ are $10^{-4}$?

8. The $(Al)(PO_4)$ solubility product of variscite is about $10^{-21}$. Calculate the $\Delta G$ of the reaction of soluble silica with variscite to form kaolinite at standard conditions and at $Si(OH)_{4aq} = 10^{-4}$.

9. Give the expression for the solubility product and complex stability constant of the hypothetical substance MA. Why do the two expressions differ? When is that assumption invalid?

10. The tendency of minerals is to go to the lowest possible energy state. Why, then, are soils, which are metastable, formed?

11. Describe the stepwise process of changes in water when an ion enters an aqueous solution. What are the two major characteristics of an ion that govern the ion's interaction with water?

12. What effect does one ion in solution have on another? How are these effects taken into account?

13. What masses of which salts would be required to produce 1 L of the solution in Problem 4 (assume the salts are water-free)?

14. Calculate (a) the activity of $Fe^{3+}$ in equilibrium with $Fe(OH)_3$ at pH 7 and (b) the activity of $FeEDTA^-$ in equilibrium with $Fe(OH)_3$ at pH 7 if 0.01 M $H_4EDTA$ was added to the solution.

15. Calculate the pH in (a) a 0.005 M solution of HCl, and (b) a 0.005 M solution of acetic acid. Explain why the pH values are different.

16. For a pesticide initially present at a concentration of 25 mg/kg of soil and having a half-life of approximately 15 days under the present field conditions, what concentration of the pesticide will remain after 110 days?

# 4

# OXIDATION AND
# REDUCTION

Oxidation and reduction dramatically change the behavior of the chemical elements. *Oxidation* is the loss or donation of electrons by an element; *reduction* is the gain or acceptance of electrons. Oxidation of one substance and reduction of another always occur together—free electrons do not exist in chemical reactions. A substance can donate electrons only if another substance can accept them. The importance of oxidation–reduction (*redox*) reactions is that energy, the energy of life, is transferred by these electron transfers. Oxygen, carbon, nitrogen, and sulfur—and to a lesser extent, iron and manganese—are the primary elements that carry out electron transfer, energy transfer, in the metabolism of living organisms and in soils.

The hydrogen ion $H^+$ and the electron $e^-$ have been called the two master variables that govern chemistry. Together, the availability of $H^+$ and $e^-$ determines the direction and rate of almost all organic and many inorganic reactions. The availability of $H^+$ and that of $e^-$ are similar conceptually but different in reality. The $H^+$ availability, or potential, is related to its concentration in water and can be measured by the familiar pH glass electrode. The glass membrane around the electrode shields the electrode from other possible reactions. The measurement is so good that some people define pH as the $H^+$ potential, even though ions cannot be measured unequivocally. The measured pH is probably very close to the $H^+$ potential, but we cannot know for certain. The pH electrode is the best example of *ion-selective electrodes*, where a selective membrane isolates the electrode from all substances but the desired substance in an aqueous solution.

In contrast, the $e^-$ availability and $e^-$ potential are measurable under some conditions, but there is no concentration of free electrons that corresponds to the $H^+$ concentration. Someone calculated that the concentration of free electrons is about $10^{-45}$ M, or about 1 free electron per galaxy. Also in contrast to $H^+$, the measurement of the $e^-$ potential is qualitative in natural systems. Because the measuring electrode is nonselective and has no shield analogous to the glass membrane of the pH electrode, many ions and molecules can donate and accept electrons from the measuring electrode. Each substance has its own degree of electron availability and hence its own potential at the electrode. The electrode's voltage is therefore a mixed potential, a weighted sum of the potentials of all the electron transfers at the electrode surface. The potential does not represent the $e^-$ potential of any ion or molecule at the electrode. When only one substance is present in the system, the electron's potential may be measurable if the kinetics of electron transfer between the substance and the electrode surface are favorable.

Virtually all biological reactions and many inorganic reactions in soils are redox reactions. Dioxygen $O_2$ is the major and final electron acceptor (*oxidizing agent*) in nature and therefore buffers the $e^-$ availability in *aerobic* systems (where $O_2$ is available). Dioxygen diffuses through soil pores to plant roots, soil microbes, and inorganic substances from the atmosphere. Until the soil becomes quite wet and/or the oxygen demand is high, the oxygen diffusion rate is usually rapid enough to maintain adequate oxygen availability. Even if only the larger soil pores are open to the atmosphere, the oxygen supply can be sufficient because gas diffusion through the gas phase is 10 000 times faster than gas diffusion through water.

If the diffusion path length from the soil surface through the soil solution is long, combined with a high oxygen demand from actively metabolizing roots and microbes, oxygen may be lacking. Oxygen deprivation (*anaerobic* conditions) slows the rates of root metabolism and ion uptake, weakens plant resistance to soil pathogens, and increases the concentration of undesirable reduced ions in the soil solution. Oxygen dissolved in water in a flooded soil can supply oxygen for about 24 hours to plants.

Much of the earth's soils are flooded or very wet for part or all of the year, subsoils have restricted water drainage and low oxygen concentrations, and the interior pores of soil aggregates in aerobic soils can have considerably lower oxygen concentrations than does the atmosphere. The ocean depths have restricted access to atmospheric oxygen. We think of aerobiosis as the ideal and common state of nature, but that is egocentrism. Most of the living earth has a limited oxygen supply.

Carbon dioxide is formed when oxygen accepts electrons from carbon compounds during metabolism. The $CO_2$ diffuses away in the same way as oxygen diffuses toward the reaction site. If diffusion rates are slow, as in flooded soils, the $CO_2$ and $H_2CO_3$ concentrations increase and they begin to buffer pH. The pH range of oxygen-deprived regions and poorly drained soils is therefore narrower than that of well-drained soils.

Agricultural practices can change the soil's ability to supply oxygen. Irrigation, cultivation, introduced crops, lower plant densities, and the shorter growing season of agricultural crops change the soil's water content and therefore the change the pore space available for gas transfer between the root zone and the atmosphere. For example, large areas of the midwestern North America have tile drains to remove the water that accumulates during the crop season and reduces oxygen availability to roots. The cultivated plants are less dense and have a shorter growing season than native plants so water accumulates more. The native plant population transpires more water and lessens water accumulation in the soil. Cultivation destroys the large soil pores through which gases and water move most rapidly. Cultivation also destroys the organic matter content, which maintains an open soil structure and improves permeability.

## 4.1    SOIL OXIDATION–REDUCTION

Redox reactions in the soil are mostly the result of a cycle started by photosynthesis. One part of the reaction is

$$CO_2 + 4e^- + 4H^+ = CH_2O + H_2O \tag{4.1}$$

where $CH_2O$ represents a carbohydrate. Carbon in $CO_2$ accepts electrons and its oxidation state changes from the $C^{4+}$ in $CO_2$ to $C^0$ in carbohydrate $(CH_2O)_n$.

Simultaneously, the $O^{2-}$ in water gives up electrons as it oxidizes to $O^0$ in $O_2$:

$$2H_2O = O_2 + 4e^- + 4H^+ \tag{4.2}$$

In these reactions $O^{2-}$ is the *electron donor*, and $C^{4+}$ is the *electron acceptor*. Equations 4.1 and 4.2 are called *half-reactions* because they describe only half of the reaction. Although half-reactions appear to imply that free electrons exist, half-reactions imply only that the other half of the reaction is unspecified. The overall reaction of photosynthesis is the sum of the half-reactions:

$$CO_2 + H_2O = CH_2O + O_2 \tag{4.3}$$

Respiration (oxidation) in plants and animals and oxidation in soils complete the photosynthetic cycle by utilizing the energy stored in the carbohydrates and organic compounds derived from the carbohydrates, by disposing of organic wastes, and by producing the $CO_2$ needed for more photosynthesis by the reaction:

$$CH_2O + H_2O = CO_2 + 4e^- + 4H^+ + energy \tag{4.4}$$

To obtain the energy and complete the reaction, organisms must find an electron acceptor to accept the electrons. If oxygen is available, the half-reaction of aerobic electron acceptance is the reverse of Eq. 4.2:

$$O_2 + 4e^- + 4H^+ = 2H_2O \tag{4.5}$$

Equation 4.4 summarizes the many steps of the intricate Krebs or citric acid cycle that organisms utilize to obtain the energy in a useful form. Equation 4.5 also oversimplifies the intricate mechanism of electron acceptance by oxygen in living organisms.

A key to obtaining the energy in organic compounds, and thus to sustain life, is to obtain an electron acceptor. Higher plants and animals can utilize only $O_2$ as an electron acceptor, but microbes in soils and elsewhere can also utilize the oxidized states of nitrogen, sulfur, iron, manganese, and other elements as electron acceptors. These electron acceptors do not release all of the photosynthetic energy and retain some of the energy in the reduced states. The energy content of these reduced states makes them reactive and potential pollutants in our concept of a healthy, that is, aerobic, environment. Reacting further with $O_2$ changes the partially oxidized compounds to higher, more benign, oxidation states.

Redox reactions of C, N, and S compounds are catalyzed by enzymes. Catalysis is necessary because most elements exchange electrons reluctantly. Enzymes lower the activation energy of electron transfer and increase reaction rates enormously. The reluctance of C, N, and S compounds to reach equilibrium creates the metastability of carbon compounds and prevents you the reader and the paper of this page from immediately oxidizing to $CO_2$. The irreversibility of electron transfer is a nuisance for physical chemists who like the simplicity of equilibrium, but is essential for life.

## 4.2  ELECTRON DONORS

The major electron donors in soils are the carbon compounds in living roots and microbes, in dead plant matter, and in soil organic matter (SOM). Table 4.1 shows the approximate C, H, and O contents of the two largest plant components, cellulose and lignin, and of typical SOM. For simplicity, Table 4.1 ignores the amounts of N, S, P, and other elements in these materials. Assuming that plant matter contains $1/3$ lignin and $2/3$ cellulose, an empirical formula for "plant matter" is approximately $C_{1.7}H_{2.2}O$. Assuming further that all the carbon in this material oxidizes to $C^{4+}$ in $CO_2$, the half-reaction of plant matter oxidation is

$$C_{1.7}H_{2.2}O = 1.7C^{4+} + H_2O + 0.2H^+ + 7e^- \tag{4.6}$$

The reaction does not go to completion immediately. Some carbon remains as SOM—microbial biomass and partially metabolized by-products.

The SOM in Table 4.1 is richer in carbon than is plant matter. SOM tends to contain more aromatic (cyclic and resonating carbon–carbon bond) compounds, and contains less oxygen, than the plant matter from which it is derived. Alternatively, because cellulose oxidizes faster than lignin, the aromatic groups may represent an accumulation of aromatic carbon from unreacted lignin. All of these materials eventually oxidize in soils, but each succeeding oxidative step is much slower. The half-

**Table 4.1. Approximate C, H, and O composition of lignin, cellulose, and soil organic matter (nitrogen, sulfur, and other elements are ignored)**

|  | C(%) | H(%) | O(%) | Empirical Formula |
|---|---|---|---|---|
| Lignin | 61–64 | 5–6 | 30 | $C_{2.8}H_{2.9}O$ |
| Cellulose | 44.5 | 6.2 | 49.3 | $C_{1.2}H_2O$ |
| Soil organic matter | 58 | 5 | 36 | $C_{2.2}H_{2.2}O$ |

reaction for the oxidation of soil organic matter is

$$C_{2.2}H_{2.2}O = 2.2C^{4+} + H_2O + 0.2H^+ + 9e^- \tag{4.7}$$

Soil organic matter contains amino (—$NH_2$) and sulfhydryl (—SH) groups, which also are electron donors.

Inorganic electron donors in soils are generally in much smaller amounts and include sulfide $S^{2-}$, sulfur $S^0$, $Fe^{2+}$, $Mn^{2+,3+}$, and ammonia $N^{3-}$. The reduced oxidation states of the trace elements Cr, Cu, Mo, Hg, As, and Se are also electron donors in soils.

## 4.3 ELECTRON ACCEPTORS

The role of soil in the oxidation of reduced C compounds is to provide electron acceptors for plant roots and microbes. Oxygen is the strongest electron acceptor in nature and yields the most energy from oxidation (Eq. 4.5). Oxygen is also the only electron acceptor that plant roots can utilize. Oxygen is made available by diffusion through soil pores and by being dissolved in the soil solution. At soil temperatures, the dissolved $O_2$ concentration is about 10 mg $L^{-1}$ so that the $O_2$ in air and water are about the same, on a volume basis.

The $O_2$ supply can be insufficient because soil pores are water-filled. The $O_2$ supply in unsaturated soils can also be less than the microbial and plant root demand due to a large supply of readily decomposable organic matter. Plant root demand for $O_2$ is relatively constant while microbial demand fluctuates widely in response to organic inputs. High oxygen demand, relative to oxygen supply, also occurs in soils affected by leaks from natural gas pipes or used for organic waste disposal. Since oxygen diffusion from the surface is relatively slow, oxygen becomes deficient.

Soil microorganisms, in contrast to higher plants and animals, can utilize other electron acceptors if $O_2$ is unavailable. The prominent secondary electron acceptors in soils and their half-reactions are

$$FeOOH + e^- + 3H^+ = Fe^{2+} + 2H_2O \tag{4.8}$$

$$2MnO_{1.75} + 3e^- + 7H^+ = 2Mn^{2+} + 3.5H_2O \tag{4.9}$$

where $MnO_{1.75}$ signifies the complex Mn(III–IV) oxides in soils.

$$SO_4^{2-} + 8e^- + 8H^+ = S^{2-} + 4H_2O \tag{4.10}$$

$$NO_3^- + 8e^- + 9H^+ = NH_3 \text{ and amino acids } + 3H_2O \tag{4.11}$$

$$NO_3^- + 2e^- + 2H^+ = NO_2^- + H_2O \tag{4.12}$$

$$2NO_3^- + 10e^- + 12H^+ = N_2 + 6H_2O \tag{4.13}$$

$$2NO_3^- + 8e^- + 10H^+ = N_2O + 5H_2O \tag{4.14}$$

The conditions that govern the endproducts of the nitrogen reactions are not yet well understood. The formation of $N_2O$ (nitrous oxide) is of interest because it is a long-lived "greenhouse" gas in the atmosphere and its concentration is increasing. Nitrous oxide is often released initially after nitrate fertilizers are added to soils.

In addition to yielding less energy than $O_2$, these secondary electron acceptors also yield products unfavorable to agriculture and aquaculture. The reduced oxidation states are more toxic than the oxidation states that are stable in the presence of $O_2$. Ammonia and nitrite, for example, are more toxic than nitrate; $H_2S$ is more toxic than sulfate. Reduction of Fe(III) and Mn(III–IV) can cause phytotoxic $Fe^{2+}$ and $Mn^{2+}$ concentrations in rice paddies. Reduction of $NO_3^-$ to gaseous $N_2$ and $N_2O$ is agriculturally undesirable because soil nitrogen is lost.

If both oxygen and secondary electron acceptors are absent, microorganisms in soils and other systems can still extract some energy from photosynthetically produced compounds by *fermentation*. Microbes can rearrange carbon compounds into more stable structures and release about 10% of the total energy in the initial compound. The products of fermentation include ethanol ($C_2H_5OH$), methane ($CH_4$), peat, and $CO_2$. In geologic time, further nonmicrobial reactions produce coal and petroleum. The fermentation products retain about 90% of the original energy and are useful fuels.

Fermentation and reduction of secondary electron acceptors are temporary expediencies for soil microbes. The resulting products are unstable in the presence of oxygen and eventually oxidize further when oxygen becomes available. SOM is a beneficial result of incomplete oxidation and fermentation. The SOM content reflects the difference of the rates of organic matter addition vs. oxidation rates. The rate of addition is essentially the rate of net photosynthesis. The oxidation rate is governed by temperature and by the rate of oxygen supply.

## 4.4   REDOX REACTIONS

The tendency of a substance to donate or accept electrons, and the measure of the electron's availability, is given by its *electrode potential*. In principle, electrode potentials can be measured directly by an electrode and a voltmeter. All chemical elements can transfer electrons and thus change their oxidation states. Table 4.2 shows the standard electrode potentials of the half-reactions of several elements. The general reaction is

$$Ox + e^- = Red \tag{4.15}$$

where Ox is an oxidized state of the element and Red is a reduced state.

High electrode potentials mean that Ox is available and readily accepts electrons. The halogen gases, for example, have high electrode potentials and thus are strong oxidizing agents; they want to accept electrons and be reduced. Low electrode potentials mean that Red is available and readily donates electrons. The alkali metals are strong electron donors, are strong reducing agents, and are avid electron donors.

**Table 4.2. Electrode potentials (reduction potentials) of selected half-reactions at 25° C. (The dashed lines show the limits of electrode potential in aqueous systems).**

| Reaction | $Eh^0$ (V) |
|---|---|
| $F_2 + 2e^- = 2F^-$ | +2.87 |
| $Cl_2 + 2e^- = 2Cl^-$ | 1.36 |
| $NO_3^- + 6H^+ + 5e^- = \frac{1}{2}N_2 + 3H_2O$ | 1.26 |
| - - - - - - - - - - - - - - - - - - - - - - - - - - - | |
| $O_2 + H^+ + 4e^- = 2H_2O$ | 1.23 |
| $NO_3^- + 2H^+ + 4e^- = NO_2^- + H_2O$ | 0.85 |
| $Fe^{3+} + e^- = Fe^{2+}$ | 0.77 |
| $SO_4 + 10H^+ + 8e^- = H_2S + 4H_2O$ | 0.31 |
| $CO_2 + 4H^+ + 4e^- = C + 2H_2O$ | 0.21 |
| $N_2 + 6H^+ + 6e^- = 2NH_3$ | 0.09 |
| $2H^+ + 2e^- = H_2$ | 0 |
| - - - - - - - - - - - - - - - - - - - - - - - - - - - | |
| $Fe^{2+} + 2e^- = Fe$ | −0.44 |
| $Zn^{2+} + 2e^- = Zn$ | −0.76 |
| $Al^{3+} + 3e^- = Al$ | −1.66 |
| $Mg^{2+} + 2e^- = Mg$ | −2.37 |
| $Na^+ + e^- = Na$ | −2.71 |
| $Ca^{2+} + 2e^- = Ca$ | −2.87 |
| $K^+ + e^- = K$ | −2.92 |

The range of electrode potentials possible in soils is limited by the stability of water with respect to oxidation and reduction. High electrode potentials can oxidize water to $O_2$. Low electrode potentials can reduce water to $H_2$. If a solution contains an oxidizing agent such as $Cl_2$ with an electrode potential greater than that of the $H_2O$–$O_2$ *couple*, or half-reaction, the oxidizing agent can oxidize water to $O_2$:

$$\frac{1}{2}O_2 + 2e^- + 2H^+ = H_2O \tag{4.16}$$

The oxidation of water to $O_2$ prevents the electrode potential from rising above the electrode potential of Eq. 4.6. Strong oxidizing agents (high electrode potential) such as hypochlorite ($ClO^-$) and $Cl_2$ are unstable and decompose in soils and water by reducing to $Cl^-$ as they oxidize water to $O_2$. Although oxidizing agents stronger than $O_2$ should not in principle be formed in an aqueous system, $N_2O$ is an apparent exception. Nitrous oxide has a high electrode potential, but nitrogen electron transfers are sluggish and irreversible. Not only is $N_2O$ formed in soils, it is somewhat stable in soils and is stable for many years in the atmosphere.

Strong reducing agents, in contrast, can reduce water to $H_2$:

$$2H^+ + 2e^- = H_2 \tag{4.17}$$

The $H^+$ ion is reduced rather than $H_2O$, but $H^+$ is always present in aqueous solutions because water dissociates. The $H^+$–$H_2$ couple (Eq. 4.17) is the lower limit of electrode potential in aqueous systems and soils.

The stability of water with respect to oxidation to $O_2$, and reduction to $H_2$, limits the range of electrode potentials and the oxidation states possible in any system containing water. These limits are the dashed lines in Table 4.2. Dissolved $F_2$ and $Cl_2$ gases are unstable because of their high electrode potentials. The reduced states $F^-$ and $Cl^-$ are the stable states in water because they have accepted an electron and thereby discharged their oxidizing power. Metals are likewise unstable in water and soils because of their tendency to oxidize, to donate electrons. Their oxidized states ($Al^{3+}$, $Ca^{2+}$, $K^+$, etc.) are stable in water. Table 4.2 implies correctly that the common metals are unstable and will corrode. Exceptions such as aluminium and zinc metals are metastable; an oxide layer that forms initially on their surfaces inhibits further oxidation. Soils and seawater catalyze the breakdown of these protective layers and speed up their corrosion (oxidation). Iron and steel do not form this protective layer.

For the redox couples between the dashed lines in Table 4.2, both states are stable in water and soil, depending on the electron availability. Under reducing conditions $Fe^{2+}$, sulfide, and ammonia are stable. If $O_2$ is available, Fe(III), $SO_4^{2-}$, and $NO_3^-$ are stable.

Soil microorganisms utilize the strongest electron acceptors available, in order to obtain the maximum energy from their food substrate. If the $O_2$ supply is insufficient, the next strongest electron acceptor available in soils is nitrate. The manganese(IV,III–II) redox couple may be stronger, but Mn(IV,III) oxides are solids, which cannot diffuse to the microbes, so their availability is low. Nitrate reduces to amino acids, $N_2$ or $N_2O$. After oxygen and nitrate have been exhausted, Fe(III) and Mn(IV,III) hydroxyoxides can be reduced, and $Fe^{2+}$ and $Mn^{2+}$ concentrations in the soil solution increase. If the rate of electron acceptance by these acceptors is less than the availability of food sources, even stronger reducing conditions result. Microbes can reduce sulfate to sulfur or sulfide, ferment organic matter to $CO_2$ and methane or peat, and in extreme cases reduce water to $H_2$.

The stepwise order of reduction in Table 4.3 is idealized. The rate of electron availability is usually much faster than the rate at which electron acceptors can dissolve and diffuse to the soil microbes. The reactions then overlap and several proceed simultaneously.

Under slower conditions in the laboratory, the order of utilization of electron acceptors follows the order of electron potentials at pH 7, as listed in the second column of Table 4.3. The standard electrode potentials of Table 4.2 are at pH 0. Most electrode potentials change similarly with pH, however, so the order of standard electrode potentials and of utilization of electron acceptors are similar. Appendix 4.1 describes how the pH dependence of electrode potentials can be calculated.

The third column of Table 4.3 lists the measured ranges of potentials over which each reaction occurs in soils. The potentials are measured with a platinum electrode whose potential responds roughly to the electron availability. Reasons for the difference between electrode and redox potentials are discussed in Appendix 4.2. In

**Table 4.3. Order of utilization of principal electron acceptors in soils, equilibrium potentials of these half-reactions at pH 7, and measured potentials of these reactions in soils**

| Reaction | $Eh$ at pH 7 (V) | Measured Redox Potential in Soils (V) |
|---|---|---|
| $O_2$ disappearance | | |
| $\frac{1}{2} O_2 + 2e^- + 2H^+ = H_2O$ | 0.82 | 0.6 to 0.4 |
| $NO_3^-$ disappearance | | |
| $NO_3^- + 2e^- + 2H^+ = NO_2^- + H_2O$ | 0.54 | 0.5 to 0.2 |
| $Mn^{2+}$ formation | | |
| $MnO_2 + 2e^- + 4H^+ = Mn^{2+} + 2H_2O$ | 0.4 | 0.4 to 0.2 |
| $Fe^{2+}$ formation | | |
| $FeOOH + e^- + 3H^+ = Fe^{2+} + 2H_2O$ | 0.17 | 0.3 to 0.1 |
| $HS^-$ formation | | |
| $SO_4^- + 9H^+ + 6e^- = HS^- + 4H_2O$ | $-0.16$ | 0 to $-0.15$ |
| $H_2$ formation | | |
| $H^+ + e^- = \frac{1}{2}H_2$ | $-0.41$ | $-0.15$ to $-0.22$ |
| $CH_4$ formation (example of fermentation) | | |
| $(CH_2O)_n = n/2\ CO_2 + n/2\ CH_4$ | — | $-0.15$ to $-0.22$ |

addition, part of the reason for the range in Table 4.3 is that the potentials were measured at soil pH values other than 7, and redox potentials are pH dependent.

Redox conditions in soils vary widely over short distances because $O_2$ must diffuse through pores of various sizes and water-filled pores. In aerobic soils the interior of soil aggregates may be partially anaerobic. The change from oxygen sufficiency to deficiency can occur within a few millimeters. In wet soils, only the largest pores are open to gas diffusion from the atmosphere.

Redox conditions in saturated and flooded soils can be more homogenous and redox measurements with the platinum electrode tend to be more reliable. Some water-loving plants such as rice, however, can conduct oxygen through the stem to the root. The soil immediately surrounding such roots is oxidized compared to the rest of the flooded soil. Convection currents in overlying water also bring oxygen downward, so that submerged sediments are often topped by a thin oxidized layer.

## 4.5  FLOODED SOILS

Ponnamperuma (1972) reviewed the chemistry of flooded soils. This research has understandably concentrated on the soil conditions of paddy rice agriculture. Only some generalizations are mentioned here. The behavior of C, N, S, Fe, and Mn generally follows that shown in Table 4.3. When rice paddies are drained before harvest, redox potentials rise, $Fe^{2+}$ and $Mn^{2+}$ concentrations decrease, and C, N, and S oxidize. When the soils are flooded again, the reactions reverse.

Phosphate apparently precipitates as Fe(III) and Al phosphates during the dry part of the rice culture cycle. Under subsequent reducing conditions, the Fe(III) phosphate is reduced to more soluble Fe(II) phosphate. This reduction can account for the rather high availability of phosphate for centuries in paddy soils. The Fe(III) phosphate may be slightly more stable than Al phosphate so the Al phosphate that precipitates initially slowly transforms to Fe(III) phosphate. Similar aerobic soils often supply inadequate phosphate to plants because Fe(III) phosphate remains insoluble.

A second noteworthy flooded soil is *acid sulfate* soil. Sediments along tropical and subtropical coastlines and river deltas may contain significant quantities of Fe(II) sulfides. When drained, these sulfides oxidize to $H_2SO_4$ and the acidic $Fe^{3+}$ ion. The soil acidity can increase to pH 2. Such conditions are highly phytotoxic and can be remedied under aerobic conditions only by extensive leaching and lime applications. If resubmerged, acid sulfate soils revert rapidly to near neutrality as the Fe(III) and sulfate are reduced back to Fe(II) sulfides.

A third example of flooded soils is the extensive areas of soils rich in organic matter, peat and muck soils, or *Histosols*. The slow rate of organic matter oxidation is due to slow $O_2$ diffusion through stagnant water; to low concentrations of mineral nutrients; and, in the most extensive areas of Histosols in Canada and Siberia, to low temperatures. Peat soils are to some degree self-perpetuating. They have a high water-holding capacity, which allows them to spread up slight slopes above the water table. In warm seasons the high specific heat of water slows the warming rate of the peat. During occasional dry seasons the high thermal resistivity of peat slows warming below the immediate surface.

## APPENDIX 4.1   ELECTROCHEMISTRY

Electrode, electron, reversible, and equilibrium potentials and electron activity are closely related terms for the equilibrium potential of the electron, where potential means availability and driving force. Equilibrium implies that all electron transfers are reversible, that a small change of electron potential will bring about a corresponding electron transfer. Table 4.2 lists such reversible potentials for half-reactions of some common ions. Reversible electron transfers, however, are rare. Irreversible reactions (Appendix 4.3), in which the electron transfer is hindered by an activation energy barrier, are the norm. This barrier can be overcome by extra energy or *overvoltage*. Enzyme catalysis and soil surface catalysis can reduce, but not completely remove, this activation energy barrier. As a result of irreversibility, redox reactions in soils respond sluggishly to changes in electron potential. The irreversibility is relatively low for iron reactions; carbon and nitrogen electron transfers are exceedingly irreversible.

Oxidation–reduction equilibrium also implies that the electrode potentials of all redox couples in the system are equal. Because of irreversibility, this condition is rare in mixtures of redox couples, especially in mixtures containing organic and nitrogen compounds such as the soil solution.

Despite these limitations, equilibrium is a convenient starting point from which to study redox reactions. A generalized redox half-reaction is

$$Ox + ne^- + mH^+ = Red + \frac{m}{2}H_2O \tag{4.18}$$

where Ox is the oxidized species and Red is the reduced species of the redox couple. The equilibrium equation for this reaction is

$$K = \frac{Red}{Ox(e^-)^n(H^+)^m} \tag{4.19}$$

where $(e^-)$ has been called *electron availability*, *electrode potential*, and *electron activity*. The electron activity has no relation to ion activities in solution because the concentration of free electrons in solution is vanishingly small.

Electron (or electrode) potential is given the symbol $Eh$, while electron activity is associated with pe. Equation 4.19 can be transformed into the Nernst equation:

$$Eh - Eh^0 = \frac{RT}{nF} \log \frac{Red}{(Ox)(H^+)^m} = -\frac{0.059}{n} \log \frac{Red}{Ox} - \frac{0.059}{n/m} pH \tag{4.20}$$

where $Eh^0$ is the standard electrode potential, $R$ is the gas constant, $T$ is absolute temperature, $F$ is the Faraday constant, and 0.059 is the quotient of the constants at 25° C. The standard potential must be defined in terms of an arbitrary reference state because, like free energy, the absolute potential cannot be defined. The $Eh^0$ is related to the standard Gibbs free energy by

$$G^0 = -nF Eh^0 = -96.48n Eh^0 \tag{4.21}$$

where $G^0$ is in joules.

The electron potential pe is an alternative expression of electron availability. From Eq. 4.19,

$$pe + pH = \log K - \log(Red) + \log(Ox) \tag{4.22}$$

when $n = m$. Adding pe and pH is advantageous because the sum can be used as one axis of a two-dimensional graph, while the activities of the reduced and oxidized species are plotted on the other axis. Adding pe + pH also makes arithmetic sense because the numerical range of pe and pH are similar. Each has approximately equal weight in the sum, in accord with their joint importance at equilibrium. Adding pe + pH, however, implies that both are at equilibrium, and that is rarely the case for pe. Nonetheless, pe + pH diagrams can summarize a great deal of thermodynamic data and provide a picture of the behavior of chemical elements in nature.

At 25°C,

$$pe = 0.059Eh \tag{4.23}$$

## APPENDIX 4.2   *Eh* AND pe

The range of electrode potentials in systems containing water is limited by the stability of water with respect to oxidation and reduction (Eqs. 4.16 and 4.17). Substituting into the Nernst equation (Eq. 4.20) yields

$$Eh = Eh^0 - \frac{0.0059}{4} \log \frac{1}{P_{O_2}} - 0.059 \, \text{pH} \tag{4.24}$$

for the upper limit of oxidizing conditions in soils and water, and for the lower limit,

$$Eh = Eh^0 - \frac{0.059}{2} \log \frac{1}{P_{H_2}} - 0.059 \, \text{pH} \tag{4.25}$$

where $P_{O_2}$ and $P_{H_2}$ are the partial pressures of oxygen and hydrogen gases. Gases are essentially ideal at low pressures so the partial pressure and activity are equal. The $Eh$ is dependent on the pH and the gas concentration but changes only 15 mV per tenfold change of $P_{O_2}$ and 30 mV per tenfold change of $P_{H_2}$. The $Eh$ limits of water, and therefore of soil and biological systems, are plotted in an $Eh$–pH diagram (Fig. 4.1a) at unit partial pressures of $O_2$ and $H_2$. Water is stable between the lines; dioxygen is stable at oxidizing conditions and electrode potentials above the upper line. Dihydrogen is stable at potentials below the lower line.

Equation 4.19 can be transformed to

$$K = \frac{(H_2O)^{1/2}}{(H^+)(e^-)(P_{O_2})^{1/4}} \tag{4.26}$$

and

$$\text{pe} + \text{pH} = 20.78 + \tfrac{1}{4} \log P_{O_2} \tag{4.27}$$

The pe + pH values range from 20.78 at highly oxidizing conditions at $pO_2 = 1$ and pH 0, to pe + pH = 0 at strongly reducing conditions when $pH_2 = 1$ (Fig. 4.1b).

Carbon, nitrogen, sulfur, and iron are elements that lend themselves to redox considerations. Interested readers should consult Pourbaix (1966) and Garrels and Christ (1965) for $Eh$–pH diagrams for other elements, or Lindsay for pe + pH plots. All of these diagrams assume equilibrium and that $Eh$ can be measured accurately. Both assumptions are questionable, but some success has been obtained measuring changes in Fe chemistry in soils. Carbon $Eh$–pH and pe + pH diagrams are not very interesting. The stability of $CO_2$ totally dominates the diagrams. Carbon compounds are stable only in a narrow strip near the $H^+$–$H_2$ boundary, that is, they are stable only under strong reducing conditions.

### A4.2.1   Nitrogen

$Eh$–pH and pe + pH diagrams can qualitatively describe nitrogen chemistry at the earth's surface, including simple interactions with biological systems. The chemistry

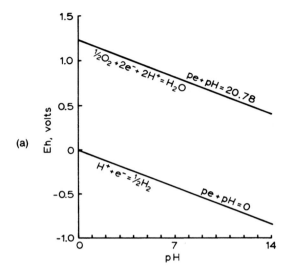

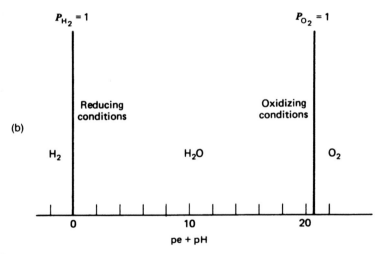

**FIGURE 4.1.** The range of oxidizing and reducing conditions as shown in (*a*) *Eh*–pH and (*b*) pe and pH diagrams. The range is defined by the breakdown of water to $H_2$ or $O_2$.

of nitrogen is usually discussed in terms of the microorganisms and the reaction mechanisms that they carry out. Microbes and enzymes are only catalysts, however, and perform only those redox reactions that electron availability permits. *Eh* and pe describe that electron availability.

Nitrogen has many oxidation states that might be stable within the redox stability range of water. By comparing the electrode potentials of all possible redox couples, the stable oxidation states can be sorted out. For example, nitrate is reduced to nitrite ($NO_2^-$) at $Eh^0 = 0.95$ volts (V), and nitrite reduces to $N_2$ at $Eh^0 = 1.25$ V. Those potentials mean that if a reducing agent is added to a nitrate solution, all of the nitrate will reduce to $N_2$ before the electrode potential increases enough to reduce nitrate to nitrite. Nitrite ions are therefore unstable and should spontaneously decompose to $N_2$. Nitrite solutions are indeed unstable but decompose only slowly, because nitrogen redox equations are invariably irreversible.

The stable oxidation states of nitrogen within the stability limits of water turn out to be nitrate, $N_2$, and ammonia. Their oxidation–reduction relationships are shown in Fig. 4.2. The stability region is dominated by $N_2$. Nitrate is stable only at strongly oxidizing conditions and pH > 3. Ammonia is stable only under reducing conditions. Although not shown, the stability regions of $NO_3^-$ and $NH_3$ are concentration dependent, increasing slightly as the their solution concentrations decrease.

Figure 4.2 also shows the stability region of the amino acid alanine ($CH_3$—$CHNH_2$—$COOH$), whose stability is typical of other amino acids. Nitrogen in alanine is more stable to oxidation than is ammonia. This accounts for the rarity of ammonia in nature compared to amino nitrogen. Ammonia is liberated when the carbon required for amino compounds disappears, such as in the composting of manure and in wastewater treatment. The alanine–$N_2$ boundary is shown as a dashed line because it is not a true equilibrium. Alanine is unstable with respect to oxidation to $CO_2$. The alanine–$N_2$ boundary is a steady state and is maintained by the continuous production of amino acid precursors by photosynthesis.

A protein–$N_2$ boundary unfortunately cannot be calculated, because the energies of formation of proteins are not yet known. Proteins are presumably more stable than amino acids, because free amino acids such as alanine are rare in nature, while proteins are common. The protein–$N_2$ boundary would represent the highest potential

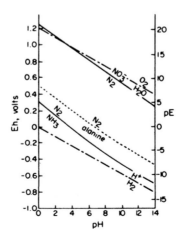

**FIGURE 4.2.** *Eh*–pH diagram of the N–$H_2O$–$O_2$–amino acid system.

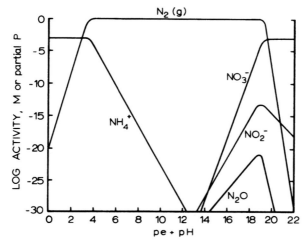

**FIGURE 4.3.** Distribution of soluble and gaseous nitrogen species. Total soluble nitrogen is assumed to be $10^{-3}$ M. (Adapted from Lindsay (1979).)

(the strongest oxidizing condition), at which $N_2$ could be converted to protein by soil microorganisms or other means during nitrogen fixation.

Figure 4.3 shows some pe + pH relations of nitrogen. This diagram has the advantage of being able to plot the concentration dependence of the redox equilibria. The vertical axis is the aqueous activity of dissolved ions or the partial pressure of gases. The concentration of soluble nitrogen is assumed to total $10^{-3}$ M and the $N_2$ partial pressure is assumed to be that of the atmosphere, 0.78. Nitrite and nitrous oxide can be present in only exceedingly low concentrations at equilibrium. Higher concentrations of $NO_2^-$ and $N_2O$ (and NO and the $NO_2^- - N_2O_4$ pair) are unstable; they will tend to degrade to $NH_3$, $N_2$, or $NO_3^-$.

The *Eh*–pH diagram shows only the "stable" nitrogen species over the range of electron potentials in aqueous systems. The pe + pH diagram shows that the "stable" species are actually only the most prominent ones. The "unstable" species are also present but in much lower concentrations.

The overall distribution of nitrogen compounds roughly follows the distribution suggested by Figs. 4.2 and 4.3. Most of the world's nitrogen exists as $N_2$. Some also exists as amino nitrogen in reduced carbon compounds, living organisms, and dead organic matter. A very small fraction exists as nitrates. If equilibrium existed, the nitrate fraction would be much larger because of oxidation of organic N, and oxidation of atmospheric $N_2$ to $NO_3^-$:

$$N_2 + 2.5O_2 + H_2O = 2HNO_3 \tag{4.28}$$

Almost all the $O_2$ in the atmosphere would be consumed. The remaining atmosphere would contain perhaps 99% $N_2$ and 1% $O_2$, and the oceans, rain, and "freshwaters" would be dilute $HNO_3$. The reaction is fortunately hindered by the general irreversibility of nitrogen reactions. Reaction 4.28 proceeds in the atmosphere to a limited extent when lightning provides sufficient activation energy.

Nitrogen reactions are generally highly irreversible, and enzymatic catalysis is necessary for nitrogen conversions in soils. Redox irreversibility is unfortunate from the standpoint of studying redox reactions, but is absolutely necessary for life. Reversibility would bring with it equilibrium, and all organisms would be transformed into $CO_2$, $N_2$, $NO_3^-$, and $H_2O$.

Irreversibility means that more energy is required to carry out, and less energy is derived from, reactions. While natural reactions are remarkably conservative of chemical energy, energy optimization is not the only criterion that determines which reactions occur. Unstable compounds, such as nitrite and nitrous oxide ($N_2O$), for example, would not be produced if soil microorganisms were concerned only with optimal utilization of available chemical energy. Spring thaw can bring about temporary nitrite accumulation in soils, apparently because microbial nitrite reducers respond slower to increased temperature than do the microbes that reduce nitrate to nitrite. In addition, nitrate fertilization or wetting of dry soil initially stimulates $N_2O$ production, so that up to 25 to 50% of the nitrogen lost by denitrification can be lost as $N_2O$. Both $N_2O$ and nitrite production appear to be temporary maladjustments during a flurry of microbial activity after a sudden environmental change.

### A4.2.2   Sulfur

Figures 4.4 and 4.5 show the equilibrium relations of the sulfur oxidation states— sulfate, elemental sulfur, and sulfide—which are stable within the stability limits of water. The stable oxidation states are similar to those of nitrogen. Elemental sulfur, however, is much less stable than is elemental nitrogen and is stable only under acid conditions.

Figure 4.4 shows the sulfur stability regions for unit activity of sulfate and sulfide. The region of sulfur stability diminishes with decreasing sulfate and sulfide concentrations. Sulfur in the amino acid cysteine ($HS—CH_2—CHNH_2—COOH$) is more stable against oxidation than is the nitrogen in alanine. The cysteine sulfate boundary is shown as a dashed line because it is unstable at $Eh^0 = 0.31$ V with respect to $CO_2$. Like alanine, the stability of cysteine is a steady state dependent on continuing production of organic compounds. The stability region of cysteine increases with increasing concentrations of sulfate compounds.

Figure 4.5 is a pe + pH diagram of sulfur at pH 7 and $10^{-3}$ M soluble sulfur species. Only $SO_4^{2-}$ and $HS^-$ are stable, with the $HS^-$ stable only under strongly reducing conditions. The soluble sulfur concentration is too low and the pH is too high to permit elemental sulfur stability. Figure 4.5 ignores organic sulfur compounds.

Sulfur redox reactions seem to be more reversible than those of nitrogen. Intermediate compounds in the reaction series from sulfate to sulfur or sulfide, and vice versa, do not appear in soils. Sulfur also differs from nitrogen in that little sulfur volatilizes from soils. Although $H_2S$ is a gas, apparently any $H_2S$ formed in soils reacts rapidly with Fe and other transition metal oxides to form sulfides. Some organic

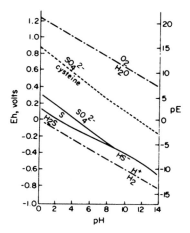

**FIGURE 4.4.** *Eh*–pH diagram of the S–H$_2$O–O$_2$–amino acid system.

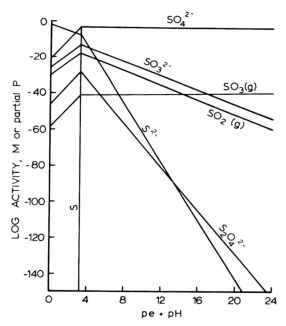

**FIGURE 4.5.** Distribution of solid, gaseous, and aqueous sulfur species at pH 7 and 10$^{-3}$ M total soluble sulfur. (Adapted from Lindsay (1979).)

gases containing —SH groups are occasionally liberated during decomposition of fresh organic matter, but the amounts are small.

### A4.2.3   Iron

Electron exchange of iron(II–III) tends to be more reversible than is electron exchange between nitrogen, sulfur, or carbon states. Iron redox reactions occur in soils without enzymatic catalysis. The Fe(II) minerals in parent material rocks oxidize spontaneously, though slowly, in aerobic soils. The electron availability for subsequent Fe redox reactions in soils is determined by microbial oxidation of carbon compounds. The reduction of Fe(III) in acid solutions is

$$Fe^{3+} + e^- = Fe^{2+} \tag{4.29}$$

The concentration of Fe(III) at soil pH levels is very low because of the insolubility of Fe(III) hydroxyoxides. As a result Eq. 4.29 consumes a negligible number of electrons. The major reaction by which Fe(III) accepts electrons in soils is the reduction of solid-phase Fe(III) hydroxyoxide:

$$FeOOH + e^- + 3H^+ = Fe^{2+} + 2H_2O \tag{4.30}$$

Substituting Eq. 4.30 into the Nernst equation (4.20) yields

$$Eh = Eh^0 - 0.059\left[\log(Fe^{2+}) + 3\,pH\right] \tag{4.31}$$

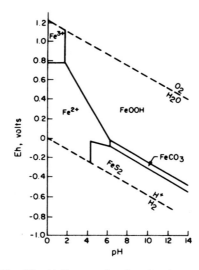

**FIGURE 4.6.** The $Eh$–pH diagram of various iron ions and compounds.

Equations 4.31 and the Nernst form of Eq. 4.29 are two of the boundaries in the $Eh$–pH diagram for Fe (Fig. 4.6). The diagram shows that FeOOH dissolves to $Fe^{2+}$ under reduced and moderately acidic conditions. The $Fe^{3+}$ ion predominates under strongly acidic and oxidizing conditions. $Fe^{2+}$ and FeOOH are the predominant states in typical well-aerated soils. Goethite (FeOOH) is less stable at equilibrium than hematite ($Fe_2O_3$) in many free-energy tabulations, but goethite seems to govern the solution chemistry of Fe in soils.

The electrode potential of Eq. 4.31 depends on the $Fe^{2+}$ concentration. The lines shown in Fig. 4.6 are for $10^{-6}$ activity of soluble Fe. Higher activities shift the equilibrium boundaries to the left. Under strongly reducing conditions FeOOH is unstable with respect to $Fe(OH)_2$, magnetite ($Fe_3O_4$), siderite ($FeCO_3$), and pyrite ($FeS_2$). The region of $FeCO_3$ stability increases with increasing $P_{CO_2}$. The region shown corresponds to $P_{CO_2} = 0.01$, which is not unreasonable for moist field soils.

## APPENDIX 4.3   REDOX POTENTIAL MEASUREMENTS

Applying models of equilibrium oxidation–reduction, such as Figs. 4.2, 4.4, and 4.6, quantitatively to soils requires that the electrode potential be known. From the electrode potential one could then calculate the soil solution concentrations of $Fe^{2+}$, $Mn^{2+}$, and $NO_3^-$ and the sulfate/sulfide ratio from Eq. 4.20. Ideally, the potential of an inert electrode in the system should equal the electrode potential, because the electrode should take on a potential corresponding to the electron availability. This measurement is called the *redox potential*.

Redox potential measurements are even simpler than pH measurements. A platinum electrode with its necessary reference electrode is inserted into a soil or suspension and the potential is measured with a sensitive potentiometer, such as a pH meter. Platinum is the preferred inert electrode material because of its greater response to change in redox conditions.

Table 4.3 shows, however, that redox potentials often differ greatly from electrode potentials. Ion activities are only qualitatively related to redox potentials, except in rare circumstances. One reason is that the Nernst equation applies only to equilibrium. Redox reactions in soils are nonequilibria, though in some cases for highly reduced soils, a steady state may be reached approximating equilibrium. Then only a few redox couples in the soil affect the platinum electrode and the result may approach a pseudo-equilibrium.

Second, equilibrium implies that the electrode potentials of all redox couples in the system are equal. Electrons would exchange between all redox couples until the potentials of all the available electrons are equal. In aerobic soils, this would mean that the potentials of all redox couples would have to equal the potential of the $O_2$–$H_2O$ couple because it would be a major factor. Measuring the redox potential would be unnecessary.

Third, redox couples do not necessarily transfer electrons equally or reversibly with platinum or other inert electrodes. Electron transfers between ions or molecules and electrodes are usually irreversible even in pure solutions. Reversibility is unlikely to be greater in a mixed system such as soils. The inert electrode responds more readily to reversible redox couples than to irreversible couples. The Fe(II–III) and $H^+–H_2$ couples are relatively reversible at platinum surfaces, so, if present, they strongly influence the potential of the electrode.

Fourth, the relative effect of a redox couple on the redox potential increases with the couple's concentration. A high concentration affects the potential more because the likelihood of electron transfer with the electrode is greater. The redox potential is a measure both of the amount of electron transfer and of the electrode potential. In aerobic soils the $O_2–H_2O$ couple is influential, despite its irreversibility, because the $O_2$ concentration is high.

The potential of the platinum electrode in a mixture of redox couples is a poorly defined, weighted average of the potentials of all the redox couples present. The contribution of each couple to the average potential is an unknown function of its concentration, irreversibility, and equilibrium electrode potential. The potential of a nonequilibrium mixture of redox couples is not the potential of any single couple and is a mixed potential.

Under aerobic conditions the redox potential deviates widely from the potentials of soil redox couples. In anaerobic soils, redox potentials may be more quantitatively related to ion activities. The $Fe^{2+}$ and perhaps $Mn^{2+}$ concentrations are high and tend to dominate the redox potential. The range of redox potentials that have been measured in soils is shown in Fig. 4.7. The envelope around those data was considered by the investigators to be the extreme limits of likely redox potentials and pH values in soils and natural waters. Redox potentials can closely approach the $H^+–H_2$ potential, because it is nearly reversible at the platinum electrode.

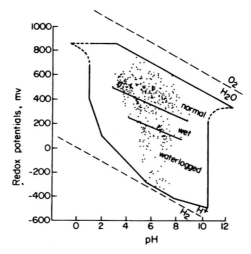

**FIGURE 4.7.** The range of redox potentials and pH values in soils. (From L. G. M. Bass Becking, I. R. Kaplan, and D. Moore. 1960. *J. Geol.* **68**:243.)

The redox potential is pH-sensitive and should be measured in conjunction with pH. To convert redox potentials to a common pH, the conversion factor of $-59$ mV per unit pH is usually employed, although the change of redox potential with pH can be as great as $-200$ mV per unit pH during rapid redox changes. The value of $-59$ mV per unit pH is probably reasonable for most measurements. Reporting both the measured redox potential and the pH avoids the conversion problem. Because redox potentials are most often mixed potentials, measurements more precise than $\pm10$ mV have little significance

The redox potential, like the pH, is an intensity measurement. In theory, it measures the availability rather than the quantity of electrons. Because the redox potential is governed by the potential rate of electron donation versus the rate of electron acceptance, however, it is also a crude estimate of the number of electron donors (reducing agents) present. The qualitative nature of redox measurements has discouraged their use. Redox measurements, however, are used frequently in wastewater treatment to indicate the extent of treatment necessary.

## BIBLIOGRAPHY

Bohn, H. L. 1970. Redox potentials. *Soil Sci.* **112**:39–45.

Garrels, R. M., and C. L. Christ. 1965. *Minerals, Solutions, and Equilibria.* Harper, New York.

Lindsay, W. 1979. *Chemical Equilibria in Soils.* Wiley, New York.

Ponnamperuma, F. N. 1972. The chemistry of submerged soils. In *Advances in Agronomy*, Vol. 24. (N. C. Brady, ed.). Academic, New York, pp. 29–96.

Pourbaix, M. 1966. *Atlas of Electrochemical Equilibrium in Aqueous Solutions* (English translation by J. A. Franklin). Pergamon, Oxford, UK.

## QUESTIONS AND PROBLEMS

1. Why do redox potentials measured in soils differ from electrode potentials?

2. What electrochemical properties for redox measurements would the ideal electrode have?

3. From the elemental analyses in Table 4.1, derive the empirical formula for SOM.

4. Write the balanced reactions for the oxidation of plant matter (Eq. 4.6) and SOM (Eq. 4.7) by oxygen.

5. Write balanced half-reactions and Nernst equations for the reduction $NO_3^-$ to $N_2$ and for $N_2$ to $NH_3$.

6. What would you expect the redox potential to be in a soil of pH 7 at field moisture capacity? If the soil were at equilibrium, what should the electron potential be? Calculate the exact potential from Eq. 4.21

7. From the hydrolysis reactions of iron given in Chapter 3 and from the Nernst equation (Eq. 4.20), derive the equation for the Fe(II–III) potential as a function of $H^+$ activity from pH 0 to pH 14.

8. Assuming equilibrium, what is the $Eh$ value of the $O_2$–$H_2O$ water couple at $P_{O_2} = 0.21$ (normal air), 0.20 (typical soil air), and 0.01 (perhaps typical of the air in an anaerobic soil aggregate)? What ranges of redox potentials might a platinum electrode measure in these three situations? Why do the redox potentials differ from the electrode potential?

9. Describe the likely chemical changes in a soil as it becomes increasingly anaerobic.

10. If oxidizing agents with electrode potentials greater than the $H_2O/O_2$ couple should not persist in aqueous solutions, why are solutions $KMnO_4$ and $NaOCl$, which are common in the laboratory and home, rather stable?

11. What redox reactions tend to "poise" the redox potentials of most soils in the "normal" region of Fig. 4.7?

# 5

# INORGANIC SOLID PHASE

Soil consists of inorganic and organic solids, the soil solution, soil air, and living organisms. Soil chemistry is primarily concerned with the solid and liquid phases and their interactions. This chapter deals with the inorganic solid phase of soils, primarily the clay fraction, and its important properties. The clay fraction ($<2$ $\mu$m in size) carries out most of soil chemistry. The larger sand (50–2000 $\mu$m) and silt (2–50 $\mu$m) fractions are much less chemically active and are composed largely of quartz ($SiO_2$), which is rather inert.

Inorganic soil particles range in size from colloidal particles ($<2$ $\mu$m) to gravel ($>2$ mm) and rocks. Inorganic components exert the major effect on most soil properties and on the overall suitability of soil as a plant growth medium. Organic components include plant and animal residues at various stages of decomposition, cells and tissues of soil organisms, and substances synthesized by the soil population. Organic components, although usually present in much smaller amounts than inorganic components, affect soil properties significantly because of their high reactivity.

Most inorganic soil components are crystalline compounds of definite structure called *minerals*. The sand and silt fractions are largely *primary minerals*, minerals formed at elevated temperatures and inherited unchanged from igneous and metamorphic rocks, sometimes after passing through one or more sedimentary cycles. Primary minerals occur in the clay fraction of weakly weathered soils but are minor constituents of the clay fraction of most agricultural soils. The most abundant primary minerals in soils are quartz ($SiO_2$) and the feldspars ($MAlSi_3O_8$), where M is a combination of $Na^+$, $K^+$, and $Ca^{2+}$. Micas, pyroxenes, amphiboles, and other primary minerals are also common, but in smaller amounts.

Minerals of the clay fraction of soils are largely secondary, that is, formed by low-temperature reactions and either inherited from sedimentary rocks or formed directly in the soil by weathering. These *secondary (authigenic) minerals* in soils commonly include the layer silicates, Al and Fe hydroxyoxides, and carbonates and sulfur compounds. Early workers assumed that clay minerals formed in soils were amorphous, noncrystalline spheres. X-ray diffraction revealed that many clay minerals are instead crystalline. The aluminosilicate clay minerals are layered, similar to mica, and have the same relative dimensions as a stack of postage stamps.

Other important constituents of the clay fraction are the so-called *free oxides*. These are Al, Fe, Mn, and Ti hydroxyoxides that accumulate in the soil as weathering removes silicon. The free oxides range from amorphous to crystalline and are often the weathered outer layer of soil particles. The hydroxyoxides, plus amorphous aluminosilicates such as *allophane*, are the most important clay-sized nonlayer minerals in soils.

The most abundant carbonate in soils is calcite ($CaCO_3$), although pure calcite rarely precipitates directly from the soil solution. Calcium carbonate is common in semiarid and arid region soils and is often present in humid region subsoils derived from calcareous parent material. Ca carbonate accumulates in loose and porous to strongly indurated and rock-like layers in semiarid and arid soils. *Gypsum* ($CaSO_4 \cdot 2H_2O$) occurs in semiarid and arid region soils. The major sulfur mineral is *pyrite* ($FeS_2$). Pyrite is frequently associated with shales and coal beds and may form in soils under reducing conditions.

## 5.1   CRYSTAL CHEMISTRY OF SILICATES

Layered aluminosilicates are the most important secondary minerals in the clay fraction of soils. When layer silicate minerals are clay or colloidal size ($<2$ $\mu$m effective diameter), their large surface area greatly influences soil properties. Most of the important clay minerals have similar silicate structures. Inasmuch as clay minerals are such important clay components, and as different clay minerals can change soil properties greatly, an understanding of soil properties begins with an understanding of silicate structures.

When atoms combine, the bond between them changes the electron distribution from that of the atomic state. The type of bond depends on the electronic structure of the combining atoms. *Ionic* or *electrostatic bonding* occurs between oppositely charged ions such as $Na^+$ and $Cl^-$. Such ions are formed by the complete loss or gain of electrons to form positive or negative ions having an electron structure like an inert gas. The formation of $Na^+$ and $Cl^-$ from Na and Cl atoms is, for example,

$$:\ddot{\underset{..}{C}}l\cdot + e^- \rightarrow :\ddot{\underset{..}{C}}l:^- \qquad Na\cdot \rightarrow Na^- + e^-$$

where the dots denote electrons in the outermost, or valence, shell. Alkali and alkaline earth metals and the halogens tend to gain or lose electrons readily and are the most likely to form ionic bonds.

Ionic bonding is strong, and ionic-bonded compounds tend to be hard solids and have high melting points. Ionic bonding is also undirected—exerted uniformly in all directions. The valence of a given ion is shared by the surrounding ions of opposite charge. The number of such neighbors is determined by their size relative to the size of the central ion. Ionic bonds predominate in many inorganic crystals, including the silicates.

*Covalent bonding* (shared electron pairs) is common between identical atoms or atoms having similar electrical properties, such as in $H_2O$, $F_2$, $CH_4$, and C (diamond). In covalent bonding the electrons are shared between atoms so that each atom attains the inert gas electronic structure. For example,

$$H\!:\!\ddot{O}\!:\qquad H\!:\!\overset{\overset{\textstyle H}{..}}{\underset{\textstyle H}{C}}\!:\!H$$

Covalent bonding is strong, but directional. Bond angles in covalently bonded structures are determined by the geometric positions of the electron orbitals (orbits) involved. Covalently bonded molecules have little tendency to ionize. Bonding within ionic radicals, or complex ions, such as $SO_4^{2-}$, is frequently covalent.

*Hydrogen bonding* occurs between $H^+$ and ions of high electronegativity, such as $F^-$, $O^{2-}$, and $N^{3-}$. The hydrogen bond is essentially a weak electrostatic bond and is important in crystal structures of oxy compounds, such as the layer silicates. Summed over many atoms, the individually weak hydrogen bonds can strongly bond adjacent structures.

The weak electrostatic force between residual charges on molecules is *van der Waals* bonding. Residual charges may result from natural dipoles of unsymmetrical molecules, polarization dipoles, or vibrational dipoles. These van der Waals forces are generally obscured by stronger ionic and covalent bonds but may dominate the properties of some molecules.

Each type of chemical bond imparts characteristic properties to a substance. If more than one type of bond occurs in a crystal, the physical properties such as hardness, mechanical strength, and melting point are generally determined by the weakest bonds. These are the first to yield under mechanical or thermal stress. Thus, the physical properties of layer silicates are determined largely by the strength of the bonds between their layers.

Although differences in the types of bonds may seem clear-cut, bonding in most crystals is somewhere in between. For example, the Si–O bond in the silicates is intermediate between purely ionic and purely covalent bonding. The degree of ionic nature of the Si–O bond is sufficient, however, to apply the rules for ionic bonding to silicate structures.

Bonding within the silicate layers is predominantly ionic. As a result, forces are undirected and ion size plays an important role in determining crystal structure. Table 5.1 shows the *crystal radii* of common ions in silicates. The distance between two adjacent ions in a crystal can be measured accurately by x-ray methods. From a series of such measurements between different ions, the effective contributing radius of each ion can be determined. An ion has no rigid boundary; an ion's radius depends on the number of its orbital electrons and on their relative attraction to the ion's nucleus. The radius of Fe ions, for example, decreases from 0.074 to 0.064 nm

**Table 5.1. Crystal ionic radii of selected cations, and coordination numbers for cations with oxygen**

| Ion | Crystal Ionic Radius[a] (nm) | Coordination Number with Oxygen | |
| | | Observed | Predicted |
|---|---|---|---|
| $Si^{4+}$ | 0.042 | 4 | 4 |
| $Al^{3+}$ | 0.051 | 4, 6 | 4 |
| $Fe^{3+}$ | 0.064 | 6 | 6 |
| $Mg^{2+}$ | 0.066 | 6 | 6 |
| $Fe^{2+}$ | 0.074 | 6 | 6 |
| $Na^+$ | 0.097 | 6, 8 | 6 |
| $Ca^{2+}$ | 0.09 | 8 | 6 |
| $K^+$ | 0.133 | 8, 12 | 8 |
| $NH_4^+$ | 0.143 | 8, 12 | 8 |
| $O^{2-}$ | 0.132 | — | — |

[a] Reprinted with permission from *Handbook of Chemistry and Physics*, 50th ed. Chemical Rubber Co. Inc. Cleveland, OH (1969–1970).

as the valence changes from Fe(II) to Fe(III). The ion radius also depends on the configuration of the ion structure.

The ion radius of oxygen ($O^{2-}$) is much larger than that of most cations found in silicates. The oxygen ion constitutes 50–70% of the mass, and over 90% of the volume, of most common silicate minerals. Silicate structures are largely determined by the manner in which the oxygen ions pack together.

An ion surrounds itself with ions of opposite charge. The number of anions that pack around a central cation depends on the ratio of the cation and anion radii and is called the *coordination number* of the central ion. Because $O^{2-}$ is virtually the only anion in soil minerals, our interest centers on the different cations. Assuming that ions act as rigid spheres, the stable arrangements of cations and anions can be calculated from the packing geometry of their crystal radii (Table 5.2).

**Table 5.2. Spatial arrangement of rigid spheres in relation to radius ratio and coordination number**

| Radius Ratio ($r_{cation}/r_{anion}$) | Arrangement of Anions Around Cations | Coordination Number of Central Cation |
|---|---|---|
| 0.15–0.22 | Corners of an equilateral triangle | 3 |
| 0.22–0.41 | Corners of a tetrahedron | 4 |
| 0.41–0.73 | Corners of an octahedron | 6 |
| 0.73–1 | Corners of a cube | 8 |
| 1 | Closest packing | 12 |

Ions are held together rigidly in a crystal structure, as determined by geometry and by electrical stability. More than one structure may meet the necessary requirements, but the most stable form will be the one having the lowest potential energy. The requirement of electrical neutrality means that the sum of positive and negative charges must be zero. Ions of opposite charge do not pair off to achieve neutrality. Instead, the cation's positive charge is divided among surrounding anions. The number of oxygen ions around each cation is determined by the coordination number, or radius ratio, of the cation and $O^{2-}$, rather than by the charge of the cation.

Table 5.2 shows the predicted and observed coordination numbers of common cations with $O^{2-}$. The $Si^{4+}$ cation occurs in fourfold or *tetrahedral coordination*. Aluminium is generally found in sixfold or *octahedral coordination* but also occurs in tetrahedral coordination in igneous minerals. Where the radius ratio is near the boundary between two types of coordination, the cation may occur in either coordination, depending on conditions during crystallization. High crystallization temperatures generally favor low coordination numbers. In high-temperature minerals Al tends to assume fourfold coordination and to substitute for Si. At lower temperatures Al tends to occur in sixfold coordination and Al substitution for Si is rare.

The tetrahedral and octahedral units formed around Si and Al are basic to the structures of silicate minerals. Crystallography developed centuries ago, before ions were known. Crystallographers very cleverly deduced that crystals were formed by the packing of simple structures like tetrahedra and octahedra. Since they did not know about ions, their concern was about shapes. The number and arrangement of ions in a structure is more fundamental than the number of faces of the structure, but the old crystallographic nomenclature persists. The tetrahedral structure (Fig. 5.1a) is four $O^{2-}$ ligands coordinated around one $Si^{4+}$, giving the unit $SiO_4^{4-}$. The electrostatic bond strength (ion charge divided by the number of bonds to the ion) for the tetrahedral unit is 1. In fourfold coordination, the hole between the four $O^{2-}$ ions is 0.225 times the $O^{2-}$ radius, or 0.030 nm, if the ions are rigid spheres. In reality, cations ranging from 0.029 to 0.052 nm radius occur in fourfold oxide coordination. The radius of $Si^{4+}$ in fourfold coordination is about 0.042 nm, indicating that the ions are not completely rigid spheres.

The octahedral (eight-sided) structure is formed by six anions coordinated around a central cation (Fig. 5.1b). The electrostatic bond strength is $1/2$ if the central ion is trivalent, or $1/3$ if the central ion is divalent. Ions commonly found in octahedral

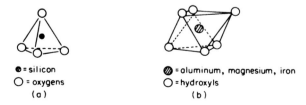

● = silicon
○ = oxygens
(a)

⦸ = aluminum, magnesium, iron
○ = hydroxyls
(b)

**FIGURE 5.1.** Diagram of (a) a silica tetrahedron, and (b) an octahedron of aluminum, magnesium, or iron.

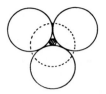

**FIGURE 5.2.** Silica tetrahedron showing position and relative size of oxygen $\bigcirc$ and silicon $\oslash$ ions.

coordination in layer silicates are $Al^{3+}$ (radius 0.051 nm in this coordination), $Mg^{2+}$ (0.066 nm), and $Fe^{2+}$ (0.074 nm). Figure 5.1 and all subsequent drawings of mineral structures greatly exaggerate bond lengths to point out structural features. Also, the size of the $O^{2-}$ ions is reduced to show the relative positions of each ion. Figure 5.2 shows a silicate tetrahedron more accurately.

Minerals vary widely in chemical composition. Substitution of one element for another in mineral structures is common; pure minerals are rare in nature. *Isomorphic substitution, isomorphism, atomic substitution,* and *solid solution* all refer to the substitution of one ion for another without changing the structure of the crystal. Such substitution takes place during crystallization and does not change afterward. Isomorphic substitution can occur between many ions of the same charge, but the size of the ions, rather than the charge, is more important than the charge. Electrical neutrality is maintained by simultaneous substitution of ions elsewhere in the structure, or by retaining ions on the outside of the structure. Isomorphic substitution generally takes place only between ions differing by less than about 10 to 15% in crystal ionic radii.

The more common isomorphic substitutions in silicate structures are $Al^{3+}$ for $Si^{4+}$ in tetrahedral coordination, and $Mg^{2+}$, $Fe^{2+}$, and $Fe^{3+}$ for $Al^{3+}$ in octahedral coordination. Substitution between ions of unequal charge in layer lattice silicates leaves negative or positive charges within the crystal that are neutralized by ions on the surface of the lattices. The common substitutions in soils produce a net negative charge and contribute to the *cation exchange capacity* of soils. Some areas of soil clay particles have a net positive charge and an *anion exchange capacity*. The anion exchange capacity is usually less than the cation exchange capacity and may result from $Ti^{4+}$ substituting for $Al^{3+}$.

A *crystal* is an arrangement of ions or atoms that is repeated at regular intervals in three dimensions. The smallest repeating three-dimensional array of a crystal is called the *unit cell*. The unit cell dimensions $a$ and $b$ (the $x$ and $y$ dimensions) are constant for a given mineral; the $c$ (or $z$) dimension is also constant, except for the special case of swelling layer silicates (Fig. 5.3). The *basal plane* is the $a$-$b$ plane. The chemical composition of layer silicate minerals is normally expressed as one-half of a unit cell, to simplify the chemical formulas.

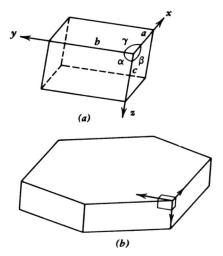

**FIGURE 5.3.** The unit cell of a crystal. It is a parallelepiped with angles $\alpha$, $\beta$, and $\gamma$ and edges $a$, $b$, and $c$. Positions of atoms in the crystal are usually given as $x$, $y$, and $z$ coordinates scaled as fractions of the corresponding cell edges $a$, $b$, and $c$. (a) Cell for kaolinite. (b) The outline of a crystal of kaolinite showing the orientation of one of its unit cells. The unit cell is $a = 0.515$ nm, $b = 0.85$ nm, $c = 0.715$ nm, $\alpha = 91.8°$, $\beta = 104.8°$, and $\gamma = 90.0°$. The crystal is about 9.0 nm wide, 10.3 nm long and 2.1 nm thick, about 10 cells by 20 cells by 3 cells, or 600 cells in all. This crystal is at the smaller end of the range of kaolinite crystals found in soils. (From D. S. Greenland and U. H-B. Hayes, eds. 1978. *Chemistry of Soil Constituents.* Wiley, New York.)

## 5.2    STRUCTURAL CLASSIFICATION OF SILICATES

The Si–O bond is so strong that the tetrahedral arrangement of four oxygen anions about the silicon cation appears to be universal in silicate structures. Different silicate structures arise from the various ways in which the $SiO_4$ tetrahedra combine with one another. Silicate structures range from single, separate tetrahedra to those in which all corners of the tetrahedron are linked through oxygen to other $SiO_4$ tetrahedra. Table 5.3 gives a structural classification of the silicates. This chapter is limited to the silicate structures distinctive to soils.

### 5.2.1    Layer Silicates

Layer silicates, sheet-like phyllosilicates such as the familiar micas, are in primary rocks and in soils. The soil minerals are often called *clay minerals*. Since other components can also be in the clay fraction, layer silicates is a more accurate term. A typical layer silicate is a combination of a layer of Al–, Mg–, or Fe(II)–O octahedra plus one or two layers of Si–O tetrahedra. The tetrahedral and octahedral sheets bond together by sharing oxygens at the corners of the tetrahedra and octahedra. Layer silicate minerals are differentiated by (1) the number and sequence of tetrahedral and octahedral sheets, (2) the layer charge per unit cell, (3) the type of interlayer bond

**Table 5.3. Structural classification of silicates**[a]

| Structural Type | Classification | Structural Arrangement | Si:O Ratio | Mineral Examples |
|---|---|---|---|---|
| Single tetrahedra | Nesosilicates | Single tetrahedra | 1:4 | Olivine, garnet |
| Disilicates | Sorosilicates | Two tetrahedra sharing one corner | 2:7 | Hemimorphite |
| Ring structures | Cyclosilicates | Closed rings of tetrahedra sharing two oxygens | 1:3 | Beryl |
| Single chains | Inosilicates | Continuous single chains of tetraheda sharing two corners | 1:3 | Pyroxene (augite) |
| Double chains | Inosilicates | Continuous double chains of tetrahedra sharing alternately two and three oxygens | 4:11 | Amphiboles (hornblende) |
| Sheet structures (layer silicates) | Phyllosilicates | Continuous sheets of tetrahedra each sharing three oxygens | 2:5 | Micas, montmorillonite |
| Framework structures | Tectosilicates | Continuous framework of tetrahedra each sharing all four oxygens | 1:2 | Quartz, feldspars, zeolite |

[a] Reprinted with permission from James D. Dana. 1959. *Manual of Minerology*, 17th ed. (C. S. Hurlbut, ed.). Wiley, New York.

and interlayer cations, (4) the cations in the octahedral sheet, and (5) the type of stacking along the $c$ dimension.

   The structural unit of the *kaolin* group is formed by superimposing a tetrahedral sheet on an octahedral sheet. Such minerals are referred to as *1:1 layer silicates*. The top oxygens of the tetrahedral sheet are shared by the octahedral sheet, forming a

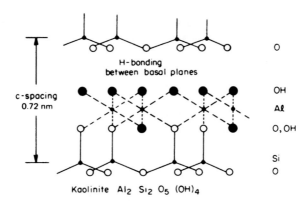

**FIGURE 5.4.** Schematic structure of kaolinite. (From F. E. Bear, ed. 1964. *Chemistry of the Soil.* ACS Monograph Series No. 160.)

common plane of oxygen ions within the structure (Fig. 5.4). In the shared plane, two-thirds of the oxygen ions are shared between Si and Al. The other one-third of the oxygen ions have their remaining charge satisfied by $H^+$. The upper surface of kaolin is a layer of closely packed OH groups. The bottom surface is composed of hexagonally open-packed oxide ions, with an OH recessed within the hexagonal (ditrigonal) oxygens. The 1:1 minerals are apparently very inflexible in their structural requirements and allow little or no isomorphous substitution. The structure is electrically neutral. The 1:1 layers are held together strongly by the many hydrogen bonds between the OH groups of one sheet and the O ions of the next sheet.

The *2:1 layer silicates* are made up of an octahedral sheet sandwiched between two tetrahedral sheets. These unit layers then stack in the *c* direction. Figure 5.5 shows the atomic arrangement of pyrophyllite, an electrically neutral 2:1 mineral. Soil 2:1 layer silicates, on the other hand, have the same structure, but have extensive isomorphic substitution, which leads to largely negative charges within the crystal. This charge must be balanced by other cations, either inside the crystal or outside the structure. The magnitude of charge per formula unit, when balanced by cations external to the unit layer, is called the *layer charge*. The 2:1 minerals are classified according to layer charge in Table 5.4.

The magnitude of layer charge plays a dominant role in determining the strength and type of bonding between the 2:1 layers. If the layer charge is zero, as in pyrophyllite, the 2:1 layers bond together by very weak van der Waals forces. If the layer charge is negative, the 2:1 sheets bond electrostatically and more strongly because cations enter between the unit layers. The greater the layer charge, the more cations, and the stronger the interlayer bond. *Smectites* (which include *montmorillonite* and *bentonite*) of low layer charge bond weakly so polar molecules, such as water, can enter between the sheets, and the minerals expand or swell as the soil becomes wet. In minerals of high layer charge, such as the micas (*muscovite* and *biotite*), the ionic bond of $K^+$ between the sheets is so strong that polar molecules cannot enter, the minerals are nonswelling (nonexpanding), and the soil does not shrink and swell with changing moisture content. *Vermiculites* are intermediate in layer charge and also in-

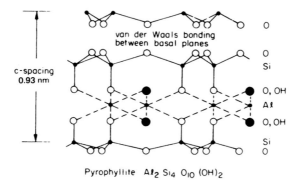

Pyrophyllite $Al_2 Si_4 O_{10} (OH)_2$

**FIGURE 5.5.** Schematic structure of pyrophyllite. (From F. E. Bear, ed. 1964. *Chemistry of the Soil.* ACS Monograph Series No. 160.)

**Table 5.4. Typical 2:1 layer silicate minerals**

| Mineral Group | Layer Charge per Formula Unit in | | Predominant Octahedral Cation | |
| | Tetrahedral Sheet | Octahedral Sheet | $Al^{3+}$ (Dioctahedral) | $Mg^{2+}$ (Trioctahedral) |
| --- | --- | --- | --- | --- |
| Pyrophyllite-talc | 0 | 0 | Pyrophyllite | Talc |
| Smectites | 0.25–0.6 | 0 | Beidellite | Saponite |
| | 0 | 0.25–0.6 | Montmorillonite | Hectorite |
| Vermiculites | 0.6–0.9 | 0 | Vermiculite | Vermiculite |
| Micas | 1 | 0 | Muscovite | Biotite[a] |

[a] $Mg^{2+}$ and $Fe^{2+}$ in octahedral coordination, $K^+$ in the interlayer position.

termediate between micas and the smectites in their swelling properties. Within a given mineral group, specific minerals are defined by the predominant ion in the octahedral coordination. Table 5.4 gives names for common minerals having $Al^{3+}$ in octahedral coordination.

The 2:1 layer silicate minerals are sometimes defined on the basis of the number of octahedral positions occupied by cations. When two-thirds of the octahedral positions are occupied, such as in pyrophyllite ($Al_2Si_4O_{10}(OH)_2$), the mineral is call *dioctahedral*; when all three positions are occupied, such as in talc ($Mg_3Si_4O_{10}(OH)_2$), the mineral is called *trioctahedral*. This difference in composition of layer silicates can be fairly easily determined by x-ray diffraction, because each substitution slightly changes the dimensions of the unit cell.

*Chlorites* are closely related to the micas and have about the same layer charge. In chlorite, the interlayer $K^+$ of mica is replaced by a positively charged octahedral *brucite* ($Mg_3(OH)_6$) sheet. The brucite sheet develops a positive charge when the $Mg^{2+}$ is partially replaced by $Al^{3+}$. Such replacement is often about one-third of the $Mg^{2+}$ positions, to give the basic unit ($Mg_2Al(OH)_6)^+$ that fits into the interlayer position of 2:1 layer silicates to yield the 2:1:1 chlorite (Fig. 5.6). Chlorites are nonexpanding, have low cation exchange capacities, and are generally trioctahedral.

## 5.2.2 Relation of Structure to Physical and Chemical Properties

The interlayer bond has a big effect on the physical and chemical properties of layer silicates. Bonding within the unit layers is much stronger than between adjacent unit layers. When the mineral is subjected to physical or thermal stress, it fractures first between the unit layers, along the basal plane. This is the reason for the flake-like shape of most macroscopic layer silicate crystals. Also, the stronger the interlayer bond, the greater the crystal growth in the *c* dimension before fracture. Hence, the size and shape of layer silicate crystals is a direct consequence of the strength of their interlayer bonds.

The surface area of layer silicates is related to their expanding properties, and may be either external only or external plus internal. *External surface* refers to the

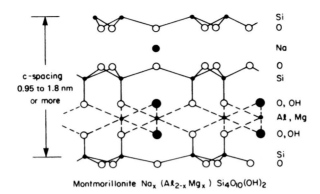

Montmorillonite $Na_x$ $(Al_{2-x} Mg_x)$ $Si_4O_{10}(OH)_2$

**FIGURE 5.6.** Schematic structure of chlorite. (From F. E. Bear, ed. 1964. *Chemistry of the Soil.* ACS Monograph Series No. 160.)

faces and edges of the whole crystal; *internal surface* is the area of the basal plane surfaces of each unit layer. Nonexpanding minerals exhibit only external surface; expanding minerals have both internal and external surface. The total surface area of montmorillonite can be as large as $800 \times 10^3$ m$^2$ kg$^{-1}$. The surface area of kaolinite, a nonexpanding and 1:1 layer silicate, is usually only 10 to $20 \times 10^3$ m$^2$ kg$^{-1}$.

The $c$ spacing of the layer silicates is determined by (1) the number of O–OH sheets per unit structure in the $c$ dimension, and (2) the presence of ions and/or polar molecules in the basal plane. This spacing is conveniently measured by x-ray diffraction. Pyrophyllite has an electrically neutral lattice and hence no interlayer cations. Adjacent units of this mineral approach one another closely and form van der Waals bonds, preventing water entry. The resultant constant $c$ spacing is 0.93 nm (Fig. 5.5). In minerals having weak interlayer bonds such as montmorillonite, cations and water or other polar molecules can enter between the basal planes, causing the $c$ spacing to increase. The expansion varies greatly with the amount and type of polar molecule. In minerals with strong interlayer bonding, such as mica, chlorite and kaolinite, water, and other polar molecules cannot enter between the basal planes.

Adsorbed cations are held by layer silicates to balance the negative charge resulting from isomorphic substitution and from unsatisfied bonds on crystal edges. The magnitude of the exchange capacity of the crystal edge is related to the number of unsatisfied bonds, and therefore is a direct function of crystal size. In 2:1 layer silicates, the larger portion of the cations balancing this mostly negative charge is in the basal plane. Where the layer charge is high, such as in mica, the bond energy is so great, the adjacent layers are so tightly "collapsed," and the K$^+$ fits so well into the hexagonal holes in the basal plane that the adsorbed cations are not exchangeable. The adsorbed cations of smectite minerals, on the other hand, are readily exchangeable. For vermiculite, Ca$^{2+}$, Mg$^{2+}$, and Na$^+$ ions in interlayer position are exchangeable, but K$^+$ and NH$_4^+$ ions, because of their good fit, are not exchangeable by ordinary procedures. Thus, exchange capacity is directly related to layer charge until the charge becomes so large that the adsorbed cations cannot be removed.

Charged areas on the mineral surface that arise from unsatisfied bonds and isomorphic substitution help to retain polar molecules. In addition, the asymmetrical distribution of orbital electrons in O and OH groups produces local negative and positive (polar) areas. Both the charged and the polar surfaces actively adsorb polar molecules by hydrogen bonding and by van der Waals forces.

## 5.3  SOIL LAYER SILICATES

Soil clay minerals often differ appreciably from those of the pure minerals. Soil clays are usually less well ordered and smaller in size than pure minerals. Soil clay minerals of different composition and structure often overlap. Neighboring particles or sheets and interleafings and interstratifications of different layer silicates are common. The mineralogy of soil clays is rarely simple or uniform because of appreciable isomorphic substitution. Coatings of Fe and Al oxides and organic matter on most layer silicates in soils further complicate the mineralogy, identification, and properties of soil clays. Such coatings can decrease the cation exchange capacity, surface area, swelling, and collapse of expansible minerals. Oxide coatings, however, increase anion exchange and other properties associated with positively charged surfaces. The minerals below dominate the clay fraction of most soils.

### 5.3.1  Kaolins

*Kaolinite* ($Al_2Si_2O_5(OH)_4$) is the 1:1 layer silicate mineral that typifies the kaolins (Fig. 5.4). Kaolinite is common in soils, as hexagonal crystals with an effective diameter of 0.2 to 2 $\mu$m. $Si^{4+}$ is apparently the only cation in the tetrahedral sheet of kaolinite, but $Al^{3+}$ or $Mg^{2+}$ may occupy the octahedral positions. When $Al^{3+}$ is in the octahedral sites, the mineral is kaolinite or one of its poorly crystallized polymorphs, dickite or nacrite. When $Mg^{2+}$ is in octahedral sites, the kaolin mineral is *antigorite* ($Mg_3Si_2O_5(OH)_4$). *Halloysite* is a form of kaolinite in which water is held between structural units in the basal plane, yielding a $c$ spacing of 1.0 nm when fully hydrated. Most kaolin units, however, are held together in the basal plane by hydrogen bonding between oxygen ions of the tetrahedral sheet and OH ions of the octahedral sheet.

Such hydrogen bonding prevents expansion (swelling, entry of polar molecules between unit layers) of the mineral beyond its $c$ spacing of 0.72 nm. Surface area is limited to external surfaces and hence is relatively small, ranging from 10 to 20 $\times$ $10^3$ $m^2$ $kg^{-1}$. Kaolinite is a coarse clay with low colloidal activity, including low plasticity and cohesion, and low swelling and shrinkage.

The unit formula for kaolinite has a Si/Al ratio of 1. This ratio is matched by its chemical composition, which suggests that soil kaolinites have little or no isomorphic substitution. Any differences from 1 could be due to surface coatings that were not removed during preparation of the sample. Most of the 10- to 100-mmol(+) $kg^{-1}$ cation exchange capacity of kaolinite has been attributed to dissociation of OH groups at clay edges. However, if only one $Si^{4+}$ of every 200 in the silica sheet were substituted by $Al^{3+}$, the net negative charge would be 200 mmol($-$) $kg^{-1}$. Chemical analysis of clays is not sufficiently sensitive to prove or disprove this extent of isomorphic substitution. The cation exchange capacity of kaolinite is highly pH dependent, suggesting that OH dissociation is the predominant source of charge rather than isomorphic substitution.

### 5.3.2 Smectites (Montmorillonite)

Smectites are 2:1 layer silicates with layer charge of 0.25 to 0.6 per formula unit. An idealized formula for smectite is $KAl_7Si_{11}O_{30}(OH)_6$, but much isomorphic substitution occurs. Because of the relatively low layer charge compared to mica and vermiculite, smectites expand freely. The $c$ spacing varies with the exchangeable cation and the degree of interlayer solvation. Complete drying yields a spacing of 0.95 to 1.0 nm, and full hydration can swell the layers up to tens of nanometers.

Depending on the predominant octahedral cation and on the location of isomorphic substitution, several names have been assigned to minerals within the smectite group (Table 5.4). The predominant one is montmorillonite in which $Al^{3+}$ is the major, and $Mg^{2+}$ the minor, octahedral cation (Fig. 5.7). A typical half-unit-cell formula is $Na_x((Al_{2-x}Mg_x)Si_4O_{10}(OH)_2)$ in which $Na^+$ is the charge-compensating exchangeable cation, Al and Mg are in the octahedral layer, and Si is in the tetrahedral layers. Soil montmorillonites often exhibit imperfect isomorphic substitution, with some $Al^{3+}$ substituting for $Si^{4+}$ in the tetrahedral sheet and $Fe^{2+}$ and $Mg^{2+}$ substituting for $Al^{3+}$ in the octahedral sheet.

Cation exchange capacities for montmorillonite range from 800 to 1200 mmol($+$) $kg^{-1}$. The cation exchange capacity is only slightly pH dependent. The low layer charge allows the mineral to expand freely, exposing both internal and external surfaces. Such expansion yields a total surface area of from $600$ to $800 \times 10^3$ $m^2$ $kg^{-1}$,

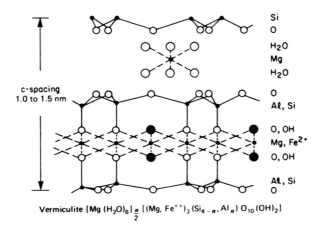

Vermiculite $[Mg (H_2O)_6] \frac{n}{2} [(Mg, Fe^{++})_3 (Si_{4-n}, Al_n) O_{10}(OH)_2]$

**FIGURE 5.7.** Schematic structure of montmorillonite. (From F. E. Bear, ed. 1964. *Chemistry of the Soil.* ACS Monograph Series No. 160.)

with as much as 80% of the total due to internal surfaces. Montmorillonite has high colloidal activity, including high plasticity and cohesion, and high swelling and shrinkage. Montmorillonite normally occurs as a fine clay with irregular crystals having an effective diameter of 0.01 to 1 $\mu$m. Smectites are common in Vertisols and in soils of alluvial plains.

### 5.3.3 Vermiculites

Vermiculites occur extensively in soils formed by weathering or hydrothermal alteration of micas. The layer structure of vermiculite resembles that of the mica from which the mineral is derived (Fig. 5.8). Both trioctahedral and dioctahedral vermiculites exist. Weathering or alteration of the precursor micas replaces the interlayer $K^+$ mostly with $Mg^{2+}$ and expands the $c$ spacing to 1.4–1.5 nm.

The name vermiculite includes several minerals. For our purposes, vermiculite refers to a 2:1 layer silicate capable of only limited expansion, having $Al^{3+}$ substituted for $Si^{4+}$ in the tetrahedral sheet to the extent of 0.6 to 0.9 per unit formula, and with $Mg^{2+}$ and $Fe^{2+}$ as the octahedral cations. An idealized half-unit-cell formula is $(Mg(H_2O)_6)_{n/2}((Mg, Fe)_3(Si_{4-n}, Al_n))O_{10}(OH)_2$, where the hydrated $Mg(H_2O)_6^{2+}$ is the exchangeable cation. An interesting property of some vermiculites is the internal balancing of layer charge originating from the parent mica by substituting $Fe^{3+}$ or $Al^{3+}$ for $Mg^{2+}$ and $Fe^{2+}$ in the octahedral sheet to yield $Na_{0.7}(Al_{0.2}Fe(III)_{0.4}Mg_{2.4})(Si_{2.7}Al_{1.3})O_{10}(OH)_2$. Isomorphic substitution in the tetrahedral sheet yields a charge of $-1.3$ per unit cell. Isomorphic substitution in the octahedral sheet yields a charge of $+0.6$ in the unit cell. The net charge of $-0.7$ per unit formula is balanced in this example by $Na^+$.

The layer charge in vermiculite gives rise to a cation exchange capacity of from 1200 to 1500 mmol(+) $kg^{-1}$, which is considerably higher than the exchange capacity of montmorillonite. As with montmorillonite, the cation exchange capacity is only slightly pH dependent. Vermiculite swells less than montmorillonite because of its higher layer charge. The mineral is nonswelling (with a $c$ spacing of 1.0 nm) when saturated with $K^+$ or $NH_4^+$ ions. Such ions are termed *fixed* because they cannot be

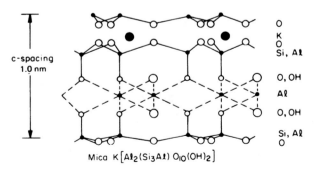

Mica $K[Al_2(Si_3Al) O_{10}(OH)_2]$

**FIGURE 5.8.** Schematic structure of vermiculite. (From F. E. Bear, ed. 1964. *Chemistry of the Soil.* ACS Monograph Series No. 160.)

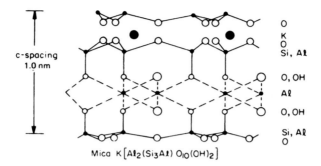

Mica $K[Al_2(Si_3Al)O_{10}(OH)_2]$

**FIGURE 5.9.** Schematic structure of mica. (From F. E. Bear, ed. 1964. *Chemistry of the Soil.* ACS Monograph Series No. 160.)

exchanged with ordinary salt solutions. The total surface area of vermiculite, when not $K^+$ or $NH_4^+$ saturated, ranges from 600 to $800 \times 10^3$ m$^2$ kg$^{-1}$.

### 5.3.4 Micas

Micas are abundant in soils as primary minerals inherited from parent materials. Micas are not known to form to any significant extent in soils. Micas are precursors for other 2:1 layer silicates, notably vermiculites. Micas are commonly present in soils as components of particles that have been partially transformed to expansible 2:1 minerals. As a result, mica is often interstratified with other minerals. Altered mica containing less $K^+$ and more water than well-ordered mica is called *hydrous mica* (formerly *illite*).

A typical half-unit-cell formula for mica is $K(Al_2(Si_3Al)O_{10}(OH)_2)$, and micas can be either dioctahedral or trioctahedral. Figure 5.9 shows the layer structure of the dioctahedral form, *muscovite*, which is clear and colorless. Isomorphic substitution of Al for Si in the tetrahedral layer creates negative charge close to the layer surface, which results in a strong coulombic attraction for charge-compensating cations. Interlayer $K^+$ is so strongly adsorbed that it is not exchanged in standard cation exchange capacity (CEC) determinations. Thus, despite the large layer charge (about $-1$ per unit cell for many micas), the CEC is only 200 to 400 mmol(+) kg$^{-1}$. Total surface area is about 70 to $120 \times 10^3$ m$^2$ kg$^{-1}$ and is restricted to external surfaces. Micas are nonswelling and are only moderately plastic. The fixed $K^+$ is released slowly during weathering and is a source of $K^+$ for plants. *Biotite* is the dark-colored mica, trioctahedral, with Fe and Mg in the octahedral sheet.

## 5.4   CHLORITES

Chlorites occur extensively in soils and are 2:1:1 layer silicates (Fig. 5.6). The positively charged and substituted brucite sheet between the negatively charged mica-like sheets restricts swelling, decreases the effective surface area, and reduces the effective cation exchange capacity of the mineral. An idealized half-unit-cell formula is $(AlMg_2(OH)_6)_x$ $(Mg_3(Si_{4-x}Al_x)O_{10}(OH)_2)$. Substitution in such classical chlorites is in the tetrahedral layer of the 2:1 portion, with the brucite sheet serving as the interlayer "cation." The layer charge of the 2:1 portion of the mineral varies but is similar to mica. Cation exchange capacities range from 100 to 400 mmol(+) $kg^{-1}$; total surface areas range from 70 to $150 \times 10^3$ m$^2$ kg$^{-1}$.

Chlorite is common in sedimentary rocks and in productive soils derived therefrom. The elemental composition of chlorites varies widely, however, with chromium and nickel occurring in *mafic* (Fe- or Mg-containing) chlorites. Serpentine-derived soils contain chlorite and many are infertile because of their high Mg and low Ca contents.

## 5.5  ACCESSORY MINERALS

As soils weather and Si, Ca, Mg, Na, and K are leached away, the soil's colloidal fraction becomes enriched with Al, Fe, Mn, and Ti oxides and hydroxyoxides. The structural organization of these hydroxyoxides ranges from amorphous to crystalline. These Al, Fe, and Ti oxides and allophane are prominent nonlayer silicate minerals in most soils and their content in soils increases with increased weathering.

Most of the crystallized secondary minerals found in soils have passed through intermediate amorphous steps during chemical weathering. *Amorphous* is defined as being nondetectable by x-ray diffraction. The distinction between amorphous and crystalline materials is vague, and various degrees of crystallinity can occur during the reorganization of hydrous gels that precipitate from the soil solution. Very small crystals can appear amorphous to many tests of crystallinity. Such materials have been termed *cryptocrystalline*.

### 5.5.1  Allophane and Imogolite

Allophane is a general name for amorphous aluminosilicate gels. These are one of the more common groups of amorphous materials in soils. The composition of allophane varies widely but includes mostly hydrated $Al_2O_3$, $Fe_2O_3$, and $SiO_2$. Only minor amounts of $Mg^{2+}$, $Ca^{2+}$, $K^+$, and $Na^+$ are generally present. The Al/Si ratio is usually between 1 and 2. Whether allophane is a mixture of individual Si or Al oxide gels or whether it is an amorphous hydrous aluminosilicate in which oxygen anions are shared between Si and Al ions is unclear. In any case, Si is apparently in tetrahedral coordination and Al is in octahedral coordination. Allophane was first noted in Japanese soils derived from volcanic parent materials that weather quickly, but it is probably present in most soils as an intermediate product of weathering.

Allophane can have a high cation exchange capacity at neutral to mildly alkaline pH, perhaps 1500 mmol(+) $kg^{-1}$, but the CEC is highly dependent on pH and degree of hydration. Values reported in the literature range from 100 to 1500 mmol(+) $kg^{-1}$, depending on pH. The CEC measurement is indefinite because exchangeable cations are loosely adsorbed and hydrolyze extensively during washing with aqueous alcohol mixtures. Allophane has a high surface area, 70 to $300 \times 10^3$ m$^2$ kg$^{-1}$, but this also varies widely with degree of crystallinity and pH.

*Imogolite* ($Al_2SiO_3(OH)_4$) was also first described in Japanese soils derived from volcanic ash. Although not highly crystalline, imogolite can be recognized by x-ray diffraction. Imogolite's and gibbsite's solubility are thought to determine the $Al^{3+}$ concentrations in the soil solution of many moderately acid soils.

### 5.5.2 Zeolites

Zeolites are three-dimensional aluminosilicate structures, like feldspars, in which tetrahedra are linked by sharing their vertices (Table 5.3). The tetrahedra are linked into 4-, 6-, 8-, and 12-membered rings, joined together less compactly than in the feldspars. *Analcime* ($NaAlSi_2O_6 \cdot H_2O$) is a representative zeolite. The open framework of zeolites leaves channels of different sizes that run in several directions through the crystal. The channels contain loosely held water molecules and charge-balancing cations that are freely exchangeable. The channels frequently interconnect and are usually larger than the diameters of the common cations, so both water and charge-balancing cations diffuse readily through the crystals. Some zeolites have smaller-sized channels that effectively prevent movement of large molecules, leading to the use of zeolites as *"molecular sieves."*

### 5.5.3 Al, Fe, Ti, and Mn Hydroxyoxides

*Gibbsite* ($Al_2(OH)_6$) is the most abundant Al hydroxide in soils, occurs in large amounts in highly weathered soils, 20–30% and more by mass, and is the stable low-temperature form of Al hydroxide. Crystalline gibbsite consists of sheets held together by hydrogen bonding between adjacent hydroxyl ions arranged directly above and below one another. In acid soils, gibbsite and the Fe hydroxides react strongly with phosphate and are responsible for keeping phosphate unavailable to plants and for adsorbing $SO_4^{2-}$ and reducing its availability to plants. Gibbsite and the Fe hydroxyoxides are responsible for much of the pH dependence of soil CEC. These hydroxyoxides have a net positive charge in acid soils due to $H^+$ adsorption. As the pH increases, the charge changes from positive to negative as the $H^+$ dissociates and the amount of phosphate and sulfate retained decreases. *Boehmite* ($AlOOH$) occurs in intensively leached, highly weathered soils. It also can be formed from gibbsite by heating to about 130° C. *Corundum* ($Al_2O_3$) is a high-temperature form rarely found in soils.

*Hematite* ($Fe_2O_3$) and *goethite* ($FeOOH$) are the most common Fe oxides found in soils. Hematite occurs in highly weathered soils and is pink to bright red. Goethite is also characteristic of strongly weathered soils and is brown to dark yellowish-brown. Both hematite and goethite occur as amorphous coatings on soil particles, imparting the red and brown colors characteristic of soils. The amorphous coatings transform to crystalline forms with aging, or as the amounts increase. As with gibbsite, the soil content can be 20–30% in highly weathered soils. Aluminium and iron ore deposits are areas of extreme soil weathering. The crystallization of amorphous Fe hydroxyoxides is responsible for the irreversible hardening, upon drying, of *plinthite* (*laterite*) of tropical soils into stonelike materials. Chemically, the Fe hydroxyoxides behave similarly to gibbsite as described above. *Magnetite* ($Fe_3O_4$) is a magnetic Fe oxide inherited from the parent rock. It usually occurs as sand-sized grains of high specific gravity. Magnetite oxidizes to *maghemite* ($Fe_2O_3$), which is also magnetic.

The Ti oxides commonly found in soils and clay sediments are *rutile* and *anatase*, both $TiO_2$ and inherited from the parent rock. Because Ti oxides resist weathering so strongly, they are often used as indicators of the original amount of parent material from which a soil has formed.

Manganese oxides are a poorly understood and amorphous mixture of Mn(III) and Mn(IV). Pure *pyrolusite* ($MnO_2$) is rare, a more accurate formula would be approximately $MnO_{1.8}$. Many transition metal ions have the same size as the Mn ions so isomorphous substitution is common. Hence, Mn hydroxyoxides can retain these other cations. For a time some workers believed that Mn hydroxyoxides were an important part of the soil's retention of trace metals. Per unit weight, this may be so. Soils, however, contain about 0.1% Mn hydroxyoxides, little compared to the amounts of Al and Fe hydroxyoxides. The *manganese nodules* that receive attention as a Mn ore are Mn-rich iron oxide nodules that are found on some ocean floors. Manganese nodules have not been found in soils.

## 5.6    CHARGE DEVELOPMENT IN SOILS

The two properties that most account for the reactivity of soils are surface area and surface charge. Surface area is a direct result of particle size and shape. Most of the total surface area of a mineral soil is due to clay-sized particles and soil organic matter. Charge development in soils is due to these same two fractions, although the sand- and silt-size fractions may contribute some cation exchange capacity if coarse-grained vermiculite is present. Charge development in soils occurs as a result of isomorphic substitution and of ionization of functional groups on the surface of solids, again primarily in the colloidal fraction, resulting in the permanent and the pH-dependent charges of soils.

### 5.6.1    Permanent Charge

Isomorphic substitution is the substitution of one ion for another of similar size within a crystal lattice. The substituting ion may have a greater, equal, or lower charge than the ion for which it substitutes. In layer silicate structures, cations can substitute for coordinating cations in either the tetrahedral or the octahedral sheets. If a cation of lower valence substitutes for one of higher valence, such as $Mg^{2+}$ for $Al^{3+}$ or $Al^{3+}$ for $Si^{4+}$, the negative charge of $O^{2-}$ and $OH^-$ ions in the mineral structure is left unsatisfied, yielding a net negative charge. Isomorphic substitution can also result in positive charge, by $Al^{3+}$ substituting for $Mg^{2+}$ in the brucite interlayer of chlorite, but negative charge tends to predominate in soil minerals.

Isomorphic substitution occurs during crystallization of layer silicate minerals in magmas and in soils. If the primary cation is unavailable as the unit cell forms, another cation can sometimes squeeze in. The resulting permanent charge is essentially independent of the soil solution composition surrounding the particle. Isomorphic substitution is the principal source of negative charge for the 2:1 and 2:1:1 layer silicates, but is of minor importance for the 1:1 minerals.

### 5.6.2 pH-Dependent Charge

The total charge of soil particles varies with the pH at which the charge is measured. Figure 5.10 illustrates *pH- dependent charge*, where some portion of the soil changes from positive charge at low pH to negative charge at higher pH. The soil's total charge is the algebraic sum of its negative and positive charges. The relative contribution of permanent and pH-dependent charge depends on the composition of soil colloids. Relatively young and weakly weathered soils characteristic of Europe and North America have a net negative charge, because of the higher pH and layer silicate and organic matter content of these soils. Highly weathered and volcanic soils, on the other hand, are dominated by allophane and hydrous oxides, may have a low pH, and may have a net neutral to positive charge (Table 5.5). Subsoils are usually lower in organic matter so the relative amount of negative charge relative to positive charge decreases. The *zero point of charge* (ZPC) is an index of the positive and negative charge on soil colloids. The ZPC is the pH at which negative and positive charges of a colloid are equal. The ZPC for the soil of Fig. 5.10 would be pH < 3.

Crystal bonding ends at the particle–soil solution interface. At the particle edge, the charge of the structural cations and $O^{2-}$ ions is not compensated by surrounding structural ions. Electrical neutrality is necessary and is maintained by interacting with $H^+$, $OH^-$, and water, and by adsorbing cations or anions from the soil solution. The primary source of pH-dependent charge is considered to be the loss of adsorbed $H^+$ and $OH^-$ on inorganic solids and $H^+$ from organic acids, phenols, and other functional groups in soil organic matter.

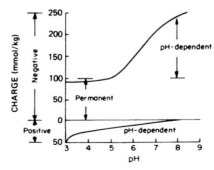

**FIGURE 5.10.** Representative change of positive and negative charges on soils with pH. (From W. D. Guenzi, ed. 1974. *Pesticides in Soil and Water*. American Society of Agronomy, Madison, WI.)

The soil solids that contain functional groups capable of developing positive pH-dependent charge include layer silicates, allophane, hydroxyoxides, and organic matter. In organic matter, the functional groups that create pH-dependent positive charge include hydroxyl (—OH), carboxyl (—COOH), phenolic (—C$_6$H$_4$OH), and amine (—NH$_2$). Equation 5.1 shows how an inorganic hydroxyoxide, an Al hydroxyoxide in this case, can change from negative to positive charge by adsorbing H$^+$:

$$(Al)-OH^{1/2+} + H^+ = (Al)-OH_2^{1/2+} \tag{5.1}$$

The other hydroxyoxides are similar. The extent of the reaction is pH–dependent. This effect is accentuated by the tendency of these hydroxyoxides to be thin coatings on soil particles, which increases their activity per unit mass. Soils containing large amounts of Al and Fe oxides have a strongly pH-dependent charge and highly variable CEC.

Figure 5.11 shows how pH-dependent charge develops at the crystal edges of kaolinite. Depending on the pH of the soil solution, the charge can be either positive or negative. Jackson suggested that the dissociation of H$^+$ occurred at p$K_a = 5.0$ for the Al–OH$_2$ group, 7.0 for (Al,Si)–OH, and 9.5 for Si–OH. The high p$K_a$ for Si–OH groups indicates that the deprotonation (H$^+$ loss) occurs only at high pH. The pH-dependent charge of layer silicates is more likely due to reversible protonation and deprotonation of Al–OH rather than Si–OH groups.

The contribution of edge OH groups to pH-dependent charge is related to the acidity of the edge groups and to the area of edge surface. For 2:1 minerals such as montmorillonite, the functional groups are apparently weakly acidic and dissociate only at high pH. In addition, the amount of edge surface for 2:1 minerals is small relative to the basal (planar) surface. Kaolinite, on the other hand, tends to stack without swelling in the $c$ dimension, increasing the edge area compared to the basal plane area. For both reasons, pH-dependent charge is more important for kaolinite than for smectites or vermiculites. As a rule of thumb, only 5 to 10% of the negative charge on 2:1 layer silicates is pH dependent, whereas 50% or more of the charge developed on 1:1 minerals can be pH dependent.

Figure 5.12 shows the pH-dependent charge in kaolinite. The Cl$^-$ anion is retained by kaolinite in acid solutions, indicating the presence of positive sites, prob-

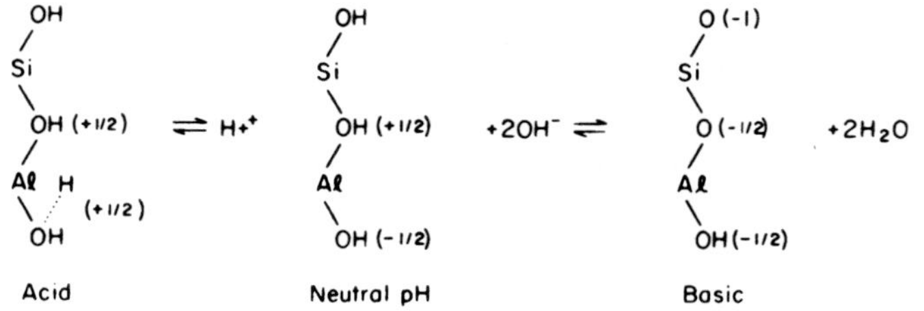

**FIGURE 5.11.** Representation of pH-dependent charge at kaolinite edges. (By permission from R. K. Schofield and H. R. Samson. 1953. *Clay Miner Bull.* **2**:45.)

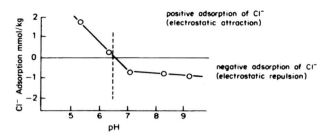

**FIGURE 5.12.** Chloride adsorption by kaolinite at various pH values. (By permission from R. K. Schofield and H. R. Samson. 1953. *Clay Miner Bull.* **2**:45.)

ably $Al–OH_2^+$. In basic solutions, the functional group changes to the negatively charged $Al–OH^-$, which repels anions. The pH at which positive and negative charges are balanced, ZPC, for this kaolinite is indicated by the vertical dashed line at about pH 6.5.

In highly weathered soils, Fe and Al oxides are abundant and can develop considerable pH dependent charge, as can Ti, Cr, and Mn oxides. For example, $Fe^{3+}$ in hematite is in sixfold coordination with $O^{2-}$. Each valence bond of an oxygen supplies $-0.5$ charge to the Fe ion. The remaining $-1.5$ charge of each O is satisfied by Fe cations in adjacent octahedra. At the soil solution interface (the edge of the crystal), however, the ion charges in the crystal have to be satisfied by $H^+$ and $OH^-$ ions in the soil solution. In effect, the ions in the crystal complete their coordination spheres by interacting with the soil solution. The result is that the crystal is coated with a layer of H or OH ions (Fig. 5.13). This charge development is similar to that developed by silicates. The sesquioxides have no permanent charge so their charge depends solely on, and varies greatly with, the pH of the soil solution. Allophane, an amorphous hydrous oxide with high surface area, also develops pH-dependent charge. Because its surface area is greater than crystalline materials, its charge is even more pH dependent.

Soil organic matter also has a strongly pH-dependent charge. The charge develops mostly by $H^+$ dissociation from carboxylic and phenolic groups. Table 5.5 summarizes the colloidal properties of the major components of the soil's clay fraction.

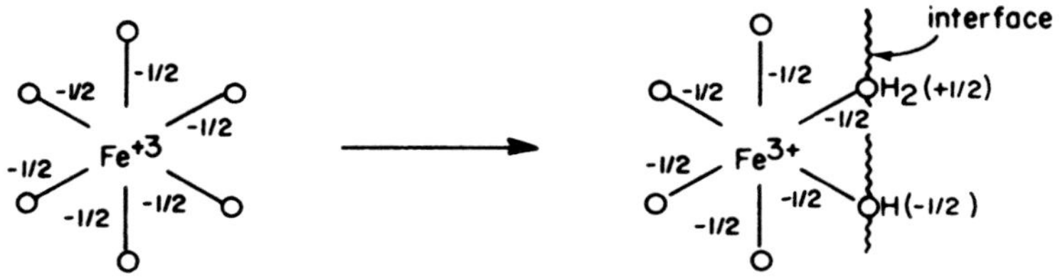

**FIGURE 5.13.** $Fe^{3+}$ and ligands in the interior and at the surface of hematite.

**Table 5.5. Summary of selected properties of inorganic solid-phase components**

| Component | Mineral Type | Chemical Formula | Layer Charge | Cation Exchange Capacity (mmol(+) kg$^{-1}$) | Surface Area $\times$ 10$^3$ m$^2$ kg$^{-1}$ | $c$ Spacing (nm) | Expansible | pH Dependency of Charge | Colloidal Activity |
|---|---|---|---|---|---|---|---|---|---|
| Kaolinite | 1:1 | $Al_2Si_2O_5(OH)_4$ | $\sim 0$ | 10–100 | 10–20 | 0.72 | No | Extensive | Low |
| Montmorillonite | 2:1 | $Na_x[(Al_{2-x}M_x)Si_4O_{10}(OH)_2]$ | 0.25–0.6 | 800–1200 | 600:800 | Variable | Yes | Minor | Extremely high |
| Vermiculite | 2:1 | $Na[Mg_3(Si_{4-x}Al_x)O_{10}(OH)_2]$ | 0.6–0.9 | 1200–1500 | 600–800 | 1.0–1.5 | No | Minor | High |
| Mica | 2:1 | $K_x[Al_2(Si_{4-x}Al_x)O_{10}(OH)_2]$ | 1.0 | 200–400 | 70–120 | 1.0 | No | Medium | Medium |
| Chlorite | 2:1:1 | $[AlMg_2(OH)_6]_x$ $[Mg_3(Si_{4-x}Al_x)O_{10}(OH)_2]$ | $\sim 1$ | 200–400 | 70–150 | 1.4 | No | Extensive | Medium |
| Allophane | — | $Si_xAl_y(OH)_{4x+3y}$ | — | 100–1500 | 70–300 | — | — | Extensive | Medium |

## APPENDIX 5.1    SURFACE AREA MEASUREMENTS

An impressive property of colloids, including layer silicate minerals, is their large area of reactive surface. Various physical and chemical properties, including water retention and cation exchange capacity, are highly correlated with the surface area of soils. Several techniques estimate the amounts of reactive surface area of soils and are briefly described below.

Colloid chemists commonly measure surface area by the adsorption of $N_2$ gas. The adsorption is conducted in vacuum and at temperatures near the boiling point of liquid nitrogen ($-196°$ C). The approach is based on the Brunauer–Emmett–Teller (BET) adsorption equation, and has been adapted to a commercially available instrument. Unfortunately, the technique does not give reliable values for expansible soil colloids such as vermiculite or montmorillonite. Nonpolar $N_2$ molecules penetrate little of the interlayer regions between adjacent mineral platelets of expansible layer silicates where 80 to 90% of the total surface area is located. Several workers have used a similar approach with polar $H_2O$ vapor and have reported complete saturation of both internal (interlayer) and external surfaces. The approach, however, has not been popular as an experimental technique.

Soil chemists more commonly measure the retention of polar liquids such as ethylene glycol or glycerol by soils. The basic procedure involves applying excess and then removing all but a monolayer from the mineral surfaces. The excess is removed under vacuum in the presence of a desiccant, to eliminate competition with $H_2O$ for retention sites. Some workers advocate a glycol–$CaCl_2$ mixture to maintain a relatively constant vapor pressure of glycol in the evacuated system, and hence to provide a more reproducible endpoint.

Glycol and glycerol retention are influenced by the species of exchangeable cation, since both the colloid surfaces and the surface cations are at least partially solvated during surface area determinations. Glycerol is preferred over glycol by some workers, because it distinguishes between vermiculitic (partially expanding) and montmorillonitic (freely expanding) surfaces under carefully controlled conditions. A single molecular layer of glycerol remains in vermiculitic interlayers, but two such layers remain in montmorillonitic interlayers.

Ethylene glycol monoethyl ether (EGME) is another polar molecule used increasingly for surface area measurements. Its results are essentially identical to the glycol method but are achieved more rapidly. It was graciously contributed to soil chemistry by a careless shipping clerk and an unknowing technician. The latter obtained unusual, but promising, results before he realized that the wrong reagent had been provided by a chemical supply firm.

Surface areas have also been measured by anion repulsion or by adsorption of certain organic solutes from aqueous solution. A particularly promising solute is cetyl pyridinium bromide, which orients differently on external and internal (interlayer) surfaces and can thus aid in distinguishing between the two types of surface.

## APPENDIX 5.2    MINERAL IDENTIFICATION IN SOILS

X-ray diffraction has probably contributed more to the mineralogical characterization of soil layer silicates than any other single technique. Other techniques being increasingly used are infrared, electron spin resonance, fluorescence spectroscopy, differential thermal analysis, and x-ray absorption spectroscopy. The simplest and most common is x-ray diffraction, which exposes material to a filtered and monochromatic beam of x-rays from an appropriate metal target. When the beam enters the sample, part of the beam is reflected by successive repeating planes of atoms. The reflected beams are reinforced (intensified) at each locus of points where the reflected beam has moved an integral number of wavelengths before being reflected by the next plane of atoms (Figure 5.14). In quantitative terms, reinforcement occurs wherever

$$n\lambda = 2d \sin \theta \qquad (5.2)$$

where $n$ is an integer, $\lambda$ is the wavelength of the x-radiation, $d$ is the "repeat" distance between successive layers of the crystal, and $\theta$ is the angle at which the radiation strikes the crystal. The loci of points can be detected either with a cylindrical photographic film placed around the irradiated sample or with a rotating detector. Radiation that has not traveled an integral number of wavelengths within the crystal emerges and strikes the film or detector out of phase with other radiation, so only minimal film darkening or detector counts are recorded. Radiation that has traveled an integral number of wavelengths within the crystal reinforces previously reflected radiation and produces strong film darkening or a peak of counts in the detector. Differences in crystal repeat distances as small as 0.01 to 0.001 nm can be detected by x-ray diffraction. The technique is particularly valuable for identifying soil colloid types, their degree of interleafing or interstratification, and variations in their interplatelet spacings resulting from pretreatments or additives.

All the techniques to identify soil minerals have difficulty coping with the heterogeneity of soils and with coatings of organic and weathered materials on soil particles, and have trouble detecting small amounts of a component in a very large

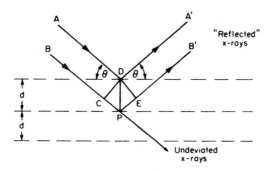

**FIGURE 5.14.** X-ray reflection from repeating mineral planes. For reinforcement, $CP + PE = 2d \sin \theta = \eta\lambda$, so that emerging radiation is in phase.

matrix of silicates and Fe and Al hydroxyoxides. If one constituent of the component has a much higher atomic weight, it can usually be more readily detected. Even Pb with atomic weight $= 207$, however, had to be $<4\%$ by mass before distinct Pb minerals could be identified in soil. A soil with that Pb content is a rich Pb ore. At normal Pb concentrations in soils, Pb and other trace elements more likely exist as isomorphous substitutes in the soil's major minerals. By going to great extremes to clean the mineral surface of coatings to reveal the minerals beneath, we may destroy the soil.

## BIBLIOGRAPHY

Feldman, S. B., and L. W. Zelasny. 1998. In *Chemistry of Soil Minerals* (P. M. Huang, ed.). Soil Science Society of America Spec. Publ. 55, Madison, WI.

Putnis, A. 1992. *Introduction to Mineral Sciences*, Cambridge University Press, New York.

## QUESTIONS AND PROBLEMS

1. Distinguish between primary and secondary minerals, and give examples of each. Which minerals are more important in determining soil properties?

2. Which minerals are commonly found in the sand and silt fractions of soil? Which are commonly found in the clay-sized fraction? Why?

3. Distinguish between ionic, covalent, hydrogen, and van der Waals bonding. Which type of bonding predominates in silicate structures?

4. What ion dominates silicate structures?

5. Calculate the theoretical range in hole size between oxygen ions in tetrahedral and octahedral coordination. Which cations "fit" in each configuration?

6. What is the dominant characteristic that determines whether ions may isomorphically substitute for one another?

7. What is a unit cell? How many unit cells are there in 1 mole of a particular mineral?

8. Why is the phrase "clay mineral" misleading, and what term is best used to describe phyllosilicate minerals of $>2$ $\mu$m effective diameter?

9. Distinguish between 1:1, 2:1, and 2:1:1 layer silicates by drawing diagram of their structures.

10. Explain how layer charge influences the following layer silicate properties:
    (a) Interlayer bonds
    (b) Crystal size
    (c) Swelling

(d) Surface area

(e) $c$ spacing

(f) Exchangeability of adsorbed cations

11. How do soil layer silicates differ from pure minerals? How are soil properties, such as cation exchange capacity, surface area, and swelling, expected to differ from the properties of pure layer silicates?

12. Write the half-unit-cell formula for a montmorillonite mineral with a layer charge of 0.3, with 90° of the substitution in the octahedral layer ($Mg^{2+}$ for $Al^{3+}$) and 10% of the substitution in the tetrahedral layer ($Al^{3+}$ for $Si^{4+}$). The saturating cation is $Na^+$. Draw a diagram of this mineral, indicating the position of $Mg^{2+}$, $Al^{3+}$, $Si^{4+}$, and $Na^+$.

13. Calculate the cation exchange capacity of the mineral in Problem 12 and express the result in millimoles(+) per kilogram. Compare this value to the value normally given for montmorillonite.

14. The $a$ and $b$ dimensions of a typical montmorillonite mineral are $a = 0.052$ nm, and $b = 0.089$ nm. Using the data in Problems 12 and 13, calculate the theoretical surface area of the montmorillonite. (Remember that montmorillonite has both internal and external surfaces.) How does this value compare with values normally given for montmorillonite?

15. Accessory minerals in the clay-sized fraction often dominate the properties of the soil solid phase. Under what conditions would you expect this to be true, what minerals are involved, and how would the soil's properties be affected?

16. What are the likely dominant sources of pH-dependent charge in

    (a) A highly weathered mineral soil low in organic matter?

    (b) A slightly weathered, montmorillonitic soil low in organic matter?

    (c) A slightly acid forest soil?

    (d) A volcanically derived soil low in organic matter?

17. What is meant by the zero point of charge?

18. Hydrous oxides are said to be amphoteric. Explain.

19. What are the functional groups responsible for pH-dependent charge in soils?

20. Calculate the proportions of mass and volume occupied by oxygen in kaolinite and goethite.

21. Would a hypothetical mineral of composition $(Al_2Mg)(Si_3Al)O_{10}(OH)_2$ be dioctahedral or trioctahedral?

# 6

# SOIL ORGANIC MATTER

The organic fraction of soil is <5% living microbes, plant roots, and soil fauna and >95% dead plant and animal residues. Per unit mass, the organic fraction is the most chemically active portion of the soil. It is a reservoir for essential elements, particularly C, N, S, and P; promotes good soil structure; is a source of cation exchange capacity (CEC) and soil pH buffering; promotes good air–water relations in soils; and is a large and active reservoir and buffer of carbon in the environment. This chapter discusses the contributions of soil organic matter to the chemical properties of soils.

Soil organic matter (SOM) is mostly (>95%) an accumulation of dead plant matter and partially decayed and resynthesized plant and animal residues. Freshly fallen leaves and dying roots rapidly decompose and the residues become part of SOM, some portions of which remain in the soil for a very long time. Crop residues, weeds, grasses, tree leaves, worms, bacteria, fungi, and actinomycetes are also part of the complex mixture. Some definitions of SOM are restricted to soil humus, omitting any undecayed organic residues and soil organisms. *Humus* is generally defined as that organic material that has been transformed into relatively stable form by soil microorganisms. We use the term SOM in its broader sense, to include all carbon-containing compounds except carbonates. Stevenson (1982) and Schnitzer and Schulten (1998) give details on the details of SOM composition and structure. This chapter is concerned with nonliving substances (particularly humus) and its chemistry. Soil microbiology texts discuss the *live microbial tissue* (*microbial biomass*) in soils.

## 6.1  SOIL ORGANIC MATTER CONTENT

Soils vary greatly in their organic matter content. A prairie grassland soil (e.g., Mollisol) may contain 5 to 6% SOM by mass to a depth of 15 cm, whereas a sandy desert

**155**

soil (Aridisol) contains little more than 0.1%. Poorly drained soils (Aquepts) often have organic matter contents greater than 10%, and peat soils (Histosols) approach 100% organic matter. Although the organic matter content of most surface mineral soils is only 0.5 to 5% by mass, the active colloidal behavior of SOM strongly affects soil physical and chemical properties.

The factors of soil formation (Section 7.3) determine the SOM content of soils. The order of importance of the factors that determine the organic matter (and nitrogen) contents of well-drained soils in the United States is climate > vegetation > topography = parent material > age.

Climate affects (1) the array of plant species, (2) the quantity of plant material produced, and (3) the intensity of soil microbial activity. Vegetation and topographic effects are difficult to separate from climatic effects. Rather, all of the factors become integrated as a soil forms and account for the generalization that forest and grassland soils usually exceed other well-aerated soils in humus content, whereas desert and semidesert soils have very little SOM.

Tropical soils, both humid and arid, were once thought to have low organic matter contents because the soils lack the dark color that characterizes SOM in temperate regions, and because temperature is inversely correlated with SOM content in temperate soils. The high soil temperature should increase microbial oxidation of SOM. The organic matter contents of similar soil orders in tropical and temperate regions, however, are quite similar (Table 6.1). The high organic matter contents of the tropical soils may be due to strong interaction of organic matter with Fe and Al hydroxy-oxides and allophane. Strong interaction could stabilize the SOM against microbial decay. The high SOM content is also due to the high rate of year-round biomass production in the humid tropics.

Topography influences the amount of SOM in two ways. North-facing slopes (in the northern hemisphere) are cooler and moister, so the organic matter content is greater than in soils on south-facing slopes. A much greater effect is topography's control of water drainage. In poorly drained and swampy soils, plant matter can be covered with water, and oxygen excluded, as soon as the plant dies. In stagnant water, oxygen diffusion from the surface is the only means of oxygen supply to the

**Table 6.1. Average organic carbon contents of several soil orders in temperate and tropical regions (to obtain approximate SOM values, multiply by 1.7)[a]**

| Soil Order (0 to 15 cm depth) | United States | Brazil | Zaire | Mean Level |
|---|---|---|---|---|
| | Carbon Contents (%) | | | |
| Mollisols | 2.44 | — | — | 2.44 |
| Oxisols | — | 2.01 | 2.13 | 2.07 |
| Ultisols | 1.58 | 1.61 | 0.8 | 1.39 |
| Alfisols | 1.55 | 1.06 | 1.30 | 1.30 |

[a] From P. A. Sanchez. 1976. *Properties and Management of Soils in the Tropics.* Wiley–Interscience, New York.

organisms carrying out the decomposition. This slow rate of oxygen supply preserves much of the plant matter from decay. Fermentation can operate in the absence of oxygen, but its effect on decay is rather small. Running water can bring enough dissolved oxygen to oxidize organic matter in streams. The turnover of lake water in spring and fall supplies some oxygen to lake bottoms.

In stagnant water, dead and partially decayed plant matter can accumulate to the water surface. Such deposits were up to 30 m thick in the Sacramento–San Joaquin Delta of California, for example, before those areas were drained. Ireland has large areas of *peat* (a type of Histosol containing 50% or more by mass of organic matter) deposits more than 10 m thick. Low temperatures also enhance peat accumulation. Major areas of peat are found in central Canada, which is both cold and swampy, with lesser amounts in low-lying areas of northern Russia and northern Europe. Smaller, but substantial, areas of Histosols are found on every continent and even in the tropics. The largest area of contiguous peat or muck soils in the continental United States is in the Everglades area south of Lake Okeechobee in southern Florida.

Peat accumulation depends on the rate of organic addition versus the rate of oxidation. The oxidation rate is a function of the rate of oxygen supply and temperature. The oxygen supply in turn depends on soil water content. When drained for cultivation, peat oxidizes as it contacts oxygen and also shrinks as it dries. Shrinkage and oxidation lowered the level of cultivated peat land in California and Florida by as much as 25 to 50 mm $yr^{-1}$. This rate has slowed in recent years as we have learned to better manage these materials. The peat islands in California are encircled by dikes to keep out the water. The peat surface is now as much as 10 m below the river level. The dikes must be continually strengthened as the peat oxidizes and shrinks. In recent years the maintenance of these dikes has become so costly that some of the islands have been abandoned after winter floods broke through the dikes. Wind erosion of the dry surface and accidental burning of the peat also contributed to the loss of these extremely productive soils.

In the formerly glaciated areas of Canada, Europe, and Siberia, peat has been accumulating for 10 000 to 15 000 years, since the glaciers retreated. This accumulation rate varies from year to year, with a rough average of about 0.1 mm $yr^{-1}$ in northern Canada. Since these areas were scraped bare by the glaciers, they represent a great transfer and redistribution of SOM and biomass carbon from more southern regions, and of $CO_2$ from the atmosphere, to the peat lands. The coal and petroleum deposits of the Carboniferous Era formed from enormous accumulations of peat soils in many areas of the world. They represent beds of SOM that were laid down under conditions similar to those that produce the peat lands of today.

Parent material influences SOM contents mainly through its effect on soil texture. In an area of similar climate and topography, SOM content tends to increase with soil clay content. The intimate association of humic substances with inorganic solids as organomineral complexes preserves organic matter. Montmorillonitic clays because of their high surface areas have particularly high adsorptive capacities for organic molecules and are particularly effective in protecting nitrogenous constituents from attack by microorganisms. This strong interaction between clays and organic matter also gives rise to important effects of SOM on soil physical and chemical properties.

As conditions change so does the SOM content. If humanity or nature lays soil bare, the SOM content can recover rapidly by geologic time standards to its previous state, but not fast enough for the human inhabitants of the area. Several years after the volcanic eruptions of Mt. St. Helens in Washington and of Mt. Pinatubo in the Phillipines, plants are reestablishing themselves on fresh-deposited pumice. Mine spoils and their revegetation also show this behavior. Plant growth is sparse in the first years as sulfides in the ores oxidize to sulfuric acid and as the rock minerals begin to weather. After the acid is leached out and some weathering has occurred, the plant increase and the rate of SOM increase follow an S-shaped curve. They start slowly, then increase exponentially, and the rate eventually levels off to a steady-state SOM content determined by climate and topography. The steady state may require 50 years for fine-textured parent material in humid climates, or as much as 1500 years for sandy soils in arid regions. These "new" soils also include flood plain deposits, road cuts, and land no longer cultivated.

Farming more subtly affects SOM. SOM levels may decrease by as much as one-half when untilled lands are cleared and cultivated. The oxidation rate is relatively fast during the first years as the cultivating tools shear soil aggregates and expose SOM that had been relatively protected from microbial attack by a coating of more resistant organic matter. The "minimum tillage" form of cultivation has the goals of minimizing the energy costs of cultivation and of increasing the SOM and nitrogen content.

## 6.2   THE DECAY PROCESS

The decay of SOM is the oxidation of organic carbon by heterotrophic organisms that utilize the energy of oxidation for their metabolism. The initial breakup of tree trunks and large objects is carried out by animals foraging for grubs, by termites, and by earthworms. Saprophytic plants such as mushrooms and snowflowers also obtain their energy from this partially decomposed plant matter. As the organic matter becomes more finely divided, the size of the decomposing organisms also decreases. Decay proceeds as long as the oxygen, water, temperature, and nutrient levels are adequate for the decomposing organisms. In the desert, the absence of water greatly hinders the oxidation rate of organic material at the soil surface. Beneath the surface of arid soils, where the moisture content is more likely to be adequate, decomposition is rapid.

The annual input rate of dead plant matter to soils in temperate regions, the *net primary productivity* of plants, is about 1 kg m$^{-2}$ of carbon or 20 Mg ha$^{-1}$ of dry matter. The annual input rate is perhaps 2 kg C m$^{-2}$ or 40 Mg ha$^{-1}$ of dry matter in humid tropical forests, and decreases to virtually zero to 0.1 kg C m$^{-2}$ in deserts and arctic tundra. This input is part of the nonhumus organic matter, which includes original plant and microbial tissue and partially decomposed material. These nonhumus substances contain carbohydrates and related compounds, proteins and their derivatives, fats, lignins, tannins, and various partially decomposed products in roots and plant tops. The portion contributed by dead animal matter is insignificant because the amount of carbon as plant biomass is 10 000 times greater than the carbon in animal biomass.

New plant material is an excellent source of food for soil microorganisms. Microbes are selective: Simple monomers of sugars, amino acids, and fats are oxidized first. The polymers—starch and proteins—follow close behind. Cellulose, the most prevalent plant component, decomposes more slowly. Lignin and hydrocarbons decompose even more slowly. Some plant materials (e.g., pine needles and oak leaves) and some synthetic organic chemicals contain inhibitory compounds and decompose slowly and by only a few microbial species. Shortly after fresh material contacts the soil, microbes begin decomposing it as a source of nutrients and energy (Fig. 6.1). The initial phase of microbial attack is a rapid loss of easily decomposable organic substances. Molds and spore-forming bacteria are especially active in consuming proteins, starches, and cellulose. Their major products are $CO_2$ and $H_2O$ plus a small amount of new microbial tissue. By-products, especially in partially anaerobic conditions, include small amounts of $NH_3$, $H_2S$, organic acids, plus other incompletely oxidized substances. In subsequent phases of decomposition, when aerobic conditions are present, these intermediate compounds and newly formed microbial tissues are attacked by other microorganisms, with production of some new biomass while a larger fraction of the organic carbon is converted to $CO_2$.

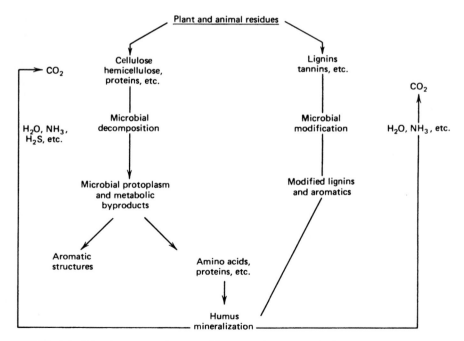

**FIGURE 6.1.** Organic matter decomposition and formation of humic substances. (From F. E. Bear, ed. 1964. *Chemistry of the Soil*, ACS Monograph Series No. 160, p. 258.)

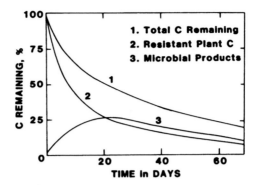

**FIGURE 6.2.** Idealized diagram for the decay of crop residues in soil under conditions that are optimal for microbial activity. (From F. J. Stevenson. 1982. *Humus Chemistry.* Wiley, New York.)

Figure 6.2 is an idealized graph of crop-residue decay with time in temperate regions. Plant residues are attacked rapidly at first, but the rate of decay soon slows. Considerable plant carbon remains in the soil at this point, but part of the residual carbon occurs as microbial by-products and part as the more resistant plant residues. Different plant components decompose at different rates. Lignin in wood is attacked much more slowly than cellulose. While most is converted into $CO_2$, the more readily decomposable constituents are also partially resynthesized by the decomposer microbes into more resistant components. Decay also slows because microbes work on the surface of particles; the resistant material can coat the underlying material and protect it from further decay.

Figure 6.3 shows the course of carbon losses from soils in a temperate climate (England) and a humid tropical climate (Nigeria). The shapes of the curves are similar, but the time scale is four times faster in the tropical climate. The approximate half-life of fresh organic matter in temperate regions is about 3–4 months, but it is as little as 3–4 weeks in the humid tropics. The flatter portion of the cube represents the second stage of organic matter decomposition, with a half-life of about 1.6 years in the humid tropics and 6.2 years in temperate regions.

The final stage of decomposition is the gradual loss of the more resistant plant parts, such as lignin, in which the actinomycetes and fungi play a major role. A small fraction of the original carbon, however, persists for a very long time. When the age of soil organic carbon is measured by [14]C dating, a small but extremely old fraction raises the average age of carbon in SOM to about 1000 years in surface soils and to several thousand years in subsoils.

Results of field experiments confirm that carbon becomes increasingly resistant to decomposition with time. This has led some investigators to conclude that the organic component exists in three major fractions when considered on a dynamic basis: (1) decomposing plant residues and the associated biomass, which turn over every few years; (2) microbial metabolites and cell wall constituents that become stabilized in soil and have a half-life of 5 to 25 years; and (3) the smallest fraction, resistant organic matter, ranging in age from 250 to 2500 years or more.

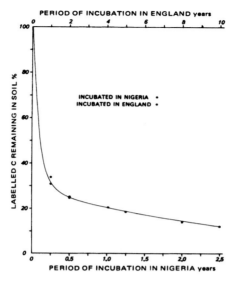

**FIGURE 6.3.** Decomposition rates of fresh organic matter in England and in Nigeria. (From D. S. Jenkinson and A. Ayanaba. 1970. *Soil Sci. Soc. Am. Proc.* **43**:912.)

Despite the stability of the resistant fractions of SOM, 50 to 80% of freshly added organic matter is lost from most temperate soils during the first year. The smaller the particle size, the faster the SOM is destroyed. Plant or animal residues must be added to soils continually if the favorable effects of organic matter on soil properties are to be maintained. To increase the SOM content by adding organic residues to soils is difficult. The decomposition rate of organic materials in soils increases in proportion to the addition rate. The more organic matter added, the more it is oxidized. An experiment in Rothamsted, England, has been adding ca. 30 tonnes ha$^{-1}$ of organic manures to a cultivated soil annually since 1843, without bringing the SOM content back to its uncultivated level.

## 6.3   EXTRACTION, FRACTIONATION, AND COMPOSITION

Soil organic matter is a complex polymer that can be studied only after it is separated from the inorganic soil fraction and after it has been broken into smaller fragments. Recent advances have lessened the degree to which SOM has to be destroyed to be studied. The separation of organic matter from the inorganic matrix of sand, silt, and clay is not physically difficult, but the extracting agent (traditionally 0.1 to 0.5 M NaOH) is harsh and alters the organic matter through hydrolysis and autoxidation. The components in such extracts can be partially fractionated by precipitation with acids or metal salts, or by taking advantage of solubility differences in various organic solvents. Students interested in the extraction and fractionation procedures should consult the works of Stevenson (1982) and Schnitzer and Schulten (1998).

The classical procedure for fractionation of extracted organic matter involves acid precipitation of some fractions from an NaOH extract, and subsequent dissolution of part of the precipitated material with alcohol (Fig. 6.4). The *humic acids* and *fulvic acids* fractions so prepared are mixture of many different chemical compounds in various stages of polymerization. Stevenson defines them as humic acid, the dark-colored organic material that is extracted from soil by various reagents and that precipitates when dilute acid is added, and fulvic acid, the colored material that remains dissolved in the extracting solution after acidification. Schnitzer and Schulten define humic acid as the fraction soluble in dilute alkali or neutral salt but coagulated by acidification, and fulvic acid as that which remains in aqueous solution after acidification, and *humin* as the fraction that cannot be extracted from soils by dilute base, neutral salts, or acids. Humin is thought to be humic acid-like material that has reacted strongly with inorganic soil components and thereby resists attack by alkalis and acids.

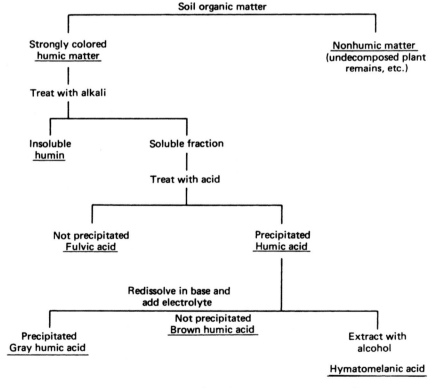

**FIGURE 6.4.** Fractionation of soil organic matter. (Modified from F. J. Stevenson. 1982. *Humus Chemistry.* Wiley, New York; and J. L. Mortenson and F. L. Hines. 1964. In *Chemistry of the Soil* (F. E. Bear, ed.). ACS Monograph Series No. 160.)

Organic materials undergo microbial enzymatic and chemical reactions in soils to form colloidal polymers called humus (Fig. 6.1). Humus is a complex and rather microbially resistant mixture of brown to almost black, amorphous and colloidal substances modified from the original plant tissues or resynthesized by soil organisms. Humus contains approximately 10% carbohydrates, 10% nitrogen components (proteins, amino acids, and cyclical N compounds), 10% "lipids" (including alkanes, alkenes, fatty acids, and esters), and 70% humic substances.

Chemists have attempted to unravel the details of humus composition for many years. Despite considerable progress in characterizing various extracts, much remains to be discovered. Humus contains primarily C, H, O, N, P, and S plus small amounts of other elements. Only a small but important portion is soluble in water, but much is soluble in strong bases. The various fractions of humus obtained on the basis of solubility characteristics are part of a heterogeneous mixture of molecules, which range in molecular weight from as low as several hundred to over 300 000 (Fig. 6.5). Carbon and oxygen content, acidity, and degree of polymerization all change systematically with molecular weight. The low molecular weight fulvic acids have higher oxygen contents and lower carbon contents than the high molecular weight humic acids. The more soluble fulvic acids are usually responsible for the brownish-yellow color of many natural waters. Humic acids precipitate with acids and polyvalent cations, thus tending to be insoluble.

Recently, SOM has begun to be analyzed by pyrolysis, which is destruction at high temperatures in the absence of oxygen, and analysis of the many volatile compounds that emanate from SOM. Carbohydrates, phenols, lignin and n-fatty acids are the

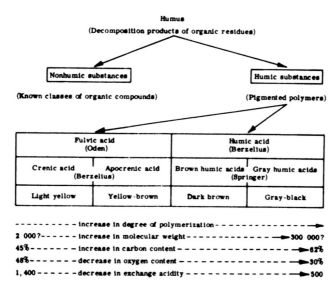

**FIGURE 6.5.** Classification and chemical properties of humic substances. (From F. J. Stevenson. 1982. *Humus Chemistry.* Wiley, New York.)

major products volatilized. Pyrolysis is a destructive technique but the compounds evolved may lead to a better understanding of humus and soil organic matter.

The composition and structure of soil humus probably varies with the material from which it was derived and with the species of microbe. Although there is no reason to believe that humus from different soils is the same, several researchers agree on a "type" structure for humic acid (Fig. 6.6). Two types of polymers, humic acid (50 to 80% by mass) and polysaccharides (10 to 30% by mass), can constitute up to 90% or more of the total humus in soils. A typical humic acid molecule probably consists of polymers of a basic six-carbon aromatic ring structure of di- or trihydroxyl phenols linked by —O—, —NH—, —N—, and —S— bonds, and containing —OH groups and quinone (—O—$C_6H_4$—O—) linkages. This structure contains a high density of reactive functional groups. Individual humic acid molecules vary in the structure and density of functional groups, but the basic structure is thought to remain approximately the same.

The chemical origin of humus components is not yet resolved. Some workers consider humus to be microbially resistant plant materials (lignin, suberin, cutins, paraffins, etc.) that are awaiting oxidation to humic acid and further oxidation to fulvic acid. Some have thought that humic materials were the largely nondegraded plant materials, which would account for the aromatic content; microbiologists thought that producing large amounts of aromatic groups was too exotic for soil microbes. A lignin origin for humus, however, would not account for the nitrogen content of humus or for the large amount of alkanes (paraffins) found in recent studies. Lignin in wood is nitrogen-free and synthesis of alkanes (aliphatic hydrocarbons) was thought to be rare. The aromatic (alkene) components have received hugh attention: They are present in lignin and make up about half of the carbon in SOM. Recent studies indicate that alkanes are also present in SOM and may play a role in binding the aromatic groups together.

## 6.4 COLLOIDAL PROPERTIES

Most of the colloidal properties of SOM are due to humus. Humus is highly colloidal and is x-ray amorphous rather than crystalline. The surface area and adsorptive capacities of humus per unit mass are greater than those of the layer silicate minerals. The specific surface of well-developed humus may be as high as $900 \times 10^3$ m$^2$ kg$^{-1}$; its exchange capacity ranges from 1500 to 3000 mmol($+$) kg$^{-1}$.

The negative charge (and hence the CEC) of humus is generally agreed to be due to the dissociation of $H^+$ from functional groups. All charge on humus is strongly pH dependent, with humic and fulvic acids behaving like weak-acid polyelectrolytes (polyprotonated weak acids). Figure 6.7 shows typical titration curves for peat and soil humic acids. Both buffer soil pH over a wide range. The slopes of humic and fulvic acid titration curves do not change as sharply as for monomeric acids, because of electrostatic effects on the high molecular weight polyacids and because of configurational changes that occur at higher pH.

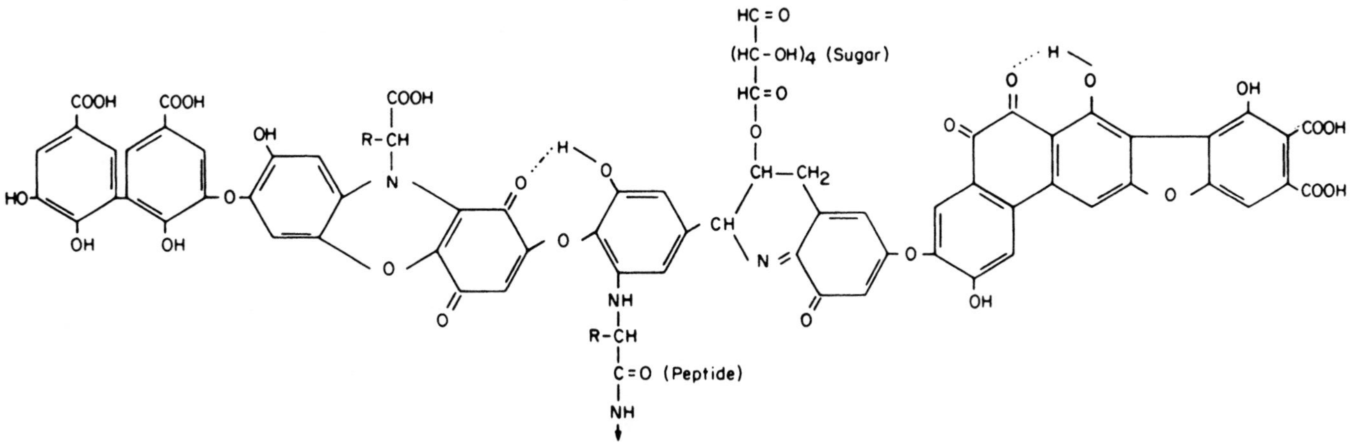

**FIGURE 6.6.** Hypothetical structure of humic acid. (From J. J. Mortvedt, P. M. Giordano, and W. L. Lindsay, eds. 1972. *Micronutrients in Agriculture.* American Society of Agronomy, Madison, WI.)

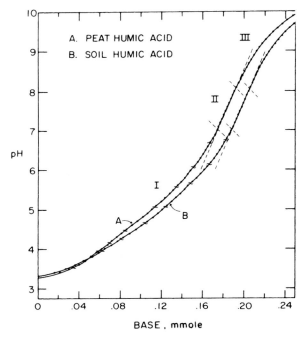

**FIGURE 6.7.** Titration curves of a soil and peat humic acid. The small wavy lines on the curves indicate endpoints for ionization of weak-acid groups having different, but overlapping, ionization constants. (From F. J. Stevenson. 1982. *Humus Chemistry.* Wiley, New York.)

The dissociation of carboxyl and phenol groups yields perhaps 85 to 90% of the negative charge of humus. Many carboxylic groups are sufficiently acid to dissociate below pH 6 (zone I, Fig. 6.7)

$$R—COOH = R—COO^- + H^+ \tag{6.1}$$

leaving a negative charge on the functional group. Here R represents any number of organic species whose differing electronegativities alter the tendency for $H^+$ to dissociate. Thus, the various R—COOH units dissociate at different pH values. As the pH of a system increases above 6, still weaker carboxylic groups and other very weak acids dissociate (zone II, Fig. 6.7). Zone III represents dissociation of phenolic OH and other very weak acids at pH > 8. Dissociation of $H^+$ from acid groups throughout the pH range adds to the total negative charge of humus. Dissociation of $H^+$ from enolic OH, imide ($=NH$), and possibly other groups also contribute to the negative charge.

No SOM fraction possessing a net positive charge has been found at normal soil pH values (Table 6.2). Protonated groups such as $(R—OH_2)^+$ and $(R—NH_3)^+$ can yield positive charges, but the overall charge on humus is negative.

Charged sites (primarily $COO^-$) enable SOM to retain cations in nonleachable but exchangeable forms that are available to plants. The bonding is primarily coulombic

**Table 6.2. Effects of pH on cation exchange capacity (CEC) for 60 Wisconsin soils**[a]

| pH | CEC (mmol(+) kg$^{-1}$, Means for 60 Soils) | | | |
|---|---|---|---|---|
| | Organic Matter | Layer Silicates | Whole Soil | % CEC from SOM |
| 2.5 | 360 | 380 | 58 | 1 |
| 3.5 | 730 | 460 | 75 | 28 |
| 5.0 | 1270 | 540 | 7 | 37 |
| 6.0 | 1310 | 560 | 108 | 36 |
| 7.0 | 1630 | 600 | 123 | 40 |
| 8.0 | 2130 | 640 | 148 | 45 |

[a]Data of C. S. Helling, G. Chesters, and R. B. Corey. 1964. *Soil Sci. Soc. Am. Proc.* **28**:517–520.

or electrostatic (e.g., —COO$^-$K$^+$). The bonding is also partly covalent, particularly when the charge is neutralized by transition-metal cations (Fe$^{2+}$, Zn$^{2+}$, Cu$^{2+}$, Ni$^{2+}$).

## 6.5 FUNCTION OF ORGANIC MATTER IN SOIL

Table 6.3 summarizes the general properties of humus and its associated effects on soil. SOM contributes to plant growth through its effects on the chemical, biological, and physical properties of soil. The SOM supplies N, P, and S for plant growth, serves as an energy source for soil microorganisms, and promotes good soil structure. Humus also indirectly affects the plant uptake of microelement and heavy metal cations, and the performance (availability) of herbicides and other agricultural chemicals. SOM is highly porous so that pesticide and other organic compounds added to soils can be enveloped by SOM. In that form they are much less active. Soils with high amounts of SOM require higher dosages to herbicides and pesticides to achieve the desired result.

The SOM supplies nearly all the N, 50 to 60% of the P, perhaps as much as 80% of the S, and a large part of the B and Mo adsorbed by plants from unfertilized, temperate region soils. Indirectly, SOM affects the supply of essential elements from other sources. The amount of N$_2$ fixation by the free-living bacterium *Azotobacter*, for example, is related to the amount of readily available energy sources in the soil, such as the carbohydrates in SOM.

In humid soils, C, N, and S are found predominantly in SOM. With increasing aridity, the amount of organic matter decreases and the fractions of the inorganic forms of the elements (carbonate, sulfate, and nitrate) tend to increase.

The mass ratio of C/N/S in the SOM of temperate region soils is roughly 100/10/1. Carbon supplies the energy for N and S reduction, as well as the matrix of compounds into which reduced N and S are incorporated and stabilized. Nitrogen

**Table 6.3. General properties of humus and associated effects in soil[a]**

| Property | Remarks | Effect on Soil |
|---|---|---|
| Color | The typical dark color of many soils is caused by organic matter | May facilitate warming |
| Water retention | Organic matter can hold up to 20 times its weight in water | Helps prevent drying and shrinking; improves moisture retention in sandy soils |
| Combination with clay minerals | Joins soil particles into structural units called aggregates | Permits gas exchange; stabilizes structure; increases permeability |
| Chelation | Forms stable complexes with $Cu^{2+}$, $Mn^{2+}$, $Zn^{2+}$, and other polyvalent cations | Buffers the availability of trace elements to higher plants |
| Solubility in water | Insolubility of organic matter results partially from its association with clay; salts of divalent and trivalent cations with organic matter are also insoluble; isolated organic matter is partly soluble in water | Little organic matter is lost by leaching |
| pH relations | Organic matter buffers soil pH in the slightly acid, neutral, and alkaline ranges | Helps to maintain a uniform reaction (pH) in the soil |
| Cation exchange | Total acidities of isolated fractions of humus range from 3000 to 14 000 mmole $kg^{-1}$ | Increases the cation exchange capacity (CEC) of the soil; from 20 to 70% of the CEC of many soils is caused by organic matter |
| Mineralization | Decomposition of organic matter yields $CO_2$, $NH_4^+$, $NO_3^-$, $PO_4^{3-}$, and $SO_4^{2-}$ | A source of nutrient elements for plant growth |
| Combination with organic molecules | Affects bioactivity, persistence, and biodegradability of pesticides | Modifies the application rate of pesticides for effective control |

[a]From F. J. Stevenson. 1982. *Humus Chemistry*. Wiley, New York.

and sulfur, in turn, are among the elements that govern the rate of plant growth and photosynthesis. The result under natural conditions is a relatively constant C/N/S ratio in SOM.

The availability of many microelement cations is strongly affected by SOM. Various low molecular weight and somewhat water-soluble components of SOM, such as fulvic acid, form stable complexes (chelates) with $Fe^{2+}$, $Cu^{2+}$, $Zn^{2+}$ and other polyvalent cations. These shield the cations from hydrolysis and precipitation reactions

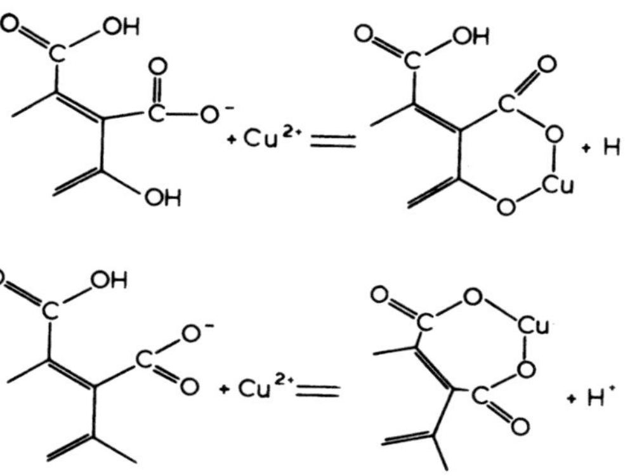

**FIGURE 6.8.** Postulated reaction between $Cu^{2+}$ and fulvic acid function groups. (After D. S. Gamble et al. 1970. *Can. J. Chem.* **48**:3197.)

and therefore increase their water solubility. A typical reaction between $Cu^{2+}$ and the functional groups of fulvic acid is given in Fig. 6.8. Inorganic precipitation, particularly in soils of high pH, greatly reduces the solubility and availability of many of the essential microelements. Organic amendments (manure, sewage sludge, etc.) can improve microelement availability in alkaline soils and correct Fe and Zn deficiencies in particular. The amendments apparently release microelements in chelated form or release (mineralize) fulvic acid compounds that chelate the inorganic Fe and Zn present in the soil. SOM also combines with toxic ions such as $Cd^{2+}$ and $Hg^{2+}$, as well as with microelement cations at high concentrations, and reduces their availability. Organic amendments to soils often decrease cation toxicities in acid soils.

Soil organic matter is involved with soil acidity. In mature Swedish forests, for example, the soil pH is perhaps 3.5. After harvest and while the succeeding trees are young, the pH is >4. The pH then decreases steadily as the forest matures, only to rise again when those trees are harvested. This pH cycle is thought to be due to organic acids formed by increasing litter fall. When litter production stops or slows and the organic acids decay or are leached away, the soil pH rises. In New England, part of the pH decline in formerly cultivated fields may be due to increased organic acids produced by greater leaf litter as trees invade the fields. In both cases, the pH is too low to be caused by $Al^{3+}$ hydrolysis, and Al hydrolysis would not account for the cyclical pH change.

Humus affects soil structure and thus soil tilth, aeration, and moisture retention. The deterioration of structure that accompanies intensive tillage is usually less severe in soils adequately supplied with humus. When humus is lost, soils tend to become hard, compact, and cloddy.

Aeration, water-holding capacity, and permeability are all improved by humus. The frequent addition of easily decomposable organic residues leads to the synthesis of complex organics (e.g., polysaccharides) that bind soil particles into aggregates. The intimate association of clay-sized particles (layer silicates) with humus via cation (e.g., calcium, magnesium, aluminum, iron) bridges also promotes aggregation. The water-insoluble salts of humic acid with polyvalent cations are called *humates*. They tend to be amorphous and glue-like. Heavy (clayey) soils, in particular, benefit from organic matter additions by promoting particle aggregation. Aggregation yields a loose, open, granular structure for good water and air permeability.

Humus also absorbs large quantities of water. The fully synthesized humus of a mineral soil contains as much as 80 to 90% water by weight. Additionally, micropores within larger soil aggregates hold available water for plants. This increase in plant-available water-holding capacity is a major benefit of organic matter additions to sandy soils.

The data in Table 6.2 point out important characteristics of the CEC of SOM. All of the charge of humus is pH dependent, even at low pH. The CEC of both organic matter and layer silicates increases with increasing soil pH, but the CEC of SOM increases faster with pH than that of the layer silicates. In one soil, for example, the SOM contributed only 1% of the total CEC at pH 2.5, but 45% at pH 8.0. In soils dominated by low-CEC layer silicates, such as kaolinite, the relative contribution of organic matter to whole-soil CEC can be even greater. A large fraction of the CEC in most soils is due to SOM.

The functional groups responsible for the high CEC of humus also buffer soil pH over a wide range. This buffering contributes significantly to the lime requirement of acid soils (Chapter 8). Total acidities of isolated fractions of humus vary from 3 to 14 mol kg$^{-1}$.

### 6.5.1   Organic Chemical Adsorption

Organic matter content is the soil factor most directly related to the sorption of most herbicides and organic compounds by soils. The manner in which organic matter sorbs organics is discussed more fully in Chapter 7. The SOM content strongly influences pesticide behavior in soil, including effectiveness against target species, phytotoxicity to subsequent crops, leachability, volatility, and biodegradability. Recommended herbicide application rates are often higher for soils high in organic matter content, to compensate for greater adsorption in these soils. The soil behavior of organic chemicals, and particularly their interaction with SOM, is an active area of current soil chemistry research.

## BIBLIOGRAPHY

Jenny, H. 1941. *Factors of Soil Formation.* McGraw-Hill, New York.

Schnitzer, M., and H.-R. Schulten. 1998. New ideas on the chemical make-up of soil humic and fulvic acids. In *Future Prospects for Soil Chemistry* (P. M. Huang, ed.). Soil Science Society of America Spec. Publ. 55, Madison, WI.

Stevenson, F. J. 1982. *Humus Chemistry.* Wiley, New York.

## QUESTIONS AND PROBLEMS

1. Distinguish between soil organic matter, humus, and soil biomass.

2. Give representative SOM contents for Mollisols, Aridisols, Oxisols, and Histosols. Justify the SOM content of each in terms of the factors of soil formation.

3. Explain why the SOM content of soils in a given climatic zone tends to be higher in fine-textured soils than in coarse-textured soils.

4. Describe the overall decay process of SOM, including discussions of the organisms involved, the decomposition products, and the time necessary for the process.

5. If increased SOM contents are so beneficial to soils, why don't farmers manage their soils to increase SOM?

6. What are the chemical properties of humus that make it special?

7. How does SOM contribute to the chemical, physical, and biological properties of soil as a medium for plant growth?

8. How may SOM alter micronutrient and trace metal availability in soils?

9. What is the buffering capacity of humus expressed in terms of $CaCO_3$ equivalent? Assume 50% dissociation of humus acidity over the pH range of 4 to 7.

10. How may SOM affect pesticide recommendations, and why?

11. Explain how only a few percent organic matter can exert a profound influence on soil properties.

12. Assuming that layer silicates and organic matter exist independently in a soil, calculate a reasonable cation exchange capacity of a soil containing 40% montmorillonite and 3% organic matter. Repeat the calculation for a soil that

    (a) Contains 40% kaolinite and 3% organic matter.

    (b) Contains 20% kaolinite and 1.5% organic matter.

    How realistic are such calculations?

13. Is SOM amphoteric? Explain.

14. What are the primary functional groups responsible for charge development in SOM?

# 7

# WEATHERING AND SOIL DEVELOPMENT

Rocks formed beneath the earth's surface are unstable when raised to the surface where they contact water, $O_2$, $CO_2$, and organic compounds. Soil is formed when the ions in rock minerals at the earth's surface change, or weather, to more stable chemical states. The change to increasing stability is slow and goes through many steps. Soil development, in the chemical sense, is roughly synonymous with weathering. This chapter discusses the general weathering reactions due to the effects of water, $O_2$, and $CO_2$ that create soil solids and the soil solution.

The particle surfaces and soil solutions created by weathering tend to be more similar chemically than the composition of the parent minerals. This relative uniformity is of great benefit to the development and maintenance of life. The small variation among different soils and soil solutions, however, affects plant growth and is often emphasized because of humanity's concern for maximum or optimum growth of agricultural plants. Compared to the vagaries of weather, pests, prices, and other production factors, soil chemical variability is relatively small and within a farmer's control.

Weathering of igneous and metamorphic rocks changes these dense solids into unconsolidated particles whose surfaces and newly formed particles often differ markedly from the chemical composition and structure of the parent minerals. The changes during weathering of sedimentary rocks is less striking. The boundary between sedimentary rocks and soil is often physically and chemically diffuse.

Table 7.1 shows the composition of common soil parent materials. The crystal structures and ion valences in rock minerals are stable at the conditions under which the rocks formed. The physical conditions of erosion, freezing and thawing, glaciation, heating and cooling, and root growth at the earth's surface break rocks apart, which exposes more surface for chemical weathering. A bigger change in the rock minerals, however, results from the new chemical conditions: exposure to water, oxygen, carbon dioxide, and organic compounds. For sedimentary rocks, weathering is

**Table 7.1. Average compositions of several parent material rocks**

| Compound | Granodiorite[a] (Granitic) (%) | Basalt[a] (%) | Shale[b] (%) | Sandstone[b] (%) | Limestone[b] (%) |
|---|---|---|---|---|---|
| $SiO_2$ | 65.1 | 49.3 | 58.1 | 78.3 | 5.2 |
| $K_2O$ | 2.4 | 1.2 | 4.3 | 1.4 | 0.04 |
| $TiO_2$ | 0.5 | 2.6 | 0.6 | 0.2 | 0.06 |
| $Al_2O_3$ | 15.8 | 14.1 | 15.4 | 4.8 | 0.8 |
| $Fe_2O_3$ | 1.6 | 3.4 | 4.0 | 1.1 | 0.5 |
| FeO | 2.7 | 9.9 | 2.4 | 0.3 | — |
| MgO | 2.2 | 6.4 | 2.4 | 1.2 | 7.9 |
| CaO | 4.7 | 9.7 | 3.1 | 5.5 | 42.6 |
| $Na_2O$ | 3.8 | 2.9 | 1.3 | 0.4 | 0.05 |
| $H_2O$ | 1.1 | — | 5.0 | 1.6 | 0.8 |
| $P_2O_5$ | 0.1 | 0.5 | 0.17 | 0.08 | 0.04 |
| $SO_3$ | — | — | 0.6 | 0.07 | 0.05 |
| $CO_2$ | — | — | 2.6 | 5.0 | 41.5 |
| Total | 100 | 100 | 100 | 100 | 100 |

[a] From G. W. Tyrell. 1950. *The Principles of Petrology*. Dutton, New York.
[b] From F. J. Pettijohn. 1957. *Sedimentary Rocks*, 2d ed. Harper & Row, New York.

due to a change in those chemical conditions. The milieu in which sedimentary rocks form is much closer to that of soil conditions than to the conditions of igneous and metamorphic rock formation. Nonetheless, the weathering of sedimentary rocks is also due to exposure to water, $O_2$, and $CO_2$, but at concentrations different from those in which the rocks formed.

The major reaction that weathers minerals is the strong tendency of ions in solids to dissolve in water (Chapter 3). In addition to the energy of hydration released by dissolving in water, an ion's entropy increases as it is freed from a rigid structure into the aqueous solutions. Entropy increases further with increasing dilution in the water. Because the Gibbs free energy decreases with increasing entropy, because dilution also increases entropy, and because substances strive to reach their lowest energy state, an ion's most stable state is at infinite dilution in water or a solid. The amount of water in soils is limited to the thin film on soil particle surfaces, so that the entropy increase due to dilution is limited to that extent.

After dissolving into the soil solution, some ions combine to create new solids that are stable under those soil solution conditions; other ions are leached away. This separation means that the composition of the soil solution changes during soil development and may change enough to cause the first-formed soil minerals to redissolve and then portions of them reprecipitate as still other minerals. Because minerals are stable over a range of composition, they change stepwise toward increased stability, while the soil solution's composition changes slowly and unidirectionally, but not continuously, during weathering.

Some minerals remain virtually unweathered despite their inherent instability, because their dissolution rate in water is exceedingly slow. Quartz particles larger than several micrometers in size (fine silt) remain in soils for so long that quartz appears to be the most stable state for soil silicon. When finely divided into clay-sized particles, however, quartz persists only slightly longer in soils than does clay-sized feldspar. Feldspar disappears from the sand and silt fractions relatively rapidly.

Slow reactions only delay the inevitable time when the unstable minerals either dissolve or form a new solid phase. New solids in soils sometimes form by recrystallization of another mineral, entirely in the solid phase. Much more frequently, new solids form by dissolution of the old mineral and subsequent precipitation of part of the solute ions, often on the surface of the old mineral. When part of the solutes reprecipitate, the overall process is called *incongruent dissolution*. *Congruent dissolution* is complete dissolution without subsequent reprecipitation of part of the original, for example, NaCl and $CaCO_3$ dissolving in water. Congruent dissolution is rare in nature. Soil weathering reactions, with the exception of limestone parent material, are incongruent dissolution because part of silicates remains behind as clay minerals and amorphous solids.

Whether an ion remains in solution or precipitates depends on the sum of the heats of formation (enthalpy $\Delta H$) of chemical bonds and the energy changes (entropy $\Delta S$) associated with the randomness of ion movement and position. The $\Delta H$ values include the changes in energy between bonds in the old and new solids and the energies of ion–water bonds. The change in randomness represents the freedom of ions to diffuse through water rather than being constrained within a crystal. Although an ion's water ligands have slightly less randomness than other water molecules, the net effect of weathering is increased randomness and hence increased $\Delta S$.

The $\Delta S$ change during ion dissolution tends to be similar for each ion and is relatively small. The $\Delta H$ of formation of new secondary minerals, in contrast, varies widely. The resulting driving force $\Delta G$, therefore, differs widely between different ions. Ions that form weak chemical bonds with other ions (slightly negative $\Delta H$, usually monovalent) tend to remain in aqueous solution, while strongly bonding ions (highly negative $\Delta H$, usually polyvalent) tend to reprecipitate. Ions remaining in the soil solution are much more easily leached from soils and are therefore considered weatherable. The chemical states of reprecipitated ions also change during weathering. Because these ions remain in the soil, however, they are considered resistant to weathering.

To illustrate, imagine a drop of pure water falling on the surface of albite ($NaAlSi_3O_8$) feldspar. This igneous mineral is unstable in water at room temperature and pressure. The first ions or molecules to dissolve achieve a high degree of randomness because they can roam in the droplet unhindered by other ions:

$$NaAlSi_3O_8 + 4H_2O + 4H^+ = Na^+ + Al^{3+} + 3Si(OH)_4 \qquad (7.1)$$

$\phantom{xxxx}$albite$\phantom{xxxxxxxxxxxxxxxxxxxxxxxxxxxxxxxxx}$soluble silica

The next ions to dissolve find successively less randomness because other ions are already present in the water droplet and restrict their freedom of movement. Hence, the $\Delta S$ change per dissolving ion progressively decreases. The $\Delta H$ of the albite bonds and of the bonds of the later minerals that precipitate are unchanged, so dissolution slows and eventually stops.

Aqueous solubility is governed by the counteracting effects of the $\Delta S$ of dissolution versus the $\Delta H$ of chemical bonding in solids. The $\Delta H$ of $Na^+$ compounds tends to be small. Therefore, $\Delta S$ prevails and $Na^+$ remains in solution. The $\Delta H$ values of Al and Si bonds with $O^{2-}$ and $OH^-$, on the other hand, are large, so the aqueous solubility of those compounds is low, and the $Al^{3+}$ and $Si(OH)_4$ concentrations soon reach their solubility maxima in water. Soil solution conditions are often such that $Al^{3+}$ and $Si(OH)_4$ precipitate as kaolinite

$$Al^{3+} + Si(OH)_4 + \tfrac{1}{2}H_2O = 3H^+ + \tfrac{1}{2}\,Al_2Si_2O_5(OH)_4 \qquad (7.2)$$
$$\text{kaolinite}$$

and $Na^+$ and some $Si(OH)_4$ remain in solution. At lower $Si(OH)_4$ concentrations, the $\Delta H$ of gibbsite predominates

$$Al^{3+} + 3H_2O = 3H^+ + Al(OH)_3 \qquad (7.3)$$
$$\text{gibbsite}$$

so that $Na^+$ and $Si(OH)_4$ remain in solution. When the water film is displaced by a second droplet of water, the $Na^+$ and $Si(OH)_4$ are removed and weathering continues. Leaching of kaolinitic soils eventually lowers the aqueous $Si(OH)_4$ concentration enough that kaolinite becomes unstable, degrades to gibbsite, and frees the remaining silicon:

$$Al_2Si_2O_5(OH)_4 + 5H_2O = 2Si(OH)_4 + 2Al(OH)_3 \qquad (7.4)$$
$$\text{kaolinite} \qquad\qquad\qquad \text{soluble} \qquad \text{gibbsite}$$
$$\text{silica}$$

Continuing this simple view of weathering and soil development, the dissolution of feldspars and the formation of secondary clay minerals (Eqs. 7.1 and 7.2), tend to control soil pH in young soils. One mole of hydrogen ions is consumed per mole of albite weathered and the soil pH is slightly alkaline. When the feldspar content or rate of feldspar weathering decreases, $Al^{3+}$ and $Fe^{3+}$ hydrolysis reactions produce the acidity characteristic of moderately weathered soils (Eq. 7.3). In the final stages of soil development, reactions like Eq. 7.4 and the leaching of $H^+$ return soil pH to near neutrality.

The albite example illustrates several general points about weathering: (1) Weathering of igneous and metamorphic minerals releases considerable alkali and alkaline earth cations during the initial transition from rock to soil. Some of these cations are retained by plant absorption and cation exchange, but most are lost immediately from the soil. (2) Weathering releases considerable silica to the soil solution. Much of the silica leaches from the soil. The remainder reacts to form the secondary minerals— kaolinites, smectites, chlorites—common in soils but transitory on a geologic time scale. (3) Aluminium (and Fe, Ti, and Mn) hydroxides are insoluble and tend to accumulate in soils. (4) Weathering initially produces some alkalinity. And (5) the second stage of weathering or soil development produces acidity.

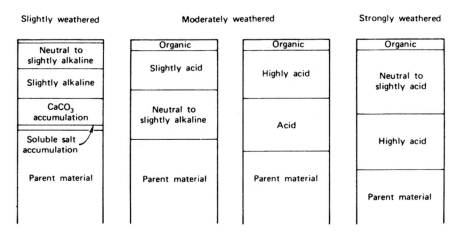

**FIGURE 7.1.** Schematic progression of basic and acidic zones through soils during soil development. This sequence also represents soil profiles from arid to humid to humid tropical regions.

Figure 7.1 represents an idealized course of weathering in a soil profile. The basicity and acidity are emphasized because pH is an easily measured indicator of the state of weathering. The zones of basicity and acidity leach down through the soil profile during development. The alkali cations released during breakdown of the parent material accumulate in a narrow, rarely noticeable zone. This is followed closely by a much more obvious band of $CaCO_3$ accumulation.

In slightly weathered soils, the surface soil pH is slightly alkaline. When the concentration of exchangeable alkali and alkaline earth cations retained on clay surfaces and in soil solution decreases, the soil becomes acidic as the hydrolysis reactions of $Al^{3+}$ and $Fe^{3+}$ become evident. When the hydrolysis reactions near completion, the surface soil pH returns to near neutrality. Ultimately, acid production ends even in the subsoil. The entire profile returns to nearly neutral pH and the residual soil is rich in the Al and Fe hydroxyoxide products of weathering.

The soil profiles in Fig. 7.1 are also somewhat typical of soil profiles found in arid (slightly weathered), humid (moderately weathered), and humid tropical (strongly weathered) regions. The course of soil weathering is the same in all climates, but in humid regions weathering rates are sufficiently faster than the rates of erosion and physical mixing to allow strongly weathered soil profiles to develop. Soils in these regions are covered with vegetation, which slows erosion rates. Because of heat and rainfall, soils in the humid tropics pass quickly through the early stages of soil development. Soils in arid regions have slow rates of soil development because of the lack of water. Many arid region soils have relatively high wind and water erosion rates, because of sparse plant cover and because the rains come in violent, infrequent storms.

The initial alkalinity from weathering is partially neutralized by the $H_2CO_3$ formed in soil pores by root and microbial respiration:

$$CO_2 + H_2O = H_2CO_3 = H^+ + HCO_3^- \tag{7.5}$$

Neutralization of the bases released by weathering and the weak acidity of carbonic acid ($pK \sim 5$) favor continued weathering. The role of $CO_2$ as the active agent of weathering has probably been exaggerated in the past. The abundance of water and the temperature more often control the rate of soil weathering than does the level of $CO_2$. Carbon dioxide is always present in soil pores, whereas water is not. Weathering in desert and frozen soils is extremely slow. Equation 7.5 is driven to the right, that is, $CO_2$'s effect on the weathering rate increases, as the $CO_2$ concentration rises to as much as several percent in soil pores during active root and microbial respiration, compared to 0.0033% v/v $CO_2$ in the atmosphere. Desert and cold soils are low in $CO_2$, because of less root and microbial activity, but still well above atmospheric concentrations. In practice, both water and high $CO_2$ levels are present during active soil weathering.

Alkali and alkaline earth cations, halides, sulfate, and silica ions tend to remain in solution, rather than precipitate as secondary minerals. Despite the soil's ability to retain ions, these ions under normal soil drainage conditions eventually reach the sea. The K, Mg, and Si move more slowly than do Na, Ca, halides, and sulfate. Under arid to semiarid conditions, much Ca precipitates as lime $CaCO_3$ and some as gypsum $CaSO_4 \cdot 2H_2O$.

The secondary minerals formed in soil from weathering products tend to be small in size and poorly crystallized to amorphous. They are primarily aluminosilicates and Al and Fe hydroxyoxides. These tiny crystals have large surface areas and tend to be charged because of unsatisfied chemical bonds within the crystals and at crystal edges. Large surface areas and unsatisfied bonds result in high surface free energies. Small particles therefore tend to dissolve and larger particles tend to grow at their expense. This aging reaction is slow, however, because of the low solubility of the ions involved. In practice, soil-formed minerals rarely grow beyond colloidal ($<2$ $\mu$m) size. Crystal growth is more evident for kaolinite. The weathering rates of smectites and chlorites, and leaching of their ions, are approximately as fast as their growth rates. Smectite and chlorite crystals larger than 1 $\mu$m are rare in soils.

Another important result of the unsatisfied bonds on the clay mineral surfaces is that they adsorb ions to balance the particle's charge. The unsatisfied charges are mostly negative, so mostly cations are adsorbed. Because they are held on the surface and not within the crystal, such ions can be exchanged for other ions. This slows the loss of ions from soils and retains the ions in states that are available for plant uptake, but ultimately the ions are lost.

Weathering continues after the formation of secondary minerals because the secondary minerals are stable only between certain concentration limits of soluble silica, alkali and alkaline earth cations, and $H^+$ in the soil solution. As these solutes are leached away, the concentration changes make the initial secondary minerals (smectites, calcite, gypsum, etc.) unstable. As weathering progresses, these intermediate minerals weather further to still more stable chemical states.

The effect of weathering reactions on the total composition of soils is illustrated by the data in Table 7.2. Three soils of increasing maturity, or degree of weathering, are compared to the average composition of igneous rocks. The comparisons are not exact because the parent materials of the soils differ from each other and from the

**Table 7.2. Composition of average igneous rocks and of three surface soils of increasing maturity**

|  | Average of Igneous Rocks | Barnes Loam (South Dakota) | Cecil Sandy Clay Loam (North Carolina) | Columbiana Clay (Costa Rica) |
|---|---|---|---|---|
| $SiO_2$ | 60 | 77 | 80 | 26 |
| $Al_2O_3$ | 16 | 13 | 13 | 49 |
| $Fe_2O_3$ | 7 | 4 | 5 | 20 |
| $TiO_2$ | 1 | 0.6 | 1 | 3 |
| MnO | 0.1 | 0.2 | 0.3 | 0.4 |
| CaO | 5 | 2 | 0.2 | 0.3 |
| MgO | 4 | 1 | <0.1 | 0.7 |
| $K_2O$ | 3 | 2 | 0.6 | 0.1 |
| $Na_2O$ | 4 | 1 | 0.2 | 0.3 |
| $P_2O_5$ | 0.3 | 0.2 | 0.2 | 0.4 |
| $SO_3$ | 0.1 | 0.1 | — | 0.3 |
| Total | 100.5% | 100.9% | 100.6% | 100.4% |

average igneous rock. The elemental contents are expressed as weight percent of the oxides, following an old geologic tradition, although only small fractions of the elements are actually present as simple oxides. Most are in more complex minerals. Expressing the elements as oxides allows the analyses to add up to 100%. More importantly, it emphasizes the importance of oxygen as *the* anion in rocks and soils. Even sulfates, phosphates, and carbonates are oxyanions in which the negative charge comes from oxygen ligands. The only other significant inorganic ligand is sulfide, and sulfide minerals are unstable in aerobic soils.

The first stage of weathering releases large amounts of Ca, Mg, Na, and K from the rock minerals, as illustrated in Table 7.2 by the change in composition from an igneous rock to the relatively young Barnes soil. With the exception of some Mg and K, these four cations are excluded from most *pedogenic* (soil-formed) minerals. Most of the alkali and alkaline earth cations remaining after the first weathering stage are in large unweathered mineral grains. Smaller fractions of Ca, Mg, Na, and K are retained by adsorption to negatively charged secondary mineral particles. These fractions are significant because they supply these essential macroelements to plants and soil microbes, they are subject to further leaching losses, and they control soil pH.

Because the efflux of cations slows markedly after initial breakdown of rock minerals, further stages are perhaps better called soil development than soil weathering. The difference in total composition between the immature Barnes soil and the rather mature Cecil soil is much less than that between the Barnes soil and igneous rock. The differences in soil maturity between the Barnes and the Cecil stages involve primarily the rearrangement of elements into secondary soil minerals. Rearrangements plus slight differences of total composition can create large differences in the availability of ions for plant growth. Phosphate is probably more available in the Barnes soil, for example, than in the Cecil soil, although the total amounts of phosphate are similar. Soluble plus exchangeable aluminium reaches phytotoxic concentrations in the Cecil soil. The slight differences in total elemental composition of the two soils can encompass a wide range of secondary mineral compositions.

The phosphate content of soils tends to remain roughly constant during soil development. Phosphate is only slowly leached from soils, at about the same rate as silica loss, so the total phosphorus content of soils varies little with soil maturity. The form of phosphate, however, changes from predominantly apatite $(Ca_5(OH,F)(PO_4)_3)$ in igneous rocks to Al(III) and Fe(III) phosphates in moderately to strongly weathered soils.

Soil sulfur, nitrogen, and carbon are associated with the soil's organic fraction and are relatively independent of weathering. Their soil contents are related to biological activity and climate. Thus, while sulfates are potentially easily leached from soils, plant and microbial uptake of sulfate and its incorporation into organic compounds tend to maintain the sulfur content of soils. Sulfur is also added to soils as fallout from natural and polluted air. In industrial regions atmospheric sulfur from coal combustion provides a large supplement to the plant's supply of natural sulfur. Plants and soils absorb sulfurous gases directly from the atmosphere and receive sulfates dissolved in rain. Nitrogen is also present as $NO$, $NO_2$–$N_2O_4$, and $NH_3$ in air and can be directly absorbed by plants and soils or from the soil solution after being washed out of the atmosphere. In addition, the reduction of atmospheric $N_2$ to amino acids and proteins by special soil and plant bacteria is an important nitrogen input to all soils.

Soil development involves a steady loss of silicon. Unfortunately, this is not readily apparent from the composition of the Barnes and Cecil soils in Table 7.2. The parent materials of these soils are apparently silica-rich. The loss of silicon is evident, however, from the low $SiO_2$ content of the highly weathered Columbiana soil. As the $SiO_2$ content of this soil decreased to less than half the average for igneous rocks, the $Al_2O_3$ and $Fe_2O_3$ contents increased threefold. The silica loss is from dissolution, not erosion. The solubilities of Fe, Al, Ti, and Mn hydroxyoxides are much lower than the solubility of silica. These hydrous oxides, or *sesquioxides*, are more stable than the secondary silicates that might have formed earlier in this soil. Assuming that the Columbiana parent material is igneous rock, the threefold increase of hydroxyoxide content means that two-thirds of the original parent rock dissolved into the soil solution and has been lost, a loss of about 30 million kg ha$^{-1}$ for each remaining meter depth of soil.

The loss of solutes during the initial stages of weathering is even more obvious from the composition of the clay fraction than from the change of the total soil. Clay particles more accurately reflect the soil's chemistry; sand and silt particles are largely vestiges of the parent material. Table 7.3 compares the composition of the average igneous rock to that of the major silicate clay minerals that form in soils; the Al, Fe, Mn, and Ti hydroxyoxides are not included. The silicate clay minerals and the hydrous oxides contain much less Si, Ca, Mg, Na, and K than do igneous rocks.

**Table 7.3. The composition of soil clay minerals[a] compared to the average of igneous rocks**

| Compound | Average of Igneous Rocks (%) | Hydrous Mica (Scotland) (%) | Montmoril- lonite (France) (%) | Kaolinite (Virginia) (%) | Allophane (Belgium) (%) |
|---|---|---|---|---|---|
| $SiO_2$ | 60 | 49 | 51 | 45 | 34 |
| $Al_2O_3$ | 16 | 29 | 20 | 38 | 31 |
| $Fe_2O_3$ | 7 | 3 | 0.8 | 0.8 | Trace |
| MgO | 4 | 1.3 | 3.2 | 0.1 | — |
| CaO | 5 | 0.7 | 1.6 | 0.1 | 2.3 |
| $Na_2O$ | 4 | 0.1 | 0.04 | 0.7 | — |
| $K_2O$ | 3 | 7.5 | 0.1 | 0.1 | — |
| $H_2O$ ($< 105°$ C) | — | 3.2 | 15 | 0.6 | 13 |
| $H_2O$ ($> 105°$ C) | — | 6 | 8 | 14 | 20 |
| Total | 99 | 100 | 100 | 100 | 100 |

[a] From E. T. Degens. 1965. *Geochemistry of the Sediments*. Prentice-Hall, Englewood Cliffs, NJ.

Aluminium, on the other hand, accumulates in the clay mineral fraction because it forms insoluble aluminosilicates and hydroxyoxides. The Al remains behind in the soil as other ions leach away. Iron also accumulates in soils but this is not apparent from Table 7.3 because the silicate clay minerals, with the exception of hydrous mica, are low in Fe. Iron precipitates in soils only as hydroxyoxides. Hydrous mica is altered parent material and is not reconstituted from the soil solution as are kaolinite, montmorillonite, and allophane. The $< 105°$ C water in Table 7.3 is, roughly speaking, adsorbed water; the $> 105°$ C water is hydroxyl ions and water within crystal structures.

Small amounts of weathered solutes reprecipitate in lower soil horizons. Examples include clay accumulation in the B horizon, "silica pans" (impermeable layers of soil particles indurated with silica), and the wide-spread "caliche" horizons of $CaCO_3$ accumulation in arid regions. Most solutes, however, reach the sea, where precipitation of other secondary minerals removes most of the weathered solutes except $Na^+$, $Cl^-$, and $Mg^{2+}$. Marine sediments, in turn, are slowly converted into igneous, metamorphic, or sedimentary rocks, which form new soil parent material. Such element recycling has circulated ions many times from land to sea during the earth's history.

Although the net effect of weathering is the loss of soluble components from the soil, the course of weathering is by no means continuous or unidirectional. The soil solution can flow upward during dry seasons. Plant uptake and decay deposit ions on the soil surface. The rates of ion absorption by plants during nutrient cycling are far greater than the rates of weathering. Because plant uptake greatly reduces weathering losses, large-scale losses of ions from soils usually occur only during periods of high rainfall or limited plant growth due to overgrazing, forest clearing, or fires. Chemical elements are also moved physically in the soil profile by "vertical mulching"; by soil fauna, such as worms, termites, and ants; and by uprooting of fallen trees. Vertical mulching occurs in montmorillonitic soils (Vertisols) that can form extensive vertical cracks as deep as 1 m in the dry season. Soil particles break from the walls of the cracks and fall to the bottom. When wetted, the fallen particles swell and force the overlying soil upward. The result is a slow, continual mixing of the surface meter of soil.

Another example of weathering reversal is salt accumulation due to impeded soil drainage or seawater inundation. Weathering under these conditions reverses in the sense that the secondary minerals formed are chemically similar to igneous and sedimentary rock minerals and are unstable under well-drained, oxidative conditions. The chemistry of soil development deals with the degradation of parent minerals and the formation of secondary minerals over a wide range of chemical conditions.

## 7.1  STABILITY OF PARENT MATERIAL MINERALS

The major minerals of igneous rocks are, in decreasing order of general abundance, feldspars ($MAlSi_3O_8$), quartz ($SiO_2$), and biotite ($K(Mg,Fe)_3AlSi_3O_{10}(OH)_2$) and muscovite ($KAl_2Si_3O_{10}(OH)_2$) micas. The feldspars include orthoclase and microcline (both $KAlSi_3O_8$) and the plagioclase series ranging from albite ($NaAlSi_3O_8$) to anorthite ($CaAl_2Si_2O_8$). Other minerals in igneous and sedimentary rocks are generally present in lesser amounts and have chemical compositions similar to those above. Granitic or acid ($>66\%$ $SiO_2$) igneous rocks are richer in silicon and potassium, and poorer in magnesium and iron, than basaltic or basic igneous (45 to 52% $SiO_2$) rocks (Table 7.1). Olivine ($(Mg,Fe)SiO_3$) is a distinctive component of basaltic rocks.

Metamorphic rocks contain the above minerals and chemically similar variants. Metamorphic rocks are occasionally rich in Mg minerals, such as the pyroxenes (approximate composition $Ca(Mg, Fe)Si_2O_6$) and the chemically similar but more complex augites. Serpentine metamorphic rocks are even richer in Mg because of antigorite ($Mg_3Si_2O_5(OH)_4$).

Sedimentary rock materials have already passed through some weathering before the rock is formed. Their composition represents depletion of weatherable elements, as in sandstone, or selective accumulation, as in shale and limestone (Table 7.1). Sandstones are mostly mechanical accumulation of quartz grains and are depleted of virtually all other elements. Shales, which are fine-grained sedimentary rocks, tend to form in regions of some chemical and particulate accumulation. They are richer than sandstones in K and Mg aluminosilicates, Fe, and Ca and Mg carbonates. The Na content of sedimentary rocks is generally much lower than that of igneous rocks, although some marine sediments contain entrapped Na.

The mineral structures of igneous minerals are varying organizations of silicon–oxygen tetrahedra and aluminium–oxygen tetrahedra and octahedra. The stability of igneous minerals also applies to the weathering of chemically and structurally similar minerals in metamorphic and sedimentary rocks.

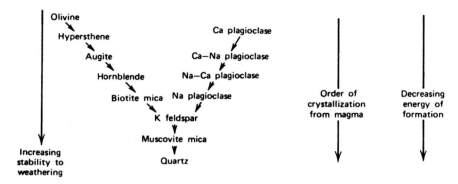

**FIGURE 7.2.** Stability to weathering of some minerals in igneous and metamorphic rocks. (Adapted from S. S. Goldich. 1938. *J. Geol.* **46**:38.)

The resistance of igneous minerals to weathering is the same as the order of crystallization from cooling magmas (Fig. 7.2). Minerals that are most stable at high temperatures, and that therefore crystallize first from the molten magma or lava, are the least stable at low temperatures. Such minerals, including calcic feldspars, olivine, and hypersthene ($(Mg,Fe)SiO_3$) tend to be rich in the water-soluble alkali and alkaline earth and Fe(II) ions. The weathering rate generally increases with increasing content of alkali and alkaline earth cations.

A second chemical factor affecting mineral weatherability is the position of ions in the structure. The tetrahedra of Ca feldspars contain half $Al^{3+}$ and half $Si^{4+}$. At room temperature, $Al^{3+}$ is more stable in octahedral coordination. The charge deficit created by the $Al^{3+}$ substitution is made up by $Ca^{2+}$ ions between the tetrahedra. The structural strain, the charge deficit in the tetrahedra, and concentrated $Ca^{2+}$ counter charge weaken the anorthite feldspar structure with respect to weathering relative to Na and K feldspars. In Na and K feldspars, only one-quarter of the tetrahedral positions are occupied by Al and that charge deficit can be locally neutralized by $Na^+$ or $K^+$. Calcium feldspars are, therefore, the least stable feldspars under soil conditions. Potassium feldspars are more stable than Na feldspars, because K fits better between adjacent tetrahedra.

A third chemical factor affecting mineral weatherability is the degree to which the tetrahedra are linked together. Increased linkage between tetrahedra means increased stability against weathering. Feldspars and quartz are three-dimensional networks of tetrahedra in which each of the four tetrahedral oxygens is a corner of another tetrahedron. This maximum sharing of oxygens produces considerable stability. Hence, quartz is very persistent in soils, if it is sand- or silt-sized. Feldspars would also be resistant to weathering if not for their Ca, Na, and K contents and the other factors described above. These are unfavorable for feldspar stability and counteract the stabilizing effect of tetrahedral linkage.

The sharing of tetrahedral oxygens in other silicates varies from the maximum sharing in quartz and feldspar to the complete independence of the tetrahedra in olivine. Olivine weathers very rapidly when exposed to water and air. Pyroxenes share two corners of each tetrahedron to form parallel single chains of tetrahedra. Amphiboles share three corners to form parallel double chains. Pyroxenes and amphiboles are appreciably more resistant to weathering than olivine. In micas the sharing is more complex and similar to the structure of smectites. Two sheets of silicon–aluminium tetrahedra are joined together by an included layer of aluminium or magnesium–iron octahedra. Charge imbalance within these three-layer sheets is balanced by $K^+$ (and occasionally $NH_4^+$) on the outer surface of the silica sheets. High-aluminium (dioctahedral) muscovite mica is more stable than high-iron and high-magnesium (trioctahedral) biotite mica, in part because of a better fit of the internal aluminium octahedral sheet. Chapter 5 gives details of mineral composition and structure.

A fourth factor of mineral stability is the Fe(II) and Mn(II) contents. In aerobic soils the presence of these ions increases weathering rates because their oxidation to Fe(III) and Mn(III,IV) creates charge imbalances. In anaerobic soils, on the other hand, minerals containing the oxidized ions are unstable.

Figure 7.2 summarizes the relative weathering rates of major minerals in igneous and metamorphic rocks. Actual weathering rates depend also on soil temperature and moisture, particle size, and planes of physical weakness (cleavage) in the crystal. The effect of moisture includes both the flow rate of soil solution past mineral surfaces and the composition of the solution. Solids dissolve more slowly if the solution already contains their constituent ions. High electrolyte concentrations, on the other hand, can maintain higher ion concentrations at equilibrium because of lower activity coefficients and because of complex-ion and ion-pair formation.

Smaller particles weather more rapidly, but the size effect is great only when the particles are less than several micrometers in size. Cleavage planes allow particles to be more easily broken apart. Feldspars and micas, for example, have clearly defined cleavage planes. Particularly in the case of feldspars, the cleavage planes hasten the rate of mineral breakdown.

## 7.2  IONIC POTENTIAL

The geochemist Goldschmidt tried to predict the fate of ions released during the weathering of igneous minerals. He found that the *ionic potential*, the ratio of crystal ion radius to ion charge, indicated fairly well whether ions eventually would leach to the sea or remain behind in soils and sediments. The crystal (dehydrated) ion size is useful for predicting solution behavior, because the size indicates how strongly it can react with water, $OH^-$, and $O^{2-}$.

Figure 7.3 plots the ratio of crystal radius versus charge for selected ions. Oxyanions—sulfate, selenate, phosphate, arsenate, borate, molybdate, carbonate, and silicate—are represented by their central cations: $S^{6+}$, $Se^{6+}$, $P^{5+}$, $As^{5+}$, $B^{3+}$, $Mo^{4+}$, $C^{4+}$, and $Si^{4+}$. The ions fall into three behavioral groups. Ions of high ionic potential, the alkali and alkaline earth cations and the halide anions, large univalent and divalent ions, are highly water soluble, easily weatherable, and leach readily from soils to the sea over geologic time.

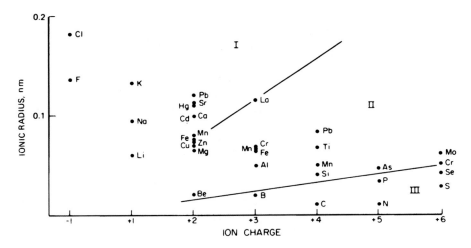

**FIGURE 7.3.** Ionic potentials of important ions.

Ions of intermediate ionic potential, such as aluminium and the transition metals, include most of the elements in the periodic table, although only Al, Fe, Ti, and Mn are present in appreciable amounts in soils. Ions in this group are of intermediate size and charge. They tend to polarize their associated water molecules and to repel $H^+$ sufficiently to precipitate as insoluble hydroxyoxides. Such ions are released during weathering of silicates but, because of the low solubilities of their hydroxyoxides, are leached only very slowly from soils. Ions of higher and lower ionic potentials migrate faster, so this middle group of ions tends to remain behind and accumulate in soils. Hence highly weathered soils, such as Oxisols or Ferrasols, contain high amounts of Al, Fe, Ti, and Mn oxides.

The ions with the lowest ionic potentials are oxyanions formed by the smallest, most highly charged cations—$SO_4^{2-}$, $NO_3^-$, $CO_3^{2-}$, and so on. These cations repel several protons from associated water molecules to form permanent oxide ligands. Such oxyanions are water soluble by geochemical standards and hence tend to be leached from soils. Phosphate and silicate are the least soluble members of this group. They lie near the boundary between soluble oxyanions and insoluble hydroxides. The loss of borates and silicates, but not so much of phosphate, is characteristic of moderate to advanced stages of soil development.

The ionic potential can be used to predict general soil chemical behavior of ions, if combined with a knowledge of the ion's oxidation–reduction behavior and of the soil chemistry of chemically related ions. The ionic potential describes best the overall process of rock weathering to form sediments, the process for which it was intended. It is somewhat less successful in explaining the soil chemical behavior of some ions. Magnesium is generally weatherable in soils. Sodium is retained much more weakly than is $K^+$. The strong preference of particular sites on layer silicate minerals for $K^+$ cannot be accounted for by ionic potential. Similarly, $Ca^{2+}$ is less strongly retained than $Mg^{2+}$. The difference is due to the incorporation of $Mg^{2+}$ into secondary minerals. Calcium released by weathering remains in the soil solution as an exchangeable cation. This difference is not accounted for by the ionic potential. Comparing the exchangeable forms only, $Mg^{2+}$ is usually adsorbed less strongly than $Ca^{2+}$. The exceptional preference of vermiculite for $Mg^{2+}$ is discussed in Chapter 8.

## 7.3 RATES OF WEATHERING AND SOIL DEVELOPMENT

Temperature and moisture are the major environmental variables affecting weathering rates. Assuming similar chronological ages and parent materials, the difference in composition of the South Dakota, North Carolina, and Costa Rican soils of Table 7.2 illustrates the effects primarily of temperature. Weathering is much faster in the warm climate of Costa Rica than in the cold winters and short summers of South Dakota. North Carolina's climate is intermediate between the two.

The rate of water movement through soils determines the rate at which weathered solutes are removed from the vicinity of soil particles. Weathering can continue even when the rate of movement is slow, such as in poorly drained soils. Lack of water, however, almost totally arrests soil development. Desert soils can be old chronologically, yet young in the sense of soil development.

Jenny (1941) proposed that soil development be regarded as the result of five *soil-forming factors*: climate, topography, biosphere, parent material, and time. In a qualitative sense, the weathering rate is related to these factors by

$$\frac{\Delta \text{weathering}}{\Delta \text{time}} = f(\text{climate, topography, parent material, biosphere}) \qquad (7.6)$$

Converting this expression into a quantitative equation, however, is beyond our present capabilities. None of the four factors has been adequately described numerically. Climate, for example, is an ill-defined integration of the intensity, duration, and seasonal distribution of temperature, moisture, and evaporation. Deposition of airborne salts and dust and parts of erosion should also be included as subtle parts of climate. The parent material factor includes chemical composition, mineral composition, crystal size, and rock fabric (structure).

The relative importance of each factor in Eq. 7.6 also varies with local and regional conditions. In a peat bog or on a steep mountain slope topography clearly has a prominent role. For a young alluvial soil, on the other hand, parent material is usually the dominant factor. In desert and polar soils, the biosphere's role is comparatively small.

The soil-forming factors also are interdependent. The biosphere clearly depends on climate. Topography may control drainage, but soil water movement is also affected by soil texture, which is derived in part from the crystal size of the parent material. Topography includes aspect (the direction of slope), which contributes to the microclimate of the soil. A north-facing slope (in the northern hemisphere) is cooler and moister than a south-facing slope. This affects the nature and distribution of plants, as well as the soil directly. The biosphere greatly affects soil–water relations, produces organic molecules that react with soil particles, dominates carbon and nitrogen chemistry and other oxidation–reduction processes of soils, and increases the $CO_2$ concentration of the soil air and soil solution. These interactions greatly increase the complexity of Eq. 7.6.

The composition of water draining from soils is an index of the rates of mineral weathering and ion leaching. The losses of several elements from the earth's continents are shown in Table 7.4. The values are the products of river water composition and river flow rates. Although they include some amounts dissolved from deeper sediments, the amounts are global indices of soil weathering and losses. The alkali and alkaline earth cations are lost mainly as dissolved ions. Silicon, aluminium, and iron are lost mainly by erosion as suspended sediments rather than by chemical dissolution, although silica dissolution is appreciable. Other estimates of the Ca, Mg, K, N, and S losses by leaching are also in the range of 2 to 20 kg ha$^{-1}$ yr$^{-1}$ (50 to 1000 mol ha$^{-1}$ yr$^{-1}$).

The rates in Table 7.4 and the soil composition data in Table 7.2 can be combined to calculate the *residence time*, or *turnover rate*, of ions in soils. Residence time is the total soil content divided by the loss rate and is the inverse of the weathering rate. The soil's sodium content is approximately 0.4%, or 50 000 kg ha$^{-1}$ m$^{-1}$. The sodium loss of 300 mol ha$^{-1}$ m$^{-1}$, or 7 kg ha$^{-1}$ m$^{-1}$, indicates that sodium's residence time is about 7000 years in the surface meter of soils. This estimate disregards atmospheric sodium input to the soil and also disregards the contribution of sediment weathering to the sodium losses in Table 7.4. If sediment weathering contributes half the sodium content of the world's rivers, sodium's residence time in soils is on the average about 15 000 years.

Similar calculations disclose that calcium's residence time in soils is about 2500 to 5000 years. The Na and Ca weathering rates seem exceedingly fast but are, of course,

**Table 7.4. Weathering losses from the continents to the sea[a] (continental land area = 1.2 x 10$^8$ km$^2$)**

| Element | Soluble | | Sediment |
| --- | --- | --- | --- |
| | $\times 10^{12}$ moles yr$^{-1}$ | moles ha$^{-1}$ yr$^{-1}$ | moles ha$^{-1}$ yr$^{-1}$ |
| Sodium | 3.6 | 300 | 130 |
| Potassium | 1.8 | 150 | 120 |
| Magnesium | 5.5 | 460 | 110 |
| Calcium | 13 | 1100 | 130 |
| Silicon | 9 | 750 | 3300 |
| Aluminum | — | — | 930 |
| Iron | — | — | 300 |
| Sulfur | 1.9 | 160 | — |
| Chlorine | 2.7 | 220 | — |
| Phosphorus | — | 2 | — |

[a] Mostly from Garrels, Mackenzie, and Hunt. 1975. *Chemical Cycles and the Global Environment*. W. Kauffman, Los Altos, CA.

counterbalanced by similar rates of input. Soils have weathered at about these rates for $4.5 \times 10^9$ years and are still far from exhausted.

The residence time can also give some idea of the rate of element cycling between soils and plants. Table 1.1 gives the soil's Na supply, relative to rate of plant uptake, as equivalent to about 5000 years. Sodium ions, therefore, cycle two to three times, on the average, through plants and soil before being leached from the soil. The 260-year plant supply of Ca in soil (Table 1.1) indicates that Ca cycles 10 to 20 times through plants before being leached from soils.

The leached ions are replenished by atmospheric fallout of sea spray entrapped in rain and by the formation of new igneous rock at the ocean floor. Assuming that the Na and Ca concentrations in the ocean are constant, their ocean residence time equals the replenishment rate of soils. Ocean residence time, in turn, is equal to the concentration divided by the rate of input from the world's rivers. The Na residence time in the oceans is about 100 million years; the Ca residence time is about 1 million years; K and Mg are intermediate. Comparing the residence times of Ca in soils and oceans shows that 200 to 400 m of soil are weathered of their Ca content during the 1 million years residence time of Ca in the oceans.

The ions are replenished mainly by formation of new igneous rock, at a rate recently estimated at $12 \times 10^9$ m$^3$ yr$^{-1}$ or 30 Tg yr$^{-1}$. The turnover rate of the entire crust of the earth is therefore about 200 million years; the entire earth's crust has been weathered and remelted into rock on the average about twenty times over the lifetime of the earth. The earth and its soil appear stable to us only because our residence time on the earth is so short.

### 7.3.1  Acidity

Increasing acidity of the soil solution hastens weathering. Discussions of weathering usually mention the $H_2CO_3$ of rainwater as both a significant input of acidity to soils and a significant factor in soil mineral weathering. The solubility of $CO_2$ in water in equilibrium with the atmosphere is about $10^{-5}$ M. An annual rainfall of 1000 mm means an input of dissolved $CO_2$ of 0.01 mole m$^{-2}$ of soil surface. The amount of $CO_2$ produced by biomass is much greater. The annual net carbon fixation by photosynthesis is about 1 kg C m$^{-2}$ of productive soil, or 80 mol C m$^{-2}$. Assuming that the amount of carbon respired by roots and soil microbes equals net photosynthesis, and that is a very conservative assumption, roots add 8000 times more $CO_2$ to the soil than does rainwater. Decay of surface plant residues adds additional $CO_2$, but most of this is liberated directly to the atmosphere.

Other atmospheric components, such as $HNO_3$ and $NH_3$, are usually too dilute in nature to affect the acidity of rain except for the first droplets, which flush out the air. The $SO_2$ and $NO_x$ ($NO+NO_2-N_2O_4$) released by coal combustion, automobiles, and ore smelting, however, can significantly increase atmospheric acidity. These gases oxidize to $H_2SO_4$ and $HNO_3$ in the atmosphere, on leaves, and at the soil surface. Acid rain of pH 4.5 or lower is common in industrial regions and may seriously affect plants, lakes, buildings, and perhaps the weathering rates of soil minerals. Acid rain has received less attention in recent years. The changes in lake acidity that caused part of the concern were due in part to changing the pH measurement technique from indicator dyes to the glass electrode. Lake and stream waters are poorly pH buffered so the slight alkalinity of the indicator dye gave a higher value than the later measurements by the glass electrode. In addition, the abandonment of agriculture in northeastern North America meant the abandonment of liming. The soils returned to their native acidity and so did the water draining from these soils. Acid rain is a real phenomenon, but air pollution controls on power plants and vehicles are reducing its intensity.

Although acid rain may increase the weathering rates of soil minerals, its effect could be rather small. Plants and microorganisms absorb sulfate and nitrate anions and simultaneously excrete an equivalent quantity of $OH^-$ to maintain internal charge balance. Plant absorption of these anions could thus tend to counteract the effect of acid rain on soil weathering. Benefits from such anion absorption by plants would accrue primarily in acid soils, where natural sulfate concentrations are relatively low. In neutral and alkaline soils, the high pH and high acid-buffering capacities would counteract the effects of acid rain. In neither case would acid rain increase soil weathering rates to the degree once feared.

### 7.3.2   Mechanisms of Mineral Decomposition

The decomposition of several common soil minerals has been examined in the laboratory and probably represents the mechanism of mineral weathering in soils as well. The decomposition rate of montmorillonite is proportional to the $H^+$ concentration of the attacking solution:

$$\frac{-\Delta(\text{mont})}{\Delta t} = k(H^+) \tag{7.7}$$

where $-\Delta(\text{mont})/\Delta t$ is the rate of disappearance of montmorillonite with time. This type of rate equation is thought to result from the free expansion of montmorillonite mineral sheets, so that the mineral's total surface is susceptible to attack. Hence, the decomposition is independent of the amount of montmorillonite, a pseudo-zero-order reaction. The rate depends only on the $H^+$ concentration as long as the $H^+$ concentration is in the range of soil solution acidities, that is, if the $H^+$ concentration is low relative to the montmorillonite concentration.

The rate equation for kaolinite decomposition has the form

$$\frac{-\Delta(\text{kaol})}{\Delta t} = k(\text{kaol})(H^+) \tag{7.8}$$

Hence, the rate of kaolinite decomposition depends on both the acidity and the kaolinite concentration. As kaolinite decomposes, its rate of decomposition decreases, and complete disappearance should theoretically require infinite time. Indeed, kaolinite is quite resistant to weathering. Kaolinite is nonexpanding so its exposed surface is small. Inasmuch as few soil minerals expand, Eq. 7.8 probably characterizes soil mineral weathering better than does Eq. 7.7. Smaller mineral particles tend to decompose first, leaving behind the larger particles. The weathering rate thus diminishes with time, so that small amounts of weatherable materials may persist even in highly weathered soils.

As the remaining particles weather to progressively smaller sizes and large surface areas, the last remnants might be expected to decompose quite rapidly. This would be true if particle surfaces remained fresh and unweathered. In practice, however, the particle surfaces become coated with residues from the weathering process and with reprecipitated secondary minerals. The residue coating hinders decomposition of weatherable soil minerals, since weathering products must diffuse through this layer before they can dissolve in the soil solution. Weathering agents must diffuse in the opposite direction to attack the mineral. The weathered surface is a protective coating. If the soil weathering rate increases, the rate of release of alkali and alkaline earth cations and silica should increase. This, in turn, would leave a thicker layer of $Al(OH)_3$ and $FeOOH$ remaining on the mineral surface. The thicker layer is a negative feedback mechanism that reduces ion diffusion rates and the weathering rate.

### 7.3.3   Time Sequence of Mineral Occurrence

Weathering involves the movement of water through the soil profile and the gradual removal of mainly silica and alkali and alkaline earth cations. The flow of water usually prevents accumulation of soluble salts but is slow enough to permit Si, Al, and some Mg to reprecipitate as secondary minerals. Under such conditions, Jackson and Sherman (1953) proposed that the change with time of the material composition of soil clays is similar in many soils (Table 7.5). The clay fraction better reflects the time, chemical composition, and environmental conditions that existed during soil

**Table 7.5. Sequence of clay mineral distribution with increasing soil development[a]**

| Relative Degree of Soil Development | Prominent Minerals in Soil Clay Fraction |
|---|---|
| 1 | Gypsum, sulfides, and soluble salts |
| 2 | Calcite, dolomite, and apatite |
| 3 | Olivine, amphiboles, and pyroxenes |
| 4 | Micas and chlorite |
| 5 | Feldspars |
| 6 | Quartz |
| 7 | Muscovite |
| 8 | Vermiculite and hydrous micas |
| 9 | Montmorillonites |
| 10 | Kaolinite and halloysite |
| 11 | Gibbsite and allophane |
| 12 | Goethite, limonite, and hematite |
| 13 | Titanium oxides, zircon, and corundum |

[a] Adapted from M. L. Jackson and G. D. Sherman. 1953. *Advances in Agronomy*, **5**:221–319.

formation than does the soil as a whole. The sand and silt fractions are usually relics of the soil parent material.

The mineral groups of Table 7.5, when present in the clay fraction, indicate progressively increasing stages of soil maturity. A given suite of minerals may not necessarily dominate the clay fraction, but its presence in detectable amounts is a fairly reliable indicator of the degree of soil weathering and development. The criterion for a mineral's presence normally is whether the mineral is detectable by x-ray diffraction. Small amounts or poor crystallinity can lead to detection problems. Jackson and Sherman's clay mineral groups that are characteristic of increasing soil maturity are as follows:

1. *Soluble salts—halite (NaCl) and gypsum (CaSO$_4$ · 2H$_2$O), as well as sulfides (pyrite, FeS$_2$) in soils reclaimed from seas or swamps.* These minerals readily dissolve in percolating water or, in the case of the sulfides, are readily attacked by oxygen. Saline and sodic ("alkali") soils are examples of this category.

2. *Calcite (CaCO$_3$), dolomite (CaMg(CO$_3$)$_2$), and apatite (Ca$_5$(F,OH)(PO$_4$)$_3$).* Carbonates of clay size are rather rapidly leached from humid soils. In arid regions, calcite accumulates. The phosphate remains in the clay phase after apatite decomposition, although in the form of Ca phosphates in alkaline soils or Al and Fe phosphates in acid soils.

3. *Olivine ((Mg,Fe)$_2$SiO$_4$) and the feldspathoids (amphiboles, mainly hornblende, and pyroxenes).* In these minerals Fe(II) can be oxidized, increasing the weathering rate.

4. *Primary layer silicates, biotite (K(Mg,Fe,Mn)$_3$Si$_3$AlO$_{10}$(OH)$_2$) from igneous and metamorphic parent materials, glauconite (K(Fe,Mg,Al)$_2$Si$_3$AlO$_{10}$ (OH)$_2$) from marine sediments, and magnesian chlorite (Mg,Fe)$_6$(Si,Al)$_4$O$_{10}$(OH)$_8$.* Aluminium is a common substitute (up to 1 in 4) for silicon in the tetrahedral sheets. This is an unstable configuration for Al at room temperature, but is more than counterbalanced by the stability imparted by the sharing of oxygens between the tetrahedral and octahedral interlayers. Iron(II) in these minerals increases their instability with respect to oxidation.

5. *Feldspars.* Albite (NaAlSi$_3$O$_8$) and anorthite (CaAl$_2$Si$_2$O$_8$) are the end members of the plagioclase feldspar continuum, covering the whole range of Na,Ca mixtures. The greater the calcium content of plagioclase, the faster its rate of weathering. Orthoclase and microcline feldspars (both KAlSi$_3$O$_8$) in the clay fraction weather at roughly the same rate as albite.

These five groups are inherited from the soil's parent material and are considered easily weatherable minerals when clay-sized. Inherited minerals that are more resistant to weathering in the clay fraction include:

6. *Quartz (SiO$_2$).* Quartz in the clay fraction is much more easily weathered than in the sand fraction, where it is the most abundant mineral.

7. *Muscovite (KAl$_2$(Si$_3$Al)O$_{10}$(OH)$_2$).* Muscovite mica is considerably more stable than biotite. Biotite contains oxidizable Fe(II) and some Mn(II), whereas muscovite does not. In addition, the dioctahedral (Al) layer of muscovite seems to fit much better between the two layers of silica–alumina tetrahedra, and hence is more stable, than the trioctahedral (Mg and Fe(II)) layer of biotite.

8. *Interstratified or intermixed layer silicates, vermiculite (M$^+$(Mg,Fe)$_3$(Si$_{4-n}$, Al$_n$)O$_{10}$(OH)$_2$) and the hydrous (slightly weathered) micas.* The M$^+$ represents an exchangeable cation. Whether the minerals of this category are inherited from the parent material or are secondary products derived from inherited minerals is uncertain in many cases.

The remaining categories of Table 7.5 contain secondary (soil-formed) minerals resulting from the weathering of primary minerals from soil parent materials. By the time these stages of weathering are reached, inherited minerals in the clay fraction have either disappeared or are present in only minor amounts.

9. *Montmorillonites or smectites (M$^+$(Al,Mg)Si$_4$O$_{10}$(OH)$_2$).* These are the secondary Mg- and Al-rich 2:1 layer silicates that can form in soils. The chemical composition can vary substantially, but the basic expansible structure remains essentially the same.

10. *Kaolinite and halloysite (Al$_2$Si$_2$O$_5$(OH)$_4$).* The 1:1 layer lattice kaolins are more stable with time than the 2:1 smectites. Kaolinite is a common component of soil clays and occurs in relatively high concentrations in moderately weathered soils.

11. *Aluminium hydroxyoxides.* These include gibbsite (Al(OH)$_3$), boehmite and diaspore (AlOOH), and allophane. The loss of silicon from soils leaves an Al- and Fe-rich residue in soil clays.

12. *Iron hydroxyoxides.* These include goethite (FeOOH), limonite (a hydrated Fe(III) hydroxyoxide), and hematite (Fe$_2$O$_3$).

13. *Titanium oxides.* These include anatase and rutile (TiO$_2$), leucoxene (hydrated, amorphous titanium oxide), and ilmenite (FeTiO$_3$), plus zircon (ZrO$_2$) and corundum (Al$_2$O$_3$). Titanium and zirconium are so immobile in soils that members of mineral group are used as indicators of the amount of parent material that has weathered to produce a given volume of soil.

The above sequence of mineral occurrence does not imply that all ions present in the soil progress from one weathering category to the next. Table 7.1 shows that only a small fraction of the alkali and alkaline earth cations are retained by soils when igneous minerals first weather. Also, when montmorillonite weathers, only a fraction of the SiO$_2$ and exchangeable cations recombine with Al(OH)$_3$ to form kaolinite. The remaining fractions are leached from the soil. Indeed, the first identifiable solid product of igneous mineral weathering is generally gibbsite. Whether the gibbsite, soluble silica, and cations remain apart or reprecipitate as smectites, kaolinite, and chlorite depends on the chemical composition of the soil solution and particularly on the cation and $Si(OH)_4$ concentrations. The sequence in Table 7.5 is primarily a time sequence, in the sense that cations and silica are increasingly lost from soils with time. The secondary minerals in each successive weathering stage of Table 7.5 are stable at lower cation and $Si(OH)_4$ concentrations, and lower soil pH, than those in the previous stage.

Recent work has shown that Table 7.5 may be too rigid. Mineral occurrence overlaps considerably. Tables 7.6 and 7.7 are a more current view of mineral occurrence in soils.

## 7.4 MINERAL FORMATION IN SOILS

The formation of secondary minerals in soils generally results from the combination and addition of ions and molecules from the soil solution to the solid phase. This mechanism was originally given little consideration, because aluminium and silicon in solution did not appear to combine during laboratory experiments. Only relatively recently has the slow kinetics of such reactions been appreciated. Experiments that take slow reactivity into account and provide nucleation centers for crystal formation have shown that secondary minerals can precipitate from solutions containing the proper constituent ions and $Si(OH)_4$.

Formerly, soil minerals were thought to form by differential migration of ions into and out of existing silicate structures. The diffusion of $Al^{3+}$ or $Mg^{2+}$ out of the lattice was supposedly balanced by the inward diffusion of other ions. Such diffusion is unlikely. In mica, for example, a cation diffusing out of octahedral coordination would leave behind a void and many unsatisfied bonds. Such diffusion would be against an enormous gradient in electrical potential. The ion would also have to break through several tetrahedra to reach the lattice surface. The cation diffusing from the soil solution would be attracted by the electrical potential but would also have to break through the tetrahedra. Furthermore, the replacing cation would have to be similar in size and charge to the vacated cation. That the ion diffusing from the soil solution would have the appropriate size is highly improbable. The common cations in soil solutions are $Ca^{2+}$, $Mg^{2+}$, $Na^+$, and $K^+$. Only $Mg^{2+}$ from this group fits into octahedral configuration. None of these cations normally occupies a tetrahedral position, and none would account for the differences in tetrahedral composition between mica and secondary minerals. Distortions during such ion diffusion would strain the crystal badly and probably cause its total rupture. The result would be more or less complete mineral breakdown before the ions recombined into a new mineral.

Despite the unlikelihood of secondary mineral formation by ion substitution into or movement within an existing solid, some secondary 2:1 layer silicates apparently are formed by solid-phase changes of mica fragments inherited from the parent material. Hydrous mica, for example, is a product of chemical weathering as well as mechanical breakdown of mica. Hydrous mica, in turn, can be modified directly to vermiculite, montmorillonite, or chlorite. The process is not completely understood, but seemingly involves the outward diffusion of $K^+$ from between the layer lattices and a subsequent or simultaneous reduction of charge within the layer lattice.

**Table 7.6. Common *primary* (residual) minerals present in soil environments in order of increasing stability[a]**

| Name | Mineral Class | Environment | Ubiquity in Soils[b] | Importance |
|---|---|---|---|---|
| Pyrite | Sulfides | Tidal marshes (reducing conditions and hard-rock mine tailings (coal and shale beds) | C | Primary mineral (oxidizing conditions) but secondary phase forms in reducing environments; large metal and acidity input to surface waters during weathering |
| Dolomite | Carbonates | Shallow, young soils formed in limestone | R | Major constituent of limestone parent material; fertilizer source |
| Pyrophyllite | Phyllosilicates | Low temp. metamorphic and hydrothermal | R | No layer charge—little chemical reactivity; unstable in soils |
| Talc | Phyllosilicates | Ultramafic rocks | R | No layer charge—little chemical reactivity; unstable in soils |
| Olivine | Nesosilicates | Basic and acidic igneous rocks | R | Source of Fe, Ca, Mg, and Mn; unstable in highly leached soil |
| Pyroxenes | Inosilicates | Igneous and contact metamorphic rocks | R | Source of Fe, Ca, Mg, and Mn; unstable in highly leached soil |
| Amphiboles | Inosilicates | Igneous and metamorphic rocks | R | Source of Fe, Ca, Mg, and Mn; vermiculite precursor |
| Chlorite (Mg-interlayer) | Phyllosilicates | Metamorphic or igneous rocks | R | Important precursor for 2:1 soil clay minerals |
| Biotite ($Fe^{2+}$-bearing micas) | Phyllosilicates | Granitic and high-grade metamorphic rocks | R | Stable in only the youngest or least weathered soils; precursor of other 2:1 soil clay minerals and Fe-oxides; source of K |

*(continued)*

**Table 7.6. (Continued)**

| Name | Mineral Class | Environment | Ubiquity in Soils[b] | Importance |
|---|---|---|---|---|
| Feldspars: | | | | |
| Calcic Plagioclase | Tectosilicates | Wide variety of igneous/metamorphic rocks; persistence in soils and geologic deposits is related to weathering intensity, and the duration of exposure to weathering | R | Source of Ca, Na, and K; alteration to secondary products is a function of microenvironment; weathering products can be kaolinite, mica, gibbsite, halloysite, smectite, or amorphous materials |
| Sodic Plagioclase | Tectosilicates | | C | |
| K-Feldspars | Tectosilicates | | C | |
| | Tectosilicates | | C | |
| Muscovite | Phyllosilicates | Granitic and high-grade metamorphic rocks | C | Vermiculite, smectite, and interstratified 2:1 precursor, K source |
| Epidote | Sorosilicates | Medium-grade metamorphic and mafic igneous rocks | R | Source of Fe, Ca, Mn; very resistant to weathering |
| Quartz | Tectosilicates | Nearly all soils and parent materials | U | Concentrated in sand- and silt-fractions; soluble in clay fraction |
| Garnets | Nesosilicates | High-grade metamorphic/acid igneous rocks | R | Source of Fe, Mg, Ca, Al, and/or Mn precursor of Fe-oxides |
| Tourmaline | Cyclosilicates | Granites and pegmatites; detrital sediments | R | Stable in soils, but may alter to secondary 2:1 clay minerals |
| Rutile | Oxides | Igneous and metamorphic rocks; detrital sediments; quartz inclusion | C | Very stable in soils, but some mobility of Ti; rarely alters to other oxides |
| Zircon | Nesosilicates | Acid and basic plutonic, metamorphic rocks | C | Very resistant; used as index minerals in pedologic studies |

[a] From Feldman and Zelazny. 1998. Soil Sci. Soc. Am. Spec. Publ. 55, Madison, WI.
[b] U, ubiquitous; C, common; R, rare.

194

**Table 7.7. Common *secondary* silicates and nonsilicates present in soil environments in order of increasing stability[a]**

| Name | Mineral Class | Environment | Ubiquity in Soils[b] | Importance |
|---|---|---|---|---|
| Halite | Halides | Arid, saline/sodic soils | C | Highly soluble; adverse osmotic effects on plants |
| Gypsum | Sulfates | Arid soils | C | Moderataly-high solubility; used to reclaim sodic soils |
| Jarosite | Sulfates | Acid sulfate soils; mine overburden, coastal wetlands | C | Product of pyrite oxidation resulting in large production of acidity and toxic metals |
| Calcite | Carbonates | Arid soils; very limited leaching | C | May act as cementing agent; high P sorption |
| Pyrite | Sulfides | Tidal marshes (reducing conditions) and hard-rock mine tailings (coal and shale beds) | C | Primary mineral (oxidizing conditions) but secondary phase forms in reducing environments |
| Allophane/Imogolite | Paracrystalline | Soils derived from volcanic ash deposits | R | Short-range order, highly sorptive |
| Sepiolite/Palygorskite | Phyllosilicates | Marine sediments, arid soils, high Si and Mg levels | R | Moderately high CEC,[c] surface area, and sorptive properties |
| Halloysite | Phyllosilicates | Volcanic ash; granitic (feldspathic) saprolite | R | Ephemeral in intensely weathered soils |
| Vermiculites (dioct.) | Phyllosilicates | Mica alteration in well-drained soils | R | Very high CEC; fixation of $K^+$; sink for solution Al |
| Smectites | Phyllosilicates | Mica and/or vermiculite alteration | C | High CEC; high surface area; high shrink-swell capacity |
| HIV[d] | Phyllosilicates | Acid, highly weathered soil surface horizons | U | Variable CEC; (degree of Al interfilling); high anion adsorption |
| Kaolinite | Phyllosilicates | Desilication of 2:1 clays/feldspathic | U | Low charge but highly pH-dependent; high anion adsorption |
| Hematite & Goethite | Oxides | Well-drained, near-surface soil | U | Low charge but highly pH-dependent; high anion adsorption |
| Gibbsite | Oxides | Old, stable soils or feldspar pseudomorphs | C | Low charge but highly pH-dependent; high anion adsorption |
| Anatase | Oxides | Dissolution of Ti-bearing parent minerals | R | Essentially chemically inert |

[a]From Feldman and Zelazny. 1998. Soil Sci. Soc. Am. Spec. Publ. 55. Madison, WI.
[b]U, ubiquitous; C, common; R, rare.
[c]CEC, cation-exchange capacity.
[d]HIV, hydroxy-interlayered vermiculite.

Kaolinite, on the other hand, has no structural counterpart among the igneous minerals. It is also the most widespread of the crystalline clay mineral. The most likely mechanism for kaolinite formation is the complete breakdown of feldspar or mica particles and the precipitation of kaolinite from $Al(OH)_3$ and $Si(OH)_4$ from the soil solution or from amorphous, less stable intermediates.

The exact chemical conditions under which soil minerals form are not known at present. Our knowledge of the thermodynamic properties of soil and igneous minerals is only sufficient to illustrate such conditions in a general way. In addition, the conditions apply only to equilibrium. Thermodynamics implies that only one energy state is the most stable state under a given set of conditions. The differences in energy between some phases are so slight, however, that kinetic factors may be more important than energy differences in determining which minerals actually form. In addition, uncertainties in the available data on free energies of formation are sometimes greater than the net free energies of potential reactions. Tabulated energies of formation for some soil minerals are little better than educated guesses. Free energies of formation also vary greatly with mineral chemical composition.

Despite such reservations, the thermodynamic data now available provide a reasonably satisfactory picture of solution conditions under which well-defined minerals form and decompose during weathering and soil development. The equilibrium between kaolinite and gibbsite, for example, can be written

$$Al_2Si_2O_5(OH)_4 + 5H_2O = 2Al(OH)_3 + 2Si(OH)_4 \qquad (7.9)$$
$$\text{kaolinite} \qquad\qquad \text{gibbsite} \quad\text{soluble silica}$$

The activities of the solids and of water are assumed to be unity. The equilibrium between the two minerals is therefore determined by the activity (concentration) of the only soluble species, $Si(OH)_4$. The equilibrium constant for the reaction is

$$K = (Si(OH)_4)^2 = 10^{-8.4} \qquad (7.10)$$

The value of this constant comes from the free energies of formation of the reactants and products of Eq. 7.10. The equilibrium constant defines the soluble silica activity at which gibbsite and kaolinite are in equilibrium:

$$(Si(OH)_4) = (10^{-8.4})^{1/2} = 10^{-4.2} \qquad (7.11)$$

At $Si(OH)_4$ activities less than $10^{-4.2}$, gibbsite is stable, kaolinite will not form, and any kaolinite present will decompose to gibbsite and soluble $Si(OH)_4$. At $Si(OH)_4$ activities greater than $10^{-4.2}$, kaolinite is the stable solid. Gibbsite is unstable at these concentrations and will react with $Si(OH)_4$ to form kaolinite. This discussion is continued in more detail in Appendix 7.1.

All of the organic matter and part of the aluminosilicates, hydroxyoxides, and silica in soil exist in structures too small or too poorly crystalline to be detectable by x-ray diffraction. These amorphous materials are not well understood, but they should logically be among the most reactive of soil components, because their structure is so open and their surface area so great. They represent a transition state between unweathered parent materials and well-crystallized secondary soil minerals.

Their high surface areas and low degrees of crystallinity suggest higher surface free energies and rapid transformation to more crystalline forms. Amorphous materials should tend to disappear from soils.

Ions in soils, however, are in constant flux because of plant and microbial uptake and subsequent organic decay. This continual input of fresh amorphous material prevents attainment of equilibrium. The presence of many different ions in soil solutions hinders recrystallization of amorphous compounds by "poisoning" the surfaces of the growing crystal. Foreign adsorbed ions prevent more favored ions from contacting the surface and allowing the crystal to grow. Furthermore, the low solubility of soil minerals allows only slow ion movement through the solution phase to larger crystals of lower surface energy. Inorganic and organic coatings inhibit ion movement to and from the particles and partially neutralize the charges and unsatisfied bonds at mineral edges. Positively charged hydroxyoxide gels and negatively charged aluminosilicates and organic gels can also neutralize each another. Such electrostatic attraction creates a greater resistance to weathering than if the phases remained separate.

Amorphous inorganic matter exists in all soils, but has been of greatest interest where it predominates in the clay fraction. Examples include soil formed from volcanic ash in which high porosity, glassy mineral structures, and chemical instability permit rapid mineral weathering. When the resulting amorphous matter has an Si/Al mole ratio of about 1, it has been termed *allophane*. Definitions of allophane vary among investigators, as might be expected from the experimental difficulty of studying amorphous substances.

The x-ray amorphous silica in soils includes opaline silica in the form of plant *phytoliths*. During plant growth, silica precipitates on the walls of plant cells. After death and organic decay of the plant, the silica phytoliths remain in the soil for many years as accurate representation of the cell wall. Phytoliths are visible under the microscope and can identify the plants in which they formed.

### 7.4.1 Soil Carbonates

In regions of limited rainfall, carbonates (particularly $CaCO_3$) accumulate in soils. Where evapotranspiration exceeds precipitation, the downward flow of water through the soil profile is sufficient to remove only the most soluble weathering products, such as $Na^+$ salts. Intermittent rains can flush out soluble salts even when the amount of percolating water is 1% or less of the total rainfall. Less soluble compounds, on the other hand, accumulate because of limited water flow. The $Mg^{2+}$ and $K^+$ form secondary aluminosilicates. Secondary silicates containing $Ca^{2+}$ as a structural ion are rare, but $Ca^{2+}$ remains instead as an exchangeable cation and precipitates as calcite, aragonite, or vaterite (all $CaCO_3$) and occasionally as the more soluble gypsum ($CaSO_4 \cdot 2H_2O$). Calcite formed in soil allows little $Mg^{2+}$ substitution into its structure, and dolomite ($CaMg(CO_3)_2$) apparently forms only under marine conditions.

Calcite, aragonite, and vaterite can also accumulate in soils when hydrostatic pressure or capillary action move $Ca^{2+}$- and $CO_2$-rich groundwaters upward in the soil profile. The loss of $CO_2$ to the atmosphere and evapotranspiration of the water lead to precipitation of $CaCO_3$. This mechanism sometimes accounts for $CaCO_3$ accumulation in soils of more humid regions.

The effect of chemical environment on $CaCO_3$ solubility in oceans and freshwaters has been considered at great length by geochemists, but under conditions that are not applicable to soils. They generally and tacitly assume (1) unlimited water content, (2) constant $P_{CO_2}$ (where $P$ is the partial pressure or concentration of $CO_2$ in the gas phase, usually assumed to be that of the atmosphere, $P_{CO_2} = 0.00035$), and (3) that $CaCO_3$ is the only source of $Ca^{2+}$. These assumptions are often invalid for soils. The water content of soils is limited and fluctuates. The $CO_2$ concentration is closer to $P_{CO_2} = 0.01$, increases to 0.2 in flooded soils and with root and microbial activity, and varies with the rate of upward diffusion of $CO_2$ to the atmosphere. The $Ca^{2+}$ inputs from weathering and exchangeable $Ca^{2+}$ maintain relatively high $Ca^{2+}$ activities in the soil solution.

Although the environmental conditions that bring about carbonate accumulation in soils are many and varied, the chemical reaction can be represented as simply

$$Ca^{2+} + H_2O + CO_2 = CaCO_3 + 2H^+ \tag{7.12}$$

Alkaline conditions favor $CaCO_3$ accumulation, by consuming $H^+$ and driving the reaction to the right. Increasing $P_{CO_2}$ causes $CaCO_3$ to react further:

$$CaCO_3 + CO_2 + H_2O = Ca^{2+} + 2HCO_3^- \tag{7.13}$$

so that $CaCO_3$ redissolves with increasing $CO_2$ concentration in the gaseous phase. This is a condition of great interest to geochemistry. In soils, however, the relatively high $Ca^{2+}$ concentrations and limited water contents tend to force reaction 7.12 to completion and to repress reaction 7.13. The effect of $P_{CO_2}$ is therefore less important than in geochemical conditions, and $CaCO_3$ precipitates in soils despite the high $P_{CO_2}$ of soil air. In acid soils, $CaCO_3$ dissolves by reversing Eq. 7.12.

When the mass of $CaCO_3$ in soils exceeds several percent, it controls both soil pH and soil solution $Ca^{2+}$ concentrations. Silicate reactions with $H^+$ and $Ca^{2+}$ in this case are relatively insignificant. The pH and $Ca^{2+}$ can then be calculated from Eqs. 7.12 and 7.13, the $P_{CO_2}$, and the solubility product of $CaCO_3$:

$$(Ca^{2+})(CO_3^{2-}) = 10^{-8.4} \tag{7.14}$$

This model has been applied successfully to irrigated arid soils after the refinements of activity coefficients, presence of ion pairs (especially $CaCO_3^0$), slight differences in solubility products between calcite and aragonite, and the inhibitory effect of organic matter on carbonate precipitation are taken into account. The $P_{CO_2}$ is also a major chemical variable in soils.

## 7.4.2   Carbon Dioxide

Dissolution of carbon dioxide from the soil air into the soil solution affects the pH of the soil solution and the solubility of soil carbonates. Carbon dioxide dissolves in water as $CO_{2(aq)}$ and rapidly establishes equilibrium with water to form the weak acid $H_2CO_3$. The solubility of $CO_2$ and other gases is governed by Henry's law, which relates the partial pressure (concentration in a gas mixture) of a gas to its concentration in aqueous solution:

$$K_H = \frac{H_2CO_3}{P_{CO_2}} = 10^{-1.5} \qquad (7.15)$$

Here $H_2CO_3$ refers to the sum of $CO_2$, $H_2CO_3$, and $CO_{2aq}$ dissolved in the aqueous solution. The $H_2CO_3$ dissociates successively to bicarbonate and carbonate ions as the pH increases:

$$H_2CO_3 = H^+ + HCO_3^- \qquad pK = 6.4 \qquad (7.16)$$

$$HCO_3^- = H^+ + CO_3^{2-} \qquad pK = 10.3 \qquad (7.17)$$

The resulting distribution of aqueous $CO_2$ species with pH is shown in Fig. 7.4. The concentrations of the various ions can be calculated from these equations by assuming equilibrium with a specific $P_{CO_2}$. Rainwater is in equilibrium with the $P_{CO_2}$ of 0.00035 in the atmosphere. In the absence of substances such as $SO_2$, $NO_x$, and $NH_3$, the acidity of rainwater is fixed at pH 5.7, as calculated from Eqs. 7.15 and 7.16.

The solubility product of $CaCO_3$ in equilibrium with water and gaseous $CO_2$ yields an acidity of

$$(H^+) = (10^{-13.5} P_{CO_2})^{1/2} \qquad (7.18)$$

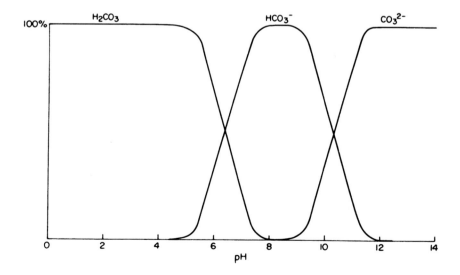

**FIGURE 7.4.** Distribution of aqueous $CO_2$ species with pH.

The pH of a solution in equilibrium with $CaCO_3$ and atmospheric $CO_2$ is 8.3. Equation 7.18 yields a value of pH 8.5 because activity coefficient corrections are ignored and because of rounding-off errors in the exponents of the equilibrium constants.

The pH of calcareous soils when measured in the laboratory is typically around 8.3, if exchangeable sodium is low. Higher pH values indicate $Na^+$, and much more rarely $K^+$, accumulation in the soil. Field pH values of calcareous soils are usually less than 8.3, however, because the $CO_2$ concentration is higher in the soil's gas phase than in the atmosphere. Root respiration and microbial decay of organic matter release $CO_2$, which must diffuse through soil pores to the atmosphere. The average $P_{CO_2}$ in the pores of agricultural soils is probably 0.003 to 0.03, 10 to 100 times that of the atmosphere. Many workers use the value of 0.01 as typical of $CO_2$ in the gas phase of soils. A lower value might be better for warmer and drier soils. The actual $CO_2$ concentration depends both on the rate of microbial and root respiration and on the diffusion rate of $CO_2$ to the atmosphere.

Gas diffusion is slower in wet and flooded soils, where soil pores are plugged by water. Gas diffusion in water is about 1/10 000th the rate of gas diffusion in air, or essentially nil in flooded soils where all soil pores are water-filled. The consumption of 1 mole of $O_2$ during respiration yields approximately 1 mole $CO_2$. In flooded soils, therefore, $CO_2$ can almost completely replace $O_2$ and reach a partial pressure of 0.2, equal to the value of $O_2$ in atmospheric air. Since the gas volume in a flooded soil is minute, it is perhaps more instructive to say that the $CO_2$ concentration in the soil solution is equivalent to $P_{CO_2} = 0.2$. At such concentrations, dissolved $CO_2$ has considerable influence on soil pH (Eq. 7.18). When soil solutions are extracted from soils, dissolved $CO_2$ is slowly lost to the atmosphere. This causes large pH increases in extracts from alkaline and flooded soils, and the possible precipitation of $CaCO_3$ and of transition and heavy metal hydroxyoxides. The loss requires several hours so immediate measurements yield pH values more representative of actual soil conditions.

### 7.4.3 Evaporites

Although calcium carbonate formation in soil is a result of high evapotranspiration rates relative to precipitation rates, the term *evaporite* is usually restricted to compounds more soluble than $CaCO_3$. Where drainage water from surrounding soils accumulates and where the amount of percolated water is small compared to the amount of water evaporated, soluble salts tend to accumulate. This subject is dealt with in more detail in Chapter 11. The present section is restricted to the extreme case of natural salt flats and playas (former and intermittent lake beds).

The high salt concentrations and accompanying high pH in these soils alter the course of soil mineral weathering toward the formation of minerals that are highly unstable under leaching conditions. The distinctive salts that form include the Na salts halite (NaCl) and trona (NaHCO3), plus smaller amounts of sulfates, borates, and similar salts of K and occasionally Li. Some secondary silicates also form, including the zeolite analcime ($NaAlSi_2O_6 \cdot 6H_2O$) and sepiolite ($Mg_4Si_6O_{15}(OH)_2 \cdot 6H_2O$).

The formation of such minerals and authigenic feldspar under highly saline conditions is sometimes termed *reversed weathering*. This is appropriate in the sense that ions weathered in other locations are thus incorporated into new minerals instead of flowing to the sea. The minerals formed, however, are considerably different from the original igneous aluminosilicates, although their chemical compositions may be similar.

## APPENDIX 7.1    STABILITY DIAGRAMS

The stability of minerals contacting aqueous solutions depends on the composition of the solution. Figure 7.5 shows the stability regions of major soil minerals at varying concentrations of $Si(OH)_4$, $Ca^{2+}$, $Na^+$, and $H^+$. The soil-formed minerals—gibbsite, kaolinite, and montmorillonite—are stable at lower Ca activity and lower pH than the igneous Ca feldspar, anorthite. Gibbsite is stable at the lowest $Si(OH)_4$ activity and montmorillonite is stable at the highest $Si(OH)_4$ activity. The stability region of Ca-montmorillonite is greater than that of Na-montmorillonite. In soils the concentrations of three aqueous species—Ca, H, and $Si(OH)_4$—are limited but the concentration of Na in the soil solution is unlimited. Precipitation of $CaCO_3$ prevents the value of $pH - \frac{1}{2}pC_a$ from increasing enough to precipitate anorthite. Anorthite is unstable under soil conditions and weathers to the soil-formed minerals. The upper limit of $Si(OH)_4$ activity is determined by the solubility of amorphous silica, which precipitates at $(Si(OH)_4) = 10^{-2.8}$.

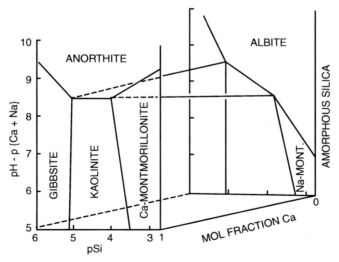

**FIGURE 7.5.** Stability diagram of the Ca–Na–Al–Si–O–$H_2O$ system at equilibrium at 25° C. The Ca system is the front vertical plane, Na is the back plane. Braces denote moles of ion charge in solution. (From H. L. Bohn and Kittrich. 1984. *Chem. Geol.* **43**:181.)

Figure 7.5 is approximate because the $\Delta G$ values of formation of Ca- and Na-montmorillonites are educated guesses. Figure 7.5 also ignores the slow kinetics of silicate reactions. In addition, the activities of $Mg^{2+}$, $K^+$, and $Na^+$ affect the stability of other soil minerals and their igneous minerals, which directly and indirectly affect the stability of Ca minerals.

Mineral stabilities often depend on the activities of $H^+$, other cations, and the $Si(OH)_4$ activity. The $H^+$ activity is usually inversely related to the activities of $K^+$, $Na^+$, $Ca^{2+}$, or $Mg^{2+}$. The $M^+/H^+$ ratio or $pH - pM$, although awkward to visualize, permits plotting both activities on one axis. High values of $pH - pK$ and of $pSi(OH)_4$ represent alkaline soils with little leaching. Low values of $pH - pK$ and $pSi(OH)_4$ represent acid soils and extensive leaching.

The gibbsite–kaolinite boundary in Fig. 7.5 is at a $Si(OH)_4$ activity of $10^{-4.2}$. Kaolinite, therefore, will not form unless the soluble silica concentration exceeds this level. Any kaolinite present at lower silica concentrations will eventually dissolve to form gibbsite and soluble $Si(OH)_4$.

At higher $Si(OH)_4$ activities, Fig. 7.5 shows that kaolinite is unstable and is transformed to montmorillonite. Assuming that $Ca^{2+}$ is the only exchangeable cation in the system, the equation for the equilibrium between kaolinite and montmorillonite can be written as

$$7Al_2Si_2O_5(OH)_4 + 8Si(OH)_4 + Ca^{2+} = Ca(Al_7Si_{11}O_{30}(OH)_6)_2 + 23H_2O + 2H^+$$

    kaolinite     soluble silica                    montmorillonite              (7.19)

where montmorillonite for simplicity is given an idealized composition. The equilibrium constant of Eq. 7.19 is

$$K_{eq} = \frac{(H^+)^2}{(Si(OH)_4)^8(Ca^{2+})} \tag{7.20}$$

The value of $K_{eq}$ can be calculated from the energies of formation of the components of Eq. 7.19. The equilibrium between kaolinite and montmorillonite depends on the $H^+$, $Ca^{2+}$, and $Si(OH)_4$ activities in solution. Rearranging Eq. 7.20 yields

$$pH - \tfrac{1}{2}pCa = 4pSi(OH)_4 - \tfrac{1}{2}\log K_{eq} \tag{7.21}$$

The equilibrium between kaolinite and montmorillonite is plotted according to Eq. 7.21 in Fig. 7.5. Montmorillonite is stable at high $Si(OH)_4$ activities and moderate $Ca^{2+}/H^+$ ratios.

The K and Na stability diagrams are similar, but the values of M/H vary. The Ca feldspars are the most unstable, and the K feldspars are the most stable, with respect to weathering. At higher M/H ratios (more basic solutions) other minerals are stable and would precipitate before the igneous feldspars, but free energy data for other silicates are lacking.

Stability diagrams such as Fig. 7.5 should not be taken too literally. Free energy data for most minerals are uncertain. Even small errors are magnified by calculating the equilibrium constant, which is the antilogarithm of a small difference between large $\Delta G$'s of formation. In addition, the activities of the solid phases are assumed to be unity, i.e, the minerals are assumed to be pure. This is probably approximately true for the major minerals and for minerals that do not form solid solutions. The activities of the minor components of solid solutions are not known with certainty but are almost surely not unity. At present, stability diagrams serve as inexact but useful illustrations of the relations between minerals in nature.

Occasionally the course of weathering can "reverse" in the sense that feldspar weathering will create secondary silicates that are unstable under the leaching and weathering conditions of well drained soils. Such reversal occurs because of the accumulation of $K^+$, $Na^+$, $Ca^{2+}$, $Mg^{2+}$, and $Si(OH)_4$ in arid and poorly drained soil solutions. These secondary silicates include zeolites, evaporites, and the authigenic feldspars. The area denoted as soil solution in Fig. 7.6 shows the extreme concentrations that have been reported in soil solutions. Within this range, several silicate minerals are stable.

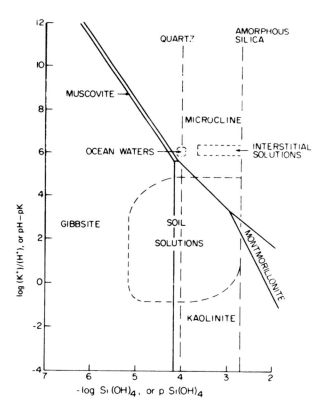

**FIGURE 7.6.** The K–Al–Si–O–H$_2$O equilibrium system at 25° C, superimposed on the ranges of compositions of soil solutions, oceans, and interstitial solutions. (From H. C. Helgeson, T. H. Brown, and R. H. Leeper. 1969. *Handbook of Theoretical Activity Diagrams Depicting Chemical Equilibrium in Geologic Systems Involving an Aqueous Solution at One Atm and 0° to 300° C.* Freeman & Cooper, San Francisco; and S. V. Mattigod. 1976. Thesis. Washington State University, Pullman, WA.)

At 0° C, the secondary minerals gibbsite, kaolinite, and montmorillonite are more stable relative to the igneous feldspars and mica than at 25° C. The stability fields of gibbsite, kaolinite, and montmorillonite are appreciably larger at 0° C. Because soil temperatures lie between 0 and 25° C, Fig. 7.5 rather conservatively illustrates the stability of soil-formed minerals. Assembling Figure 7.5 requires comparing the stabilities of many likely minerals and eliminating those that are unstable relative to other minerals. Muscovite mica is thereby found to be stable at 25° C only within a narrow region of solution compositions between the stability fields of microcline, a mineral identical in composition to orthoclase feldspar, and orthoclase. At 0° C, the stability region of muscovite disappears.

Figure 7.6 shows the solubilities of quartz and amorphous silica in relation to the minerals of Fig. 7.5. The solubilities of substances having the empirical formula $SiO_2$ or $SiO_2 \cdot nH_2O$ have been studied for decades. These studies are much more complicated than they appear, because of the reluctance of soluble silica to reach equilibrium or even metastable equilibrium with its solid phases. Soluble silica tends to polymerize slowly in supersaturated solutions rather than to precipitate cleanly. In addition, the solid phase that precipitates is often amorphous silica instead of the most stable phase, quartz or its close relative chert. Amorphous silica is metastable and much more soluble than quartz. The solubilities of amorphous silica and quartz are often assumed to be the upper and lower limits of silica solubility in soils. Viewed from the range of soil solution compositions shown in Fig. 7.6, silica concentrations can be less than the equilibrium solubility of quartz even though quartz is almost always present in the sand fraction of soils. The slow kinetics of silica reactions and the slow release of $Si(OH)_4$ during weathering create wide deviations from equilibrium.

Figure 7.6 also shows the concentration ranges of ocean waters and of the interstitial water in minerals. The composition of interstitial water is not well known, but the region shown is a reasonable guess. Ocean and interstitial waters are as influential as the soil solution in affecting mineral transformations at the earth's surface. The differences in composition of these three solutions also give rise to wide variations in the nature and distribution of soil clay minerals.

## BIBLIOGRAPHY

Feldman, S. B., and L. W. Zelazny. 1998. The chemistry of soil minerals. In *Future Prospects for Soil Chemistry* (P. M. Huang, D. L. Sparks, and S. A. Boyd, eds.). Soil Science Special Publ. 55, American Society of Agronomy, Madison, WI, pp. 139–152.

Jackson, M. L., and G. D. Sherman. 1953. Chemical weathering of minerals in soils. In *Advances in Agronomy*, vol. 5 (A. G. Norman, ed.). Academic, New York, pp. 221–317.

Jenny, H. 1941. *Factors of Soil Formation*. McGraw-Hill, New York.

## QUESTIONS AND PROBLEMS

1. Show how to calculate the pH of rainwater, assuming equilibrium with $CO_2$ and the absence of other acidic or basic solutes.

**2.** Derive Eq. 7.18.

**3.** Describe the major minerals and states of calcium, magnesium, potassium, and sodium as they cycle from rock to soil to sea to rock.

**4.** From Tables 7.2 and 7.4, calculate the average residence times of calcium, magnesium, and potassium in soils.

**5.** Show the equations for the equilibria between anorthite and gibbsite kaolinite and calcium montmorillonite.

**6.** If the $Na^+$ residence time in the oceans is about $10^8$ years, what is the $Na^+$ residence time in soils? What assumptions are necessary? How many times on the average has $Na^+$ recycled from rock to soil to oceans?

**7.** For $Ca^{2+}$, $Mg^{2+}$, $K^+$, $P^{5+}$, $Si^{4+}$, or other selected chemical elements, trace all the possible pathways that the ion might follow from its state in igneous rock through soils to its most stable state at the earth's surface.

**8.** Explain why soil feldspar particles tend to be larger than soil kaolinite particles and why both tend to be larger than montmorillonite particles?

**9.** Calculate the $Ca^{2+}$ activity in equilibrium with $CaCO_3$ and $P_{CO_2} = 0.2$ and 0.01 at (a) pH 8.0 and (b) pH 5.0.

# 8

# CATION RETENTION (EXCHANGE) IN SOILS

Probably the most important and distinctive property of soils is that they can retain ions and release them slowly to the soil solution and to plants. The retention prevents concentrations that are too high and too low. The evolution of plants has taken advantage of this buffered range of ion concentrations that soils make available in the soil solution. Over most of the earth's surface, the availability of these ions in the soil solution is adequate, but not necessarily ideal, for plants. Crop and horticultural plants and a desire for maximum yield place greater demands on the soil and may require adjusting the native soil solution. Adjustments by fertilization, liming, and salt removal are usually temporary. The soil and climate tend to return the soil to its native state.

Ion retention is actually ion exchange. Soils give up other ions, $H^+$ or $OH^-$ and $HCO_3^-$, in equal amounts to those retained. When trace ions are removed from the soil solution, the ion exchange to the soil solution is often unnoticed. The retention of organic, nonionic substances usually results in their degradation by soil microbes and conversion to $CO_2$ and water. This chapter is concerned with the exchange, the retention and release, of cations between soil particles and the soil solution.

Soil chemistry has stressed cation retention and exchange and has almost ignored anion retention and exchange. This unfortunate bias is because the clay particles of most soils of Europe and North America have a net negative change. The amount of cation exchange is therefore greater than anion exchange. Had soil chemistry begun in Australia, in soils of volcanic parent material, or in highly weathered tropical soils the bias might be toward anion exchange and retention. Soils have both a negative charge that retains cations and a positive charge that retains anions. We usually measure the soil's net charge and that is usually negative.

Cation retention by soils can be roughly divided into the weaker electrostatic interaction of soil particles with the alkali and alkaline earth cations and the soil's stronger chemical bonding with trivalent and transition metal cations. Chemical bonding is the interaction of polyvalent cations with $O^{2-}$ and $OH^-$ ligands of aluminosilicates, hydroxyoxides, and phosphates, plus retention of weak Lewis acids by soil organic matter. Chemical bonding, also called precipitation or strong adsorption, is discussed in Chapter 3 and is related to the dissolution–precipitation reactions of classical chemistry. The weaker, electrostatic retention of ions is distinctive to soils and colloidal systems and creates the major reservoir of the essential macroelement cations for plants and all living organisms. Some generalizations can be made about the attraction, exchange, and retention of cations by soils:

1. Relatively weak (electrostatic) attraction—alkali and alkaline earth cations (mainly Ca, Mg, K, and Na)

   Nonspecific, depends mostly on the concentration ratios on the solid vs. the soil solution and on the ion charge ratio. Some clay minerals prefer one ion over others.

   Reactions are fast and reversible; time scale is seconds and minutes.

   Amount of retention depends on soil's cation exchange capacity, the negative charge of soil particles.

   Largely due to aluminosilicate clay minerals plus soil organic matter.

2. Strong (chemical bonding) attraction—H, Al, Be, Ti, transition metal, and "heavy metal" cations

   Specific, that is, the strength of attraction depends mostly on the cation's water solubility and the amount of that cation on the surfaces of soil particles.

   Reaction time scale is rapid at first, but continues at ever-slower rates for long periods.

   Amount of retention depends on soil pH rather than on the charge properties of soil clays.

   Aluminosilicates are less important, Fe and Mn oxides are more important, in this retention than in electrostatic cation retention. The retention is generally much stronger than that predicted by aqueous solubility products.

   Mechanisms of retention are complex and grade gradually from one to another.

   Organic matter increases the range of sorptivity, possibly by adding soft Lewis base character.

   Amorphous materials retain more than crystalline.

## 8.1 ELECTROSTATIC CATION RETENTION (CATION EXCHANGE)

*Exchangeable ions* are those ions replaced by neutral salt solutions flowing through soils. *Soluble salts* are removed by water alone. Salt solutions also exchange anions from soils, but because soil colloids are mostly negatively charged, generally more cations exchange than anions. The major exchangeable cations are, in order of decreasing amounts, $Ca^{2+}$, $Mg^{2+}$, $K^+$, and $Na^+$. The retention helps prevent leaching losses during weathering; plants exchange for these cations and absorb them by releasing $H^+$.

In humid and temperate region soils, most of the plant-available alkaline earth and alkali metal cations neutralize the soil's negative charge. A typical agricultural loam soil contains about 20 000–30 000 kg ha$^{-1}$ of exchangeable cations in its root zone (0.5 m depth). Roughly 80% is Ca, 15% is Mg, 4% is K, and 1% is Na. As soil acidity increases, $Al^{3+}$ and $H^+$ are also exchangeable on soil surfaces. Arid soils may contain salts considerably in excess of the charge-neutralizing cations. Exchangeable cations and soil solution salts are very important to plant productivity and are easily manipulated by liming, irrigation, and fertilization. Hence, cation exchange has long been an important part of soil chemistry research.

Thompson and Way conducted the first recorded studies of cation exchange in Rothamsted, England, in 1850. They showed that passing an ammonium sulfate solution through soil columns leached calcium sulfate out of the soil. The predominant cation in the aqueous solution had changed from ammonium to calcium because of cation exchange in the soil. Thompson and Way showed that the exchange was very fast and reversible, and that the amount of ammonium ion retained equaled the amount of calcium released. Subsequent work has refined and supported these findings and has measured the cation exchange capacities of soil and soil components, the relative affinities of soils and their components for various cations, and the effects of changing soil pH on exchange reactions.

Cations in soils are roughly in three major categories: solid phase, exchangeable, and soluble. Weathering and organic decay release cations that vary in charge, size, and polarizability, so they respond differently to the soil surfaces and other ions they encounter in the soil solution. Small, polyvalent ions tend to reprecipitate/adsorb in soils by forming strong chemical bonds with aluminosilicate and hydroxyoxide surfaces. Larger, lower-charge cations (mainly $Ca^{2+}$ and $Mg^{2+}$) instead associate more weakly with surfaces of the solid phase and are the exchangeable ions. The largest, lowest-charge cations ($K^+$ and $Na^+$) are weaker competitors for surface charge neutralization and tend to dominate in the *bulk soil solution* away from the charged surfaces (the soluble ions). Weathering tends to remove the soluble and exchangeable ions from soils.

The distribution of major exchangeable cations in productive agricultural soils is generally $Ca^{2+} > Mg^{2+} > K^+ > NH_4^+ \approx Na^+$. The composition of the exchangeable cations in different soils tends to be much more uniform than the composition of the parent material rocks from which the soils are derived. The general effect of soil reactions is to smooth out the differences between soil parent materials and inputs, and to create a relatively uniform distribution of exchangeable ions for plant growth.

The soil-forming factors can modify the distribution of exchangeable cations from this desired state. Table 8.1 shows examples of exchangeable cations found in a wide variety of soils. The Merced soil has high exchangeable Na because it is poorly drained in an arid climate, it has no drainage to the sea, and upwelling water from a

**Table 8.1. CEC values and major exchangeable cations of selected soils[a]**

| Soils | pH | CEC (mmol kg$^{-1}$) | Exchangeable Cations (% of Total) | | | | |
|---|---|---|---|---|---|---|---|
| | | | Ca$^{2+}$ | Mg$^{2+}$ | K$^+$ | Na$^+$ | H$^+$ (Al$^{3+}$)[b] |
| Average of agricultural soils (Netherlands) | 7.0 | 383 | 79.0 | 13.0 | 2.0 | 6.0 | — |
| Average of agricultural soils (California) | 7.0 | 203 | 65.6 | 26.3 | 5.5 | 2.6 | — |
| Chernozem or Mollisoll (Russia) | 7.0 | 561 | 84.3 | 11.0 | 1.6 | 3.0 | — |
| Sodic Merced soil (California) | 10.0 | 189 | 0.0 | 0.0 | 5.0 | 95.0 | 0.0 |
| Lanna soil, unlimed (Sweden) | 4.6 | 173 | 48.0 | 15.7 | 1.8 | 0.9 | 33.6 |
| Lanna soil, limed (Sweden) | 5.9 | 200 | 69.6 | 11.1 | 1.5 | 0.5 | 17.3 |

[a] From F. E. Bear, ed. 1964. *Chemistry in the Soil*, 2d ed. American Chemical Society, Washington, DC, p. 167.

[b] Probably includes some titratable acidity (Chapter 10.)

high water table evaporates at the surface. Under those conditions Ca precipitates as $CaCO_3$, Mg, and K can form secondary soil minerals, leaving Na as the major cation. Exchangeable Na may exceed K in the Netherlands soils because of atmospheric inputs of NaCl from the nearby ocean. The high $Mg^{2+}$ content of the California soils may reflect the high-Mg content of rocks found in volcanic and geologically active regions. Exchangeable $Al^{3+}$ is present in appreciable quantities in acid soils (pH < 5.5), such as the Lanna soil in Sweden. This soil is formed from granitic rocks in dense forest under conditions of slowly weatherable parent material, high rainfall, good drainage, and organic acids from organic matter decomposition. This strongly acid soil (pH < 4) contains considerable exchangeable Al and some exchangeable $H^+$.

Productive agricultural soils are characteristically Ca dominated. One important part of sustainable agriculture is to maintain Ca dominance. Despite the wide range of soil-forming factors in Table 8.1, Ca is the predominant cation in all but the extremely sodic (and barren) Merced soil.

The sum of exchangeable $Ca^{2+}$, $Mg^{2+}$, $K^+$, $Na^+$, and $Al^{3+}$ generally equals, for practical purposes, the soil's *cation exchange capacity* (CEC). The CEC varies from 10 mmol(+) kg$^{-1}$ for coarse-textured soils to 500 to 600 mmol(+) kg$^{-1}$ for fine-textured soils containing large amounts of 2:1 layer silicate minerals and organic matter.

In 1850, Thompson and Way found cation exchange to be reversible, *stoichiometric* (the amount released, as moles of ion charge, equals the amount retained), and rapid. Since then some refinements have been studied and some exceptions have been found, but their results are still generally valid. Although cations are preferred in varying degrees by soil colloids, even strongly adsorbed cations can normally be replaced by manipulating solution conditions. An exception to this generalization of reversibility is the preferential retention of many polyvalent cations (especially trace metals, weak Lewis acids) by soil organic matter. Such cations, which are thought to be partially covalently bonded, can be displaced only by other polyvalent cations capable of forming even stronger covalent bonds. Other exceptions include cation fixation reactions, described later in this chapter; cases where large organic cations, such as the pesticides paraquat and diquat, are physically prevented (steric hindrance) from approaching certain interlayer exchange sites; and cases where multivalent cations are preferentially adsorbed because they can simultaneously balance several closely adjacent exchange sites.

Because cation exchange reactions are stoichiometric, the sum of all exchangeable cations present at a given pH and CEC varies little or not at all with cation species. For example, consider the exchange reaction

$$CaX + 2NH_4^+ = (NH_4)_2X + Ca^{2+} \tag{8.1}$$

where X designates a cation exchanger. Two ammonium ions replace one calcium ion to preserve the stoichiometry of the reaction. Exchangeable cation composition and CEC values normally are expressed as moles of ion charge $kg^{-1}$, formerly meq/100 g (milliequivalents per 100 g).

Exchange reactions are also rapid. The exchange step itself is virtually instantaneous. The rate-limiting step often is ion diffusion to or from the colloid surface. This is particularly true under field conditions, where ions may have to move through tortuous pores or through relatively thick, stagnant water films on soil colloid surfaces to reach an exchange site. The need for diffusion can produce *hysteresis* (the extent or speed of reaction depends on direction of the reaction) for some ion exchange reactions. Under laboratory conditions, samples normally are shaken during exchange reactions, to speed ion movement and to minimize the thickness of stagnant water layers on soil particle surfaces.

Because of their reversibility, cation exchange reactions can be driven forward or reverse by manipulating the relative concentrations of reactants and products. In the laboratory, common techniques for driving the reactions toward completion are to use high ($\geq 1$ M) concentrations of exchanging cations and to maintain low concentrations of product cations by leaching or repeated washings:

$$CaX + 2Na^+ \text{ (high concentration)} = Na_2X + Ca^{2+} \text{ (low concentration)} \tag{8.2}$$

to form insoluble precipitates

$$CaX + Na_2CO_3 = Na_2X + CaCO_3 \text{ (precipitate)} \tag{8.3}$$

or to form volatile gases

$$NH_4X + NaOH = NaX + NH_4OH = NaX + H_2O + NH_3 \text{ (gas)} \tag{8.4}$$

For exchange between cations of differing charge, diluting the solution favors retention of the more highly charged cation. For example, Eq. 8.1 has a reaction coefficient $k$:

$$k = \frac{[(NH_4)_2X][Ca^{2+}]}{[CaX][NH_4^+]^2} \tag{8.5}$$

where brackets indicate concentrations (mol $L^{-1}$ or mol $kg^{-1}$) rather than activities. Rearranging Eq. 8.5 gives a typical cation exchange equation:

$$\frac{[(NH_4)_2X]}{[CaX]} = k\frac{[NH_4^+]^2}{[Ca^{2+}]} \tag{8.6}$$

Because of the squared term on the right of Eq. 8.6, the ratio of ammonium to calcium in the colloid's double layer changes with total, as well as relative, salt concentration of the bulk solution. This dependence of cation exchange on cation valence has been termed the *valence dilution effect*. As an example, consider a solution having $[NH_4^+] = [Ca^{2+}] = 1$ mmol $L^{-1}$. The ratio $[NH_4^+]^2/[Ca^{2+}]$ in this case equals $1^2/1$, or 1 mmol $L^{-1}$. Upon tenfold dilution, the ratio is $[0.1]^2/[0.1]$, or 0.1 mmol $L^{-1}$. Hence, the ratio of ammonium to calcium on the colloid decreases with dilution (Table 8.2). The total quantities, but not the concentrations, of replacing ions remained constant. Table 8.2 shows the absence of a dilution effect for cations of the same valence. The percentage of calcium replaced by barium remained virtually constant, but that replaced by ammonium decreased with decreasing salt concentration.

Exchanging one cation for another in the presence of a third (*complementary*) cation also becomes easier as the retention strength of the third cation increases. For example, replacing calcium by ammonium is easier from a $Ca^{2+}-Al^{3+}$ soil than from $Ca^{2+}-Na^+$ soil. The fraction of the CEC satisfied by the tightly bound Al is in effect blocked off, and the Ca and ammonium ions compete for a smaller number of exchange sites.

**Table 8.2. Replacement of exchangeable calcium from 1 mmol of montmorillonite exchange sites by a constant amount (1 mmole of ion charge) of barium or ammonium, at varying replacing-cation concentrations[a]**

| Solution Added | | Percent $Ca^{2+}$ Replaced by: | |
|---|---|---|---|
| liters | mol charge $L^{-1}$ | $Ba^{2+}$ | $NH_4^+$ |
| 0.025 | 0.04 | 49.7 | 29.8 |
| 0.100 | 0.01 | 50.2 | 20.8 |
| 0.200 | 0.005 | 50.8 | 16.6 |
| 0.400 | 0.0025 | 52.7 | 15.2 |

[a] Adapted from P. Schachtschabel. 1940. *Kolloid-Beihefte.* **51**:199–276.

## 8.1.1 Exchange Selectivity

The attraction of cations for negatively charged colloid surfaces is qualitatively described by electrostatic attraction and repulsion, following Coulomb's law (Appendix 8.1). A major limitation of this simple electrostatic approach, however, is its failure to predict differences in preference or selectivity of colloid surfaces for cations of the same valence. Such preference is related to the relative hydrated sizes or to the relative energies of hydration of the various cations. Ions of smaller dehydrated radius have a greater density of charge per unit volume. Hence, they attract waters of hydration more strongly and have a larger hydrated radius. An ion of larger hydrated radius is held less tightly by coulombic attraction. Partially dehydrated ions can approach the surfaces more closely and generally are retained quite tightly by soil colloid particles.

These generalizations arise from data such as in Table 8.3. The data were generated by saturating a montmorillonite suspension with a given ion and then measuring the quantity of that ion released when a symmetry (amount equal to the CEC) of either $NH_4Cl$ or $KCl$ was added.

The most important factor determining the relative extent of adsorption or desorption of a given ion is its valence. Divalent ions in general are retained more strongly than monovalent ions, trivalent ions are retained even more strongly, and quadrivalent ions such as thorium $Th^{4+}$ are essentially unreplaced by an equivalent amount of $KCl$.

Within a given valence series, the degree of replaceability of an ion decreases as its dehydrated radius increases. An apparent exception is the "$H^+$" ion. Monovalent

**Table 8.3. Relation of ion charge and ion size to ion retention**[a]

| Ion | Crystallographic (Dehydrated) Radius (nm) | % Released by $NH_4^+$ or $K^+$ |
|---|---|---|
| $Li^+$ | 0.068 | 68 |
| $Na^+$ | 0.097 | 67 |
| $K^+$ | 0.133 | 49 |
| $NH_4^+$ | 0.143 | 50 |
| $Rb^+$ | 0.147 | 37 |
| $Cs^+$ | 0.167 | 31 |
| "$H^+$" ($Al^{3+}$) | (?) | 15 |
| $Mg^{2+}$ | 0.066 | 31 |
| $Ca^{2+}$ | 0.099 | 29 |
| $Sr^{2+}$ | 0.112 | 26 |
| $Ba^{2+}$ | 0.134 | 27 |
| $Al^{3+}$ | 0.051 | 15 |
| $La^{3+}$ | 0.102 | 14 |
| $Th^{4+}$ | 0.102 | 2 |

[a] Modifed from H. Jenny and R. F. Reitemeier. 1935. Reprinted with permission from *J. Phys. Chem.* **39**:593–604. Copyright by the American Chemical Society.

"hydrogen" in this case behaves more like trivalent lanthanum. Work with acid soils and clays since the early 1950s has demonstrated that "hydrogen" clays are unstable and rapidly decompose to produce aluminium-saturated clays. Hence, the "$H^+$" entry of Table 8.3 probably represents $Al^{3+}$.

Relative ion replaceability, or ease of removal from specific colloids, has been called the *lyotropic series*. For example, the data of Table 8.3 could be written as

$$Li^+ \approx Na^+ > K^+ \approx NH_4^+ > Rb^+ > Cs^+ \approx Mg^{2+} > Ca^{2+} > Sr^{2+}$$

$$\approx Ba^{2+} > La^{3+} \approx H(Al^{3+}) > Th^{4+}$$

in order of increasing strength of retention by montmorillonite. The order of the lyotropic series is explainable if the cations at the colloid surface include a layer of specifically adsorbed or partially dehydrated cations, the so-called Stern layer (Appendix 8.1). The composition of the Stern layer can be estimated from coulombic calculations if individual ion characteristics (e.g., hydrated radius and polarizability) are considered.

Soil colloids of high charge density, that is, of high charge or CEC per unit surface area, generally have the greatest preference for highly charged cations. For example, vermiculite normally retains more Ca than does montmorillonite from a mixed $Na^+$–$Ca^{2+}$ solution. Hence montmorillonite has a higher exchangeable Na percentage than vermiculite at the same bulk-solution Na and Ca concentrations. The monovalent cations $NH_4^+$ and $K^+$ are often exceptions to this generalization, because of their unusually strong preference by mica and vermiculite (discussed in greater detail below). Partially covalent bonding and/or complex formation may contribute to a similar preference of soils high in organic matter for many polyvalent cations. Raising soil pH can also change cation selectivity by increasing soil CEC and thus increasing the preference for polyvalent versus monovalent ions.

In addition to coulombic preferences related to ion size, certain colloids exhibit unusually high preferences for specific cations. An example is the high exchangeable magnesium content of vermiculite. Hydrated $Mg^{2+}$ apparently fits so well into the water network between partially expanded sheets of vermiculite that Mg is preferred over a wide concentration range (Fig. 8.1). The dashed line in the figure indicates no ion preference by the colloid, and the dotted line shows a more typical case of $Ca^{2+}$–$Mg^{2+}$ exchange with an exchange coefficient of 1.5. At low $Mg^{2+}$ concentrations, vermiculite prefers Ca over Mg because the hydrated $Mg^{2+}$ is larger than hydrated $Ca^{2+}$. As soon as enough Mg ($>40\%$) is present in solution to exert a significant effect on the interlattice water network, the curve shifts to a pronounced preference for Mg. Although normal soil solutions have relatively high $Ca^{2+}/Mg^{2+}$ ratios, the crossover on the figure occurs at $Ca^{2+}/Mg^{2+}$ ratios that are attainable under some natural conditions. These conditions can occur when calcium carbonate is precipitating, when former marine sediments are contributing soluble salts, or when high Mg micas are weathering to vermiculite. Vermiculite then is an excellent scavenger of Mg ions and becomes nearly saturated with Mg even when exposed to appreciable Ca concentrations and/or monovalent cations. To replace the Mg from natural

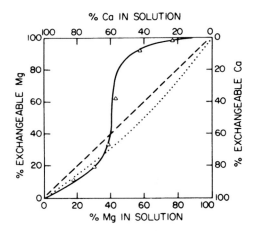

**FIGURE 8.1.** Ca–Mg exchange in a vermiculite suspension. The dashed line represents an exchange constant of 1.0 (no preference); dotted line represents a Ca–Mg exchange constant of 1.5. (From F. F. Peterson et. al. 1965. *Soil Sci. Soc. Am. Proc.* **29**:327.)

vermiculites, the mineral must be repeatedly leached with high concentrations of a replacing ion. This procedure lowers the relative Mg concentration below the point of preferential adsorption.

Another case of high preference for a particular ion is the preference of vermiculites and of weathered edges of trioctahedral micas for K and similar exchangeable cations. Mica weathers to a vermiculite-like mineral with a decrease in layer charge accompanying the weathering process. Traditionally, the preferential adsorption of K, NH₄, cesium, and rubidium by such minerals has been attributed to the excellent fit of the ions in the hexagonal or ditrigonal holes on vermiculite surfaces. The process is believed to be activated by the dehydration of large, weakly hydrated ions as adjacent silicate sheets approach one another during thermal motion or drying.

Alternatively, the affinity has been explained by the relative hydration energies of various ions, plus the relative hydration energies of individual cation exchange sites on different minerals. The relatively small hydration energies of K, NH₄, Rb, and Cs result in easy dehydration and strong retention. The hydration energy theory explains how $Ba^{2+}$, with essentially the same crystallographic radius as NH₄, is not fixed by trioctahedral micas or vermiculite. Barium ions, with their greater energy of hydration, apparently are not readily dehydrated and entrapped by adjacent mineral lattices. Barium also readily rehydrates, forcing the lattices apart when mineral surfaces are rewetted after drying.

Preferential retention of K and NH₄ by vermiculite and by weathered mica edges is sufficiently dramatic that a sizeable literature has accumulated on this so-called *fixation* reaction. Fixation generally decreases with soil acidification and increases with soil liming. This is attributed to the formation of Al and Fe hydroxide interlayers between mica and vermiculite layer lattices under acid conditions. Such interlayers prevent the lattices from collapsing completely. Lattice collapse is theoretically necessary to retain fixed cations against exchange by various extracting solutions. Fixation is accentuated by drying.

## 8.1.2  Cation Exchange Equations

To predict the effects of, for example, irrigation, liming, weathering, fertilization, and acid rain on soils, it is necessary to predict the exchangeable cation composition in equilibrium with this new input. The exchangeable cation chemistry can also provide valuable clues about plant elemental deficiencies or imbalances, rates of toxic metal movement and attenuation, and tendencies toward soil dispersion. Cation exchange equations predict those effects with varying precision.

Several equations describe cation exchange processes. Each has its own set of characteristics and merit. The choice of a particular equation often seems to be subjective, however, and may be based as much on the investigator's background as on any other factor. Lack of familiarity with the units, and of the numerical values of exchange coefficients for other equations are major deterrents to adoption of a more widespread and uniform approach to cation exchange.

Certain limitations are inherent in most cation exchange equations:

1. Cation and anion exchange are considered separately; acknowledging their simultaneous presence is rare.
2. The cation or anion exchanger is assumed to possess constant exchange capacity. Often, however, the capacity varies with the exchangeable ion, with salt concentration, and with pH.
3. Stoichiometric (1 to 1) ion exchange is generally assumed. Apparent exceptions are usually explained by simultaneous adsorption of molecules or by complex ion formation.
4. Complete reversibility is usually assumed.

The most general type of cation exchange relationship is a mass action equation:

$$CaX + 2Na^+ = 2NaX + Ca^{2+} \tag{8.7}$$

resulting in the reaction coefficient

$$k = \frac{(NaX)^2(Ca^{2+})}{(CaX)(Na^+)^2} \tag{8.8}$$

where X denotes the exchangeable form of the cation and parentheses denote activities of soluble or exchangeable cations. The major problem in all exchange equations is evaluating the activity of the exchangeable cations, since their activities cannot be measured or calculated precisely. Equation 8.8 can be rearranged to

$$\frac{(NaX)^2}{(CaX)} = k\frac{(Na^+)^2}{(Ca^{2+})} \tag{8.9}$$

This has been termed a *Kerr*-type exchange equation. Kerr used ion concentrations in place of ion activities, thus tacitly assuming concentrations and activities to be directly proportional. Nevertheless, the equation often holds fairly well over narrow concentration ranges. The activity coefficient of the divalent cation is more concentration dependent than that of the monovalent cation, but the monovalent cation activity coefficient is squared, offsetting much of this variation. The ion activity ratio in the soil solution, therefore, may be roughly proportional to the concentration ratios in the soil solution over an appreciable range.

Ion activities in aqueous solution can be estimated from Debye–Hueckel theory (Eq. 3.16) or approximated by measurements with specific-ion electrodes. The more difficult problem of estimating the activities of adsorbed cations is still unanswered. Different assumptions for estimating the activities of exchangeable ions have resulted in the several cation exchange equations that are commonly used for exchange between ions of different valence. All of these exchange equations reduce to the Kerr equation (8.9) when ions are the same valence. The goal of each equation is to provide a relatively uniform exchange "constant" (more correctly, *exchange coefficient*) over a wide range of exchangeable cation compositions. The difficulty of this, even for ions of the same valence, is apparent from the Kerr equation "constants" plotted in Fig. 8.2. The exchange coefficient is approximately constant only over a limited concentration range for Na–K exchange and over even more restricted ranges for Na–Rb exchange. Such variability limits the practical usefulness of most exchange equations to relatively small ranges.

The *Gapon* equation, proposed in 1933, has found considerable use:

$$\frac{[NaX]}{[Ca_{1/2}X]} = k_G \frac{[Na]}{[Ca]^{1/2}} \qquad (8.10)$$

where exchangeable-cation concentrations are in mmoles of charge per gram (or kilogram), and soluble-cation concentrations are in millimoles (or moles) per liter.

The Gapon equation also uses concentrations rather than activities for the soluble ions, and writes the mass action equation with chemically equivalent quantities both

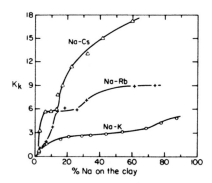

**FIGURE 8.2.** Selectivity coefficient $K_k$ versus exchangeable ion composition for an attapulgite clay. (From C. E. Marshall and G. Garcia. 1959. *J. Phys. Chem.* **63**:1663.)

for colloid exchange sites and exchanging cations. The Gapon equation corresponds to the following chemical reaction:

$$(Ca)_{1/2}X + Na^+ = NaX + \tfrac{1}{2}Ca^{2+} \tag{8.11}$$

It differs from the square root of the Kerr equation by including the term $[Ca_{1/2}X]$ rather than $[CaX]^{1/2}$. As was discussed for Eq. 8.9, the successful use of concentration ratios in solution, instead of activity ratios, is fortuitous over fairly narrow (though important) ranges of soluble-ion composition. One example, Na–Ca exchange, is important in irrigated regions. Dispersion and physical deterioration of many soils occur if exchangeable Na becomes too high. The Gapon equation is unsatisfactory if applied over the entire range of Na–Ca compositions, but works fairly well over the range of compositions of most interest to irrigated agriculture. The Gapon exchange coefficient is fairly uniform from 0 to 40% exchangeable sodium for many irrigated soils of the western United States, at $k_G = 0.010$ to $0.015$ $(Lmmol^{-1})^{1/2}$.

Other cation-exchange equations are discussed in Appendix 8.2. Most of these have a better theoretical basis than the Gapon equation, but few are as simple to apply or visualize. Many workers are willing to sacrifice a little theoretical rigor to gain some simplicity in a cation-exchange equation. The Gapon and the Kerr equations are the simplest ion-exchange equations. Gapon adequately predicts cation-exchange behavior over practical ranges for many soil systems.

### 8.1.3  Diffuse Double Layer

In air-dry soils, the exchangeable ions can be considered to reside directly on the surface of the colloid (Fig. 8.3). The negative charges of soil clays and the layer of exchangeable cations make up two slightly separated layers, called a *Helmholtz double layer*. When water is present, however, the cations are no longer so tightly held on the surface. The electrostatic attraction of cations is counteracted somewhat by diffusion into the aqueous solution. Diffusion tries to equalize the concentration throughout the aqueous phase. Figure 8.4 shows the net result of electrostatic attraction versus diffusion of cations at two bulk solution salt concentrations, ignoring any anion affects. The cation concentration decreases with distance from the negatively charged surface. The colloid's negative charge is neutralized by a swarm of positive ions in the aqueous phase, the diffuse double layer (DDL).

Increasing the salt concentration from $C_1$ to $C_2$ reduces the tendency for diffusion away from the surface and thus shrinks the DDL. The thickness of the DDL is loosely defined as the distance over which the solution concentration is affected by the colloid's charge. The solution outside the DDL is termed the *bulk solution*.

Anion repulsion within the DDL also neutralizes the colloid's negative charge by increasing the net positive charge within the DDL. Figure 8.5 shows how anions are repelled by the colloid's charge. Assuming for the moment that cations do not affect the anion distribution, the anion concentration outside the DDL, $C_3$, is then higher than if no repulsion occurred, $C_0$. Since ion charges in the bulk solution must

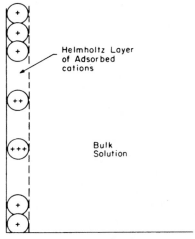

**FIGURE 8.3.** Distribution of monovalent cations and anions near the surface of a typical montmorillonite particle according to the Helmholtz model. (Adapted from D. R. Neilsen et al. 1972. *Soil Water*, p. 45, by permission of the American Society of Agronomy and Soil Science Society of America.)

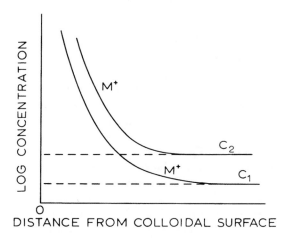

**FIGURE 8.4.** The distribution of cations away from a negatively charged soil surface at two cation concentrations, with effects of anions disregarded. The cation exchange capacity is proportional to the area between the curves and their corresponding dashed lines.

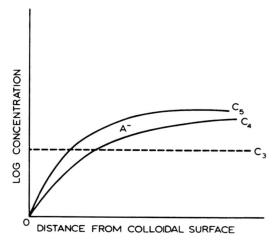

**FIGURE 8.5.** Distribution of anions near a negatively charged surface at two anion concentrations, $C_4$ and $C_5$, disregarding cation effects.

balance, this means that anion repulsion increases the salt concentration in the bulk solution. Increasing the anion concentration to $C_4$ shrinks the DDL, as it did for cation attraction, though in this case the effect arises from increased anion diffusion toward the surface. The effect of anion repulsion is mostly interesting in the laboratory, because the contribution of anion repulsion to soil behavior is small except when bulk solution salt concentrations are $>1$ M, where few organisms can survive.

Figure 8.6 shows the combined result of cation attraction, anion repulsion, and ion diffusion on the cation and anion distribution next to a negatively charged particle. The solution near the surface has an excess of cations and a deficit of anions. The thickness of the DDL decreases with increasing cation or anion charge. The effect of anion charge is less significant, because fewer anions are in the DDL. The DDL shrinks with increasing cation charge because fewer ions are necessary for charge neutralization and the more highly charged cations are attracted more strongly to the colloid. Diffusion, on the other hand, results from ion concentrations rather than ion charge.

Table 8.4 gives values for an arbitrary double-layer "thickness" for monovalent and divalent cations at three bulk solution salt concentrations. Double-layer thickness varies inversely with the square root of the bulk solution salt concentration or the valence of the exchangeable cation. These thicknesses are small compared to the diameters of soil pores, which are on the order of 1000 to 50 000 nm, but they are of the same magnitude as water film thicknesses in relatively dry soils.

If the DDL contained only those cations necessary to neutralize the colloid charge, the anion concentration would be zero within the DDL. Because diffusion continually drives anions toward the colloid surface, however, the total negative charge within the DDL is that of the anions plus the colloid's charge. Cations within the DDL must neutralize both sources of negative charge. The cations that neutralize

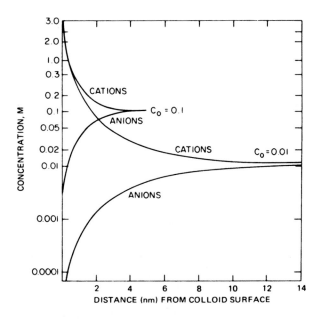

**FIGURE 8.6.** Distribution of monovalent cations and anions near the surface of a montmoril-lonite particle. (Adapted from D. R. Nielsen et al. 1972. *Soil Water*, p. 45, by permission of the American Society of Agronomy and the Soil Science Society of America.)

the colloid charge are "exchangeable"; the remainder are "soluble," because they neutralize the anions that have diffused into the DDL. The total positive charge in the DDL must exactly equal the total negative charge of that region. For positively charged colloids, the behavior of cations and anions in the DDL and bulk solution are reversed. Anions are attracted to the surfaces and cations are repelled.

**Table 8.4. Approximate "thickness" of a typical soil colloid double layer as a function of electrolyte concentration[a]**

| Bulk-Solution Concentration of Cations (mol charge $L^{-1}$) | "Thickness" of the Double Layer (nm) for: | |
| --- | --- | --- |
| | Monovalent Cations | Divalent Cations |
| $10^{-5}$ | 10 | 5 |
| $10^{-3}$ | 1 | 0.5 |
| $10^{-1}$ | 0.1 | 0.005 |

[a] Adapted from H. Van Olphen. 1963. *An Introducton to Clay Colloid Chemistry*. Interscience, New York. Reprinted by permission of John Wiley & Sons, Inc.

## 8.2  STRONGLY-RETAINED CATIONS

The strongly retained cations in soils include many of the essential microelements and also the "toxic" cations. The concentrations of these ions in the soil solution are low and they are apparently retained by two means. One group is the cations that in aqueous solutions precipitate as insoluble oxides and hydroxyoxides. The root zone of a typical agricultural soil might contain as much as 300 000 kg ha$^{-1}$ of Fe and Al, but their plant availability is only a few kg ha$^{-1}$.

The second and smaller group is those cations that also are insoluble in pure aqueous solutions, but that tend to associate in soils with soil organic matter and sulfide. These cations are the weak and intermediate Lewis acids of Table 3-7—$Cu^{1-2+}$, $Cd^{2+}$, $Hg^{1-2+}$, $Ni^{2+}$, $Zn^{2+}$, and $Fe^{2+}$. Soil retains these cations by hydroxyoxide precipitation/adsorption and by soil organic matter adsorption in varying degrees. The amounts of the weak Lewis acids in soils are small; the root zone contains on average about 300 kg ha$^{-1}$ of Pb and $<1$ kg ha$^{-1}$ of Cd and Hg but the variation is large. The amounts of these ions absorbed by plants is a tiny fraction of the total.

Soil retention of strongly retained ions generally increases with pH. Above pH 7, the effect of increasing pH on ion movement, plant availability, and chemical extractability lessens. Molybdenum is an exception: $MoO^{2-}$ reacts strongly with $Ca^{2+}$ and precipitates at higher pHs. If these ions are added to soils, as in municipal and industrial wastes, contaminated water, fly ash, and so on, most tests have shown that the amounts retained and the strength of that retention increases with time. The amounts retained and the strength of retention increase rapidly at first; the rate then slows over periods ranging from days to months.

The generalization that heavy metal retention increases with time may be wrong for weak Lewis acids applied in organic wastes to soils. One recent experiment with Mo, Zn, and Cd added to soils in sewage sludge showed that their plant availability remained unchanged for many years after application. The plant uptake of these ions was linearly proportional to soil content even after 23 years since the last addition. The amount of Zn available was 0.5 to 3% of that added initially. The amount of available Cd was 4 to 18% of that added. The reasons for the long-term and high availability in this case is uncertain.

### 8.2.1  Oxide-Retained Cations

The relative retention of divalent ions by amorphous Fe hydroxyoxides is

$$Pb > Cu > Zn > Ni > Cd \geq Co > Sr > Mg$$

Retention of the divalent cations by Al hydroxide is slightly different:

$$Cu > Pb > Zn > Ni > Co \geq Cd > Mg > Sr$$

Retention by silica is

$$Pb > Cu > Co > Zn > Ni = Cd > Sr > Mg$$

Although differing in detail, the retention by these three major soil components is rather similar. The measurements were done in the absence of organic matter or sulfide ions, so only the attraction to $O^{2-}$-dominated surfaces was determined.

The retention by silicate and hydroxyoxide minerals and surfaces is by $O^{2-}$ ions. Since the oxide ion interacts strongly with $H^+$, the cation retention is strongly pH dependent. Figure 8.7 illustrates both the pH and time dependence of the retention of $Cd^{2+}$ by Fe hydroxyoxide. At pH $< 6$, the amount of $Co^{2+}$ retained by Fe hydroxyoxide increased with pH and time. At pH $> 6.5$, $Co^{2+}$ retention was complete within 2 weeks.

The initial retention shown by the black triangles is low as acidity decreased to pH 7, whereupon retention increased sharply, the so-called *adsorption edge*. When the solid was treated later with acid, the amount of $Co^{2+}$ released back to the solution was less than the amount added, and the release decreased as time of aging the Co–Fe hydroxyoxide mixture increased. Other ions react similarly with silicates and hydroxyoxides. The pH of the adsorption edge and the effect of time differ somewhat, but the behavior is rather general. Plant deficiencies of these essential microelements, such as Fe chlorosis and Zn deficiency, may occur above soil pH 8. The deficiencies have been noted in fruit and nut trees and in some varieties of sorghum. The deficiencies are in arid regions under irrigation in which the faster growth puts a greater stress on soil availability. Native plants growing under native conditions do not show deficiency symptoms. Plant deficiencies are more complicated than simple soil unavailability, but the deficiencies can usually be overcome by acidifying the soil.

Conversely, when the concentrations of these cations are too high—Al toxicity in acid soils or Fe and Mn toxicity in rice paddies—raising the soil pH by liming is ef-

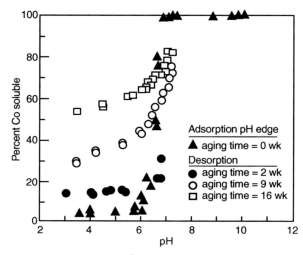

**FIGURE 8.7.** Fractional adsorption of $Co^{2+}$ to hydrous Fe-oxide (HFO) as a function of pH and HFO-$Co^{2+}$ aging time. (From C. C. Ainsworth, J. L. Pilou, P. L. Gassman, and W. G. vander Aluys. 1994. *Soil Sci. Soc. Am. J.* **58**:1615.)

fective. Since $O^{2-}$ ligands are so prevalent in soils, they can retain/adsorb/precipitate large amounts of these ions if pH > 6. Changing the soil pH or foliar spraying to circumvent the soils is generally more effective than adding these microelements as soil fertilizers. That also means that soil pollution by these ions is readily overcome by increasing the soil pH. The ions are immobile and unavailable for plant or microbial uptake. Liming contaminated soils, covering them with clean soil, or mixing them with clean soil underneath, is often sufficient to overcome any hazards. Regulatory agencies realize that the mere presence of an ion in soils does not necessarily constitute a hazard to humans. Soils have a high capacity to reduce high additions of these elements to their plant availability in native soils.

The retention of this group increases, that is, mobility and plant availability decreases, rapidly at first and the rate slows with time. This behavior is similar to diffusion and suggests that the mechanism is the slow transfer of surface ions into the weathered layer on soil particles. The initial rate is rapid because the surface concentration is relatively high and the diffusion path length is short. As diffusion inward proceeds, the surface concentration decreases and the diffusion path length increases. The shallower concentration gradient slows the rate of further cation diffusion.

Although the cations retained primarily by aluminosilicates and hydroxyoxides in soils, that is, the hard Lewis acids, are controlled by interaction with $O^{2-}$ and $OH^-$ ligands on soil particle surfaces, the aqueous solubility of these cations is usually much less than that predicted by the solubility products of their pure hydroxyoxides. The big differences between solubility products and ion activity products (IAP) indicate that soils retain the cations more strongly than their own pure hydroxyoxides. How soils can accomplish this is still uncertain. Although the cations fit better into their own hydroxyoxide structures than into aluminosilicate or major Al, Fe, Mn, and Ti hydroxyoxide structures, soils nonetheless retain the ions very strongly.

One school of thought maintains that certain "sites" on soil surfaces can retain these cations strongly; radiographs show the cations are both bunched and spread out on soil surfaces. Another school suggests that these adsorption sites are where the cations can mix as solid solutions with the other ions on the surfaces. The free energy of mixing on the surfaces (Appendix 3.2) is responsible for the strong retention rather than any uniquely favorable adsorption spots on soil particle surfaces. In any case, soil retention can reduce the aqueous solubility of these ions to well below that in equilibrium with their pure hydroxyoxides.

The models that predict soil solution concentrations by their oxide/hydroxyoxide or other solubility products ($K_{sp}$) have usually been only qualitatively successful. Predictions of aquated Al and Fe concentrations have sometimes been successful, probably because their hydroxyoxides were a major component of the solid phase and therefore unaffected by the effects of solid solution mixing. For trace metals, the models have been less successful. In some cases the measured IAP has been similar to the $K_{sp}$ of a pure solid that has not yet been identified in soils. Pb solubility in the soil solution has been linked to the solubility of Pb phosphate, for example. This must still be tested by measuring the Pb response to added phosphate to see if the agreement of IAP and $K_{sp}$ is coincidental rather than causal.

Another approach is simply to statistically relate the ion concentration in the soil solution to the total amount in the soil. In 100 British soils, the Pb concentration in the extracted soil solution closely followed the following equation:

$$\log Pb_{total} - \log Pb_{soilsolution} = 1.30 + 0.55pH \tag{8.12}$$

In 30 British soils, the soil Cd concentrations followed

$$\log Cd_{total} - 1.09Cd_{soilsolution} = 1.11 + 0.38pH \tag{8.13}$$

Such relations depend on the analytical procedure employed. The relations do not support one retention mechanism over another but are useful for regulatory purposes. The soil solution concentrations ranged from 3.6 to 3600 $\mu$g Pb $L^{-1}$ and 2.7 to 1280 $\mu$g Cd $L^{-1}$. The amounts in the extracted solutions were less than 1% of the total amounts in the soils, indicating how effective soils are in reducing availability of these toxic ions. Although several retention mechanisms may be involved for Pb and Cd, all the mechanisms apparently respond to pH. Neither equation supports the generalization of increasing retention with time.

### 8.2.2 Cations Retained by Soil Organic Matter

The soft Lewis acids (Section 3.5)—Cd, Cu, Zn, Hg, $Pb^{2+}$, and to a lesser extent $Fe^{2+}$ and $Mn^{2+}$—tend to be more affected by soil organic matter and sulfide. Some generalities about these cations are the following:

1. Soil retention is less pH sensitive than for hard Lewis acids.
2. Retention may be less time dependent.
3. Fraction retained, and strength of retention, decreases as concentration increases.
4. The ion activity products of the cation's hydroxyoxide are less than their hydroxyoxide solubility products.

Although soft Lewis bases are associated with reducing (anaerobic, low oxygen) conditions, normal (aerobic) soils contain sufficient amounts of organic matter to retain the low amounts of these trace metal ions in soils.

The soft Lewis acids and bases are among the very toxic ions. The soft Lewis acid–base situation that has created considerable concern is in municipal and industrial landfills. The worry is that the strong reducing conditions will create soluble soft Lewis acid–base ions, complex ions and molecules that will leach out of the landfills and into groundwater. To counteract this problem, clay and plastic liners are being installed beneath landfills and water-impermeable caps are being placed above the landfills to entomb the wastes. The slow migration of ions beneath natural swamps, another strongly reducing condition, suggests that the severity of the problem may be exaggerated. The problem, however, has made people more aware of proper and improper use of soil as a disposal/recycling medium. The idea of entombing our wastes in "secure" landfills seems naive; wise treatment would seem to be a better alternative.

## APPENDIX 8.1   DIFFUSE DOUBLE-LAYER THEORY

The *Guoy–Chapman* theory, derived concurrently by them in the early 1900s, is the basis of describing the DDL on charged colloid surfaces. Their assumptions are similar to those used later and more successfully by Debye and Hueckel to describe ion activities in the much simpler case of aqueous solutions. Gouy–Chapman theory assumes that (1) exchangeable cations exist as point charges, (2) colloid surfaces are planar and essentially infinite in extent, and (3) surface charge is distributed uniformly over the entire colloid surface. These assumptions inaccurately describe actual systems, but the theory of the DDL works surprisingly well for soil colloids. Apparently, many of the errors inherent in the assumptions tend to cancel each other.

Cations are attracted toward, and anions are repelled from, negatively charged soil colloids. Such interactions follow Coulomb's law:

$$F = \frac{qq'K}{Dr^2} \tag{8.14}$$

where $F$ is the force of attraction or repulsion (newtons), $q$ and $q'$ are the electrical charges (coulombs), $K$ is a proportionality constant ($= 8.9 \times 10^9$ for these SI units), $r$ is the distance of charge separation (meters), and $D$ is the dielectric constant ($= 78$ for water at 25° C). The strength of ion retention or repulsion increases with increasing ion charge, with increasing colloid charge, and with decreasing distance between the colloid surface and either the source of charge or the soluble ion.

The increased cation concentration in the DDL develops a countertendency for cation diffusion away from the surface. The diffusion tends to equalize cation concentrations throughout the solution phase. Combining the equations for cation attraction and diffusion yields the Boltzmann equation:

$$\frac{C}{C_0} = \exp\left(-Ze\frac{\psi}{kT}\right) \tag{8.15}$$

where $C$ is the concentration of an ion at a specified distance from the charged surface, $C_0$ is the concentration of the ion in the bulk solution, $Z$ is the valence of the ion, $e$ is the unit of electronic charge, $\psi$ is the electrical potential of the colloid at the specified distance, $k$ is the Boltzmann constant (the gas constant per molecule), and $T$ is the absolute temperature. Equation 8.15 describes the distributions of both cations and anions in the double layer, provided that $\psi$ is made negative because of the net negative charge of most soil colloids.

The treatment of double-layer phenomena is straightforward when the change of electric potential with distance from the colloid surface can be adequately estimated. This distribution can be considered to arise from the termination of individual lines of force from the colloid when they encounter cations in the double layer. Solutions of the equations describing electric potential distribution are, however, mathematically complex.

The DDL is generally treated quantitatively in either of two ways. The more standard approach is to regard soil colloids as having constant surface charge but variable surface potential (constant charge colloids). The distribution of potential thus varies with bulk solution salt concentration and with average valence of the counter (exchangeable) ions, but the charge (CEC) of the soil remains constant.

A second approach is to treat the colloid as having constant surface potential but variable surface charge (*constant potential colloids*). This behavior is common for colloids such as gold sols or glass surfaces and for soils having predominantly pH-dependent charge. In most soils, however, the main potential-determining ions are H and OH ions. Hence, the charge remains virtually constant as long as the pH is held constant, unless the salt concentration or the exchangeable cation composition changes markedly. Unlike for variable potential colloids, the charge of constant potential colloids varies appreciably and predictably with the salt concentration of the bulk solution. Adsorption of certain anions can also change the colloid charge.

Hydroxyoxide surfaces often appear to have concentration-dependent anion exchange capacities, but such behavior can also be explained by the collapse of the double layer at high salt concentrations. Under these conditions, positively charged sites are no longer masked by the DDLs of the predominantly negatively charged soil matrix.

Fully expanded double layers are rare in field soils. Double-layer expansion normally is restricted to thin water films on colloid surfaces or by interactions with double layers on adjacent soil particles within aggregates. Figure 8.8 represents such

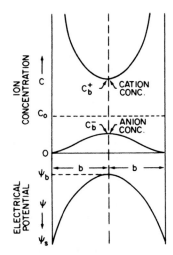

**FIGURE 8.8.** Electrical potential and ion concentration between interacting negatively charged platelets.

a restricted double layer. The distributions of cations, anions, and electrical potential in Fig. 8.8 are assumed to be symmetrical between two vertical colloid particles. The cation concentration decreases to $C_b^+$, and the anion concentration increases to $C_b^-$, the solute concentrations at the midplane between adjacent particles. Similarly, the electrical potential varies from $A_s$, the *surface potential* of the colloid, to $\psi_b$, the *midplane electric potential*. The excess of cations at the midplane compared to the bulk solution causes an osmotic gradient. This, in turn, causes water imbibition, or swelling, of the colloid. Water imbibition continues until the tendency to swell is balanced by interparticle bonds; that is, until the osmotic potential at the midplane equals that in the bulk solution, or until swelling is retarded by the lack of additional water. For most soils, obvious swelling of the entire matrix is uncommon, but it can be pronounced for highly montmorillonitic soils, such as Vertisols.

In theory, one can calculate the distribution of electric potential within the double layer for any combination of colloid charge, salt concentration, counter-ion valence and interparticle distance. The Boltzmann equation (8.15) can then be used to calculate cation and anion distributions. From such distributions, cation exchange, colloid swelling, and anion repulsion can be inferred but the calculations are complex, tedious, and often only approximate.

In practice, a set of curves developed by Kemper and Quirk (Fig. 8.10), yields approximate electric potentials as a function of distance from the colloid surface. Such potentials can then be substituted directly into the Boltzmann equation to infer cation and anion distributions. $Y_b$ is the scaled electric potential (equal to $-Ze\psi/kT$ of Eq. 8.15) at the midplane between interacting colloids, $\Gamma$ is the surface charge density in coulombs $m^{-2}$ (96.5 times the ratio of CEC, in mmoles charge $kg^{-1}$, divided by the specific surface, in $m^2$ $kg^{-1}$), $Z$ is the valence of the exchangeable cation, $C_0$ is the molar salt concentration in the bulk solution, and $x$ is the distance (in nm) from the midplane between colloids to the plane at which the ion concentration is to be calculated.

Figure 8.9 is designed for partially expanding (interacting) double layers. A value of $Y_b = 0.01$, however, is normally a satisfactory approximation for noninteracting colloid surfaces. To estimate the thickness of the water film, move across the figure from right to left on the line representing $\Gamma_s/(C_0)^{1/2}$ axis to obtain the distance from midplane to surface $(x)$. When the distance to the midplane is known and $Y_b$ is to be estimated, on the other hand, move from right to left across the figure on a $\Gamma_b/(C_0)^{1/2}$ line to the appropriate value of $Zx(C_0)^{1/2}$. Then estimate $Y_b$ by interpolating between $Y_b$ lines.

In either of the above cases, the intersection of horizontal and vertical lines, when projected horizontally onto the $Y$ axis, determines the surface potential $Y_s$. Values of the electric potential for the system can vary only between $Y_b$ and $Y_s$. The $Y_b$ line then yields values of $Y$ corresponding to selected $x$ values. The cation and anion concentrations can then be calculated as functions of distance from the colloid surface or from the midplane between interacting colloids by using these $Y$ values and Eq. 8.15. When estimating anion distributions from the Boltzmann equation, $Y$ must first be multiplied by $Z_{(-)}/Z_{(+)}$ to obtain values that decrease appropriately with proximity to the colloid surface.

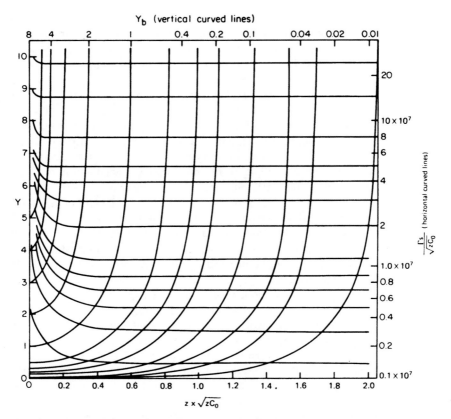

**FIGURE 8.9.** Scaled potential ($Y_b = Ze\psi/kT$) as a function of solution concentration and of distance from the midplane, charge density, and soil colloid surfaces. (From W. D. Kemper and W. P. Quirk. 1970. *Soil Sci. Soc. Am. Proc.* **34**:347–350.)

To derive Fig. 8.9, Kemper and Quirk assumed symmetric electrolytes (e.g., NaCl and $CaSO_4$, but not $CaCl_2$). Such an assumption is unreasonable for many soil solutions, but causes relatively minor errors in most cases. The effects of symmetric versus nonsymmetric electrolytes on electrical potential distribution are considerably less than the effects of ion valence, total salt concentration, or surface charge density.

*Stern* improved the Gouy–Chapman theory of the DDL by assuming that some ions are tightly retained immediately next to colloid surfaces in a layer of specifically adsorbed or Stern- layer cations. The double layer is diffuse beyond this layer. A satisfactory approximation of the Stern model can be made by assuming that the specifically adsorbed ions quantitatively reduce the surface density of the colloid. The diffuse portion of the double layer then is assumed to develop on a colloid surface of correspondingly reduced charge density. Sample Stern-modification calculations for a series of monovalent cations are shown in Fig. 8.10. Relatively few of the

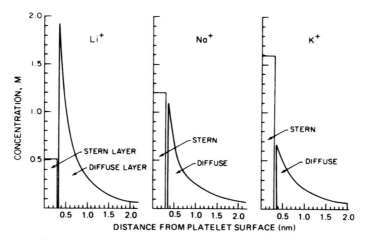

**FIGURE 8.10.** Calculated cation distributions near a mineral surface. (From I. Shainberg and W. D. Kemper. 1966. *Soil Sci. Soc. Am. Proc.* **30**:707–713.)

strongly hydrated lithium ions are strongly adsorbed in the Stern layer. Most lithium ions are in the diffuse layer instead. The opposite trend is evident for the weakly hydrated potassium ions. Shainberg and Kemper treated the implications and applications of the model, and consequences of its assumptions, during the mid 1960s.

## APPENDIX 8.2    CATION EXCHANGE EQUATIONS

As indicated in Section 8.4, the primary difference between various cation exchange equations is their differing treatment of the activities of exchangeable cations. Vanselow, for example, assumed that the activities of exchangeable cations were proportional to their mole fractions. This is equivalent to saying that ions on soil colloid surfaces behave as if in ideal solution (Appendix 3.2). The mole fraction of an ion in a binary system is

$$\text{mole fraction of species } a = \frac{n_a}{n_a + n_b} \tag{8.16}$$

where $n$ is the number of moles per unit volume or per unit mass. Substituting this assumption into a Kerr-type expression (Eq. 8.8), yields the *Vanselow–Argersinger* equation for monovalent–divalent cation exchange

$$\frac{[\text{NaX}]^2}{[\text{CaX}][\text{NaX} + \text{CaX}]} = K_v \frac{(\text{Na}^+)^2}{(\text{Ca}^{2+})} \tag{8.17}$$

Here brackets indicate exchangeable ion concentrations (mmol $\text{kg}^{-1}$) and parentheses denote soluble ion activities (mmol $\text{L}^{-1}$). Since the left-hand side of the equation is dimensionless, $K_V$ has units of L mmol$^{-1}$. The Vanselow–Argersinger equation has been used extensively to characterize cation exchange on simple, relatively uniform exchangers. In the surface chemistry literature, this equation has been used to calculate so-called *thermodynamic exchange constants*. However, these constants generally are simply averaged values of Vanselow coefficients over a range of exchangeable and soluble ion compositions. They rarely describe the ion distribution at any specific composition precisely.

*Davis* developed an equation similar to the Vanselow equation from statistical thermodynamics. Electrostatic forces between colloid surfaces and adsorbed cations were calculated for various surface configurations of charge sites. These sites were assumed to be neutralized by individual adsorbed ions. Hence, the model resembles most closely the Helmholtz model of the double layer with the charge of cations on the surface assumed to be just equal to the number of colloid charges. The resultant equation is

$$\frac{[NaX]^2}{[Ca][NaX + q_{Ca}CaX]} = K_d \frac{(Na^+)^2}{(Ca^{2+})} \tag{8.18}$$

where

$$q_{Ca} = Z_{Ca} - \frac{2Z_{Ca}}{Y} + \frac{2}{Y} \tag{8.19}$$

Here $Y$ is the number of nearest-neighbor (closest) charge sites, and $Z$ the cation valence. The main difference between the Davis equation (8.18) and the Vanselow equation (8.16) is the specific ion factor $q_i$ for the divalent cation. For monovalent ions, $q_i$ is unity.

Despite the entirely different theoretical bases of the Davis and Vanselow equations, each produces essentially the same expression for ion exchange. Values for the specific ion factor ($q_i$) include 1.0 for a linear array of cation exchange sites (two nearest-neighbor exchange sites), and 1.67 for a close-packed array (six nearest-neighbor sites) (Fig. 8.11). The linear array gives results that are numerically equal to the Vanselow equation. Krishnamoorthy and Overstreet tested several configurations and concluded that a large number of soils behaved as if the exchange sites were in

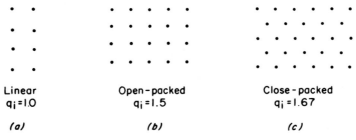

Linear  
$q_i = 1.0$  

*(a)*

Open-packed  
$q_i = 1.5$  

*(b)*

Close-packed  
$q_i = 1.67$  

*(c)*

**FIGURE 8.11.** Surface arrays of cation exchange sites used in the Davis (statistical thermodynamics) equation for cation exchange.

open-packed array ($q_i = 1.5$). Because Davis delayed publishing his work, Eq. 8.18 has sometimes been called the *Krishnamoorthy–Overstreet* equation instead.

Eriksson and Bolt used another approach to develop the Eriksson, or *double layer, exchange equation*:

$$\frac{\Gamma_1}{\Gamma} = \frac{r}{\Gamma\sqrt{\beta}}\, \sinh^{-1} \frac{\Gamma\sqrt{\beta}}{r + 4v_d\sqrt{M_2}} \qquad (8.20)$$

where $\Gamma$ is the colloid charge density (mmol($-$) m$^{-2}$); $\Gamma_1$ is the portion of the charge neutralized by the monovalent cation; $r$ is the reduced ratio ($= M_1/(M_2)^{1/2}$), where $M_1$ and $M_2$ are the activities of monovalent and divalent cations in the bulk solution; $\beta$ is a constant equal to $1.08 \times 10^{20}$ mmol$^{-1}$ for aqueous systems at 25° C; and $v_d = \cosh Y_b$ (the hyperbolic cosine of $\exp -Ze\psi/kT$) from Eq. 8.15 at the midplane between adjacent colloid particles). In practice, $v_d$ commonly is set equal to unity. This is equivalent to no interaction between adjacent particles or to infinite interparticle distance (realistically, to a particle separation of a few tens of nanometers).

An advantage of the double-layer equation is that it calculates ion distributions entirely from physically measurable parameters, such as CEC, surface area, and bulk solution solute concentrations. Disadvantages are that these measurements are rather involved, and that the inverse hyperbolic sine makes it difficult to visualize the effects and implications of changes on cation exchange. Generally, experimental surface charge densities for soils and clays must be multiplied by a factor of 1.2 to 1.4 to make experimental results and this equation's predictions agree satisfactorily. The need for this modification may result either from errors in the surface area estimates or because different portions of the surface affect the exchange capacity and surface area measurements differently. Nonetheless, Eq. 8.20 permits a wide variety of calculations almost entirely from measurable soil properties. The Kerr, Vanselow, and Davis equations, in contrast, require measurement of empirical exchange coefficients for different sets of experimental conditions to make such predictions.

Workers at the U.S. Salinity Laboratory substituted the sum of Ca plus Mg for the exchangeable and bulk-solution Ca concentrations in the Gapon equation (8.10). This yields

$$\frac{[NaX]}{[CaX + MgX]} = k\frac{[Na+]}{[Ca^{2+} + Mg^{2+}]^{1/2}} \qquad (8.21)$$

The $[Ca^{2+} + Mg^{2+}]$ term was necessary because many early water analyses did not distinguish between the two ions. The left side of Eq. 8.21 was termed the ESR (exchangeable sodium ratio). The solution concentration ratio on the right side was termed the SAR (sodium adsorption ratio). The SAR is written as $[Na^+]/([Ca^{2+} + Mg^{2+}]/2)^{1/2}$ when the concentration units are millimoles of charge per liter. The reduced ratio ($r$) of the double-layer exchange equation (8.19) is equal to the SAR divided by $(1000)^{1/2}$.

From analyses of saturation extracts and exchangeable cation concentrations for a large number of soils from the western United States, the statistical relation of ESR and SAR was found to be

$$ESR = -0.01 + 0.015 \ (SAR) \qquad (8.22)$$

This is equivalent to the Gapon equation, with an exchange constant of 0.015, except for the small negative intercept. For many applications, the intercept is negligible. Soils outside the principal irrigated portions of the western United States, such as the irrigated tropics, may have Gapon constants appreciably different from 0.015.

Another early approach to ion exchange was that of the *Donnan* equilibrium. This concept described a system in which a solution and suspension were separated by a membrane that is permeable to ions but impermeable to the exchanger or clay. An example is filter paper separating a soil suspension and its extract. In a "micro-Donnan" system, each soil colloid particle with its ion swarm is regarded as being separated from the bulk solution by an imaginary membrane. Basic Donnan equilibria apply to homovalent exchange in soils and can also be used to explain dilution effects during exchange between ions of different valence.

*Eriksson* applied Donnan equilibrium calculations to heterovalent exchange, reasoning that clays in a salt solution could be thought of as an ion species restricted from free diffusion. His equation was

$$\frac{(B)_i}{(A)_i^2} = \frac{(B)_o}{(A)_o^A} \qquad (8.23)$$

where i and o refer to ions inside the clay phase and outside (in the bulk solution). This is equivalent to a Kerr equation with $K_K = 1$. Concentrations were multiplied by activity coefficients for the solution phase, and calculated from the amount of ad-

**Table 8.5.  A comparison of exchange coefficients for several cation exchange equations, as calculated from the ammonium–calcium exchange data of Table 8.2**

| Equation | Exchange Coefficient at an Ammonium Concentration of: | | | |
| --- | --- | --- | --- | --- |
| | 0.04 M | 0.01 M | 0.005 M | 0.0025 M |
| Kerr (using concentrations) | $1.91 \times 10^{-3}$ | $1.81 \times 10^{-3}$ | $1.58 \times 10^{-3}$ | $2.30 \times 10^{-3}$ |
| Kerr (using solution activities) | $1.33 \times 10^{-3}$ | $1.51 \times 10^{-3}$ | $1.37 \times 10^{-3}$ | $2.07 \times 10^{-3}$ |
| Gapon (using concentrations) | 1.77 | 1.07 | 0.97 | 1.17 |
| Vanselow | 2.22 | 2.50 | 2.35 | 3.60 |
| Davis ($q_{ca} = 1.5$) | 1.75 | 1.88 | 1.73 | 2.63 |
| Davis ($q_{ca} = 1.67$) | 1.63 | 1.73 | 1.59 | 2.41 |
| | Exchangeable $NH_4^+$ Percentage | | | |
| Erikkson | | | | |
| Predicted[a] | 26.1 | 20.2 | 17.9 | 14.4 |
| Measured | 29.8 | 20.8 | 16.6 | 15.2 |

[a] Assuming a surface area of $800 \times 10^3$ m$^2$ kg$^{-1}$.

sorbed ions divided by the volume of exchanger for ions in the adsorbed phase. This fixed the activity coefficients for the adsorbed-phase ions as well. Today, Donnan equilibrium is used only as a first approximation to DDL theory. The main objection to Donnan theory is the large error involved in predicting concentrations at varying distances from the particle surface with a single ion ratio. Donnan theory predicts only an average activity ratio, which may err by a factor of two or more. Other criticisms have arisen from its inability to adequately predict the properties of clay suspensions.

Table 8.5 compares different exchange coefficients calculated from the data for ammonium–calcium exchange in Table 8.2. The simple Gapon equation (8.10) yields the most uniform exchange coefficient for this set of data; the Eriksson equation's predictions also agree well with the measured values. Bond and Verburg (1997) applied the various ion equations to the more complicated case of ternary (Ca–K–Na). Their slight modifications of the 1918 work by Rothmund and Kornfeld yielded the most consistent exchange coefficients in their study. Snyder and Cavallaro (1997) applied a single-phase mixture approach to $NH_4^+$–$Ba^{2+}$–$La^{3+}$ exchange on clays.

## APPENDIX 8.3   DETERMINATION OF CATION EXCHANGE CAPACITY AND EXCHANGEABLE CATIONS

When cation exchange relations are measured, both the total and the relative quantities of the exchangeable cations are required. To determine the total quantity of exchangeable cations, the cation exchange reaction is normally forced toward completion by either of two approaches. In one, the soil sample is exhaustively leached with a solution that contains a replacement, or index, cation to be placed on all exchange sites. Ions removed by the leaching are then analyzed to determine the initial exchangeable cation composition. A second approach involves repeated batch washings (several cycles of adding replacement cation, shaking, centrifuging, and decanting the supernatant solution). Analysis of the combined supernatant solutions yields the amounts of exchangeable ions initially present. The large excess of replacing ion in each batch-washing step drives the reaction toward completion.

In determining the CEC, a soil saturated with a single index ion is washed free of soluble salts, often with alcohol to keep the soil flocculated and to prevent loss of the index cation by hydrolysis:

$$NaX + H_2O = HX + Na^+ + OH^- \tag{8.24}$$

where X represents the clay and $Na^+$ is the index cation. The $H^+$ clay is unstable and rapidly breaks down:

$$HX + Al(OH)_3 = Al(OH)_2X + H_2O \tag{8.25}$$

Hydrolysis would yield a low estimate of the CEC. The index cation is then extracted from the soil with still another salt solution and measured to give the CEC.

The salt used to furnish the index cation should be relatively soluble in the alcohol used for sample washing. The low solubility of NaCl in ethanol has been a frequently overlooked source of error. The salt then dissolves during the subsequent extraction step and yields an anomalously high CEC value. Soils containing large quantities of hydroxyoxide or amorphous minerals may also retain salts in particle micropores, so that washing does not completely remove the salts. This also yields high CEC values.

To eliminate the problems associated with the washing step, Okazaki, Smith, and Moodie proposed a CEC procedure in which salts are not removed between the index cation saturation and extraction steps. Rather, the anion of the salt providing the index cation is analyzed in the final extract. In accordance with electrical neutrality, the CEC is then equal to the total quantity of index cations removed during extraction minus the quantity of index anions removed simultaneously. The main potential source of error from this procedure arises from anion repulsion, if the quantity of index salt remaining after saturation is merely calculated from the weight of solution retained and its initial (or average) concentration. This error minimized if the index solution is lowered to approximately 0.1 M during the final two saturation washes. The error is eliminated if the quantities of index salt are analytically determined instead.

The concentration of the index salt solution should not be high. Early measurements using 1 M salt solutions, to insure complete replacement and flocculation, yielded low CEC measurements because anion repulsion is significant at these concentrations and neutralized a significant portion of the colloid's charge. The CEC measurement is one of many examples in soil chemistry of the complexity of a seemingly simple experiment.

## BIBLIOGRAPHY

Bond, W. J., and K. Verburg. 1997. Comparison of methods for predicting ternary exchange from binary isotherms. *Soil Sci. Soc. Am. J.* **61**:444–454.

McBride, M. B. 1990. Reactions controlling heavy metal solubility in soils. *Adv. Agron.* **10**:1–59.

Snyder, V. A., and Cavallaro, N. 1997. The thermodynamic theory of ion exchange: a single-phase mixture formulation. *Soil Sci. Soc. Am. J.* **61**:36–43.

Thompson, H. S. 1850. On the absorbent power of soils. *J. R. Agr. Soc.* **11**:6874.

## QUESTIONS AND PROBLEMS

**1.** The following distribution of cations and anions exists near a soil colloid surface:

| Distance | 4.0 nm | 3.0 nm | 2.0 nm | 1.0 nm | 0.5 nm | 0.25 nm |
|---|---|---|---|---|---|---|
| Cation concentration $(mol(+) L^{-1})$ | 0.10 | 0.12 | 0.17 | 0.35 | 1.0 | 2.0 |
| Anion concentration $(mol(-) L^{-1})$ | 0.10 | 0.08 | 0.06 | 0.04 | 0.01 | 0.00 |

Assuming that the excess of cations reported for each increment represents the entire increment (e.g., that the cation concentration is 2.0 mol charge $L^{-1}$ from the colloid surface to 0.375 nm from the surface, etc.), estimate the CEC for a colloid having $800 \times 10^3$ m$^2$ kg$^{-1}$ of reactive surface (Ans. = 12.0 mmol charge kg$^{-1}$).

2. Based on the data of Table 8.4, what proportion of the cross-sectional area of a cylindrical soil pore of radius 15 $\mu$m is influenced by the electric double layer if monovalent ions predominate at a salt concentration of $10^{-1}$ M?

3. If all water of an unsaturated soil at 20% water content is spread uniformly over $100 \times 10^3$ m$^2$ kg$^{-1}$ of reactive surface, what proportion of that water is influenced by the electric double layer for the chemical conditions specified in Problem 2?

4. A soil is equilibrated with a solution of SAR = 20. Based on the Gapon equation, what would be its equilibrium exchangeable sodium percentage (ESP)? If the soil had instead been equilibrated with the same solution diluted fivefold with salt-free water, what would have been the corresponding SAR and ESP values?

5. Generate a selectivity diagram similar to Figure 8.1 for two cations (A and B) having Kerr-type coefficients of

   (a) 0.5
   
   (b) 1.0
   
   (c) 2.0

6. A vermiculitic surface soil has the ability to fix 25 mmol kg$^{-1}$ of $K^+$ or $NH_4^+$. What rate of $(NH_4)_2SO_4$ or KCl fertilizer (in kg ha$^{-1}$) would be required to saturate this fixation capacity for a 30-cm depth of soil?

7. The CEC is being estimated by the Okazaki, Smith, and Moodie procedure. If 5 g of soil retain 3 g 0.1 M(+) index solution after centrifugation and decanting, and if the total index cation retained is subsequently determined to be 1.6 mmol(+), what is the CEC of the sample? Based on the anion distribution of Problem 1, what percentage error is contributed in this case by anion exclusion, if the soil has a reactive surface area of $200 \times 10^3$ m$^2$ kg$^{-1}$?

8. Based on layer lattice thickness estimates from Chapter 4, what is the relative attraction for a dehydrated $K^+$ ion residing directly on the mineral surface of a tetrahedrally substituted 2:1 mineral when compared to an octahedrally substituted 2:1 mineral?

9. For a mineral of CEC = 100 mmol(+) kg$^{-1}$ and surface area = $800 \times 10^3$ m$^2$ kg$^{-1}$, saturated with a monovalent cation at a salt concentration of 0.001 M, use Fig. 8.8 to estimate the distance from the mineral surface to the midplane if $Y_b = 0.01$, as well as the values of $Y (= Ze\psi/kT)$ at 0.5, 2, 5, 10, and 30 nm from the mineral surface.

10. Using the $Y$ values from Problem 9, calculate the corresponding cation and anion concentration at the specified distances from the mineral surface.

11. Verify the calculations of Table 8.5.

12. Explain in your own words the differences between the Helmholtz, Guoy–Chapman, and Stern models of the double layer.

13. Explain the valence dilution effect.

14. For a CEC procedure that uses $Na^+$ as the index cation, $H_2O$/ethanol as the wash solvent, and $Mg^{2+}$ as the displacing cation, discuss the effect of each of the following on CEC measurements:

    (a) Hydrolysis due to excess washing

    (b) Presence of large amounts of lime or gypsum in the soil

    (c) Incomplete index–cation saturation

    (d) Precipitation of an insoluble $Na^+$ salt in the ethanol

    (e) Incomplete removal of the index cation by $Mg^{2+}$

15. For a solution of SAR $= 40$ and a total salt concentration of $0.01$ mol(+) $L^{-1}$, calculate ESP if $k = 0.015$.

# 9

# ANION AND MOLECULAR RETENTION

Soil particles remove anions and molecules from the soil solution, and release others to the soil solution, in varying degrees. The mechanisms of retention and release are electrostatic and chemical bonding. The mechanism of ion retention is actually exchange but the $H^+$, $OH^-$, and other ions released are usually unnoticed. These retention mechanisms are combinations of adsorption, absorption, precipitation, and solid solution mixing; distinguishing between the mechanisms is difficult. Soil–anion interaction varies from slight repulsion to weak to very strong attraction and retention (Table 9.1).

Differences in retention between the groups is illustrated by Fig. 9.1. A soil naturally high in chloride and borate was leached successively with 40 pore volumes of water. (A pore volume of water is that amount which fills all the voids between the soil particles.) The $Cl^-$ concentration in the effluent was 1100 mM initially and decreased rapidly to almost zero after 5 pore volumes of water had passed through the soil. The $H_3BO_3/H_2BO_3^-$ concentration was about 15 mM initially and decreased much more slowly than $Cl^-$.

The leaching was stopped for 30 days and then was continued for 10 more pore volumes of water. The initial chloride concentration, the "rebound," was 0.5 mM

**Table 9.1. Anion and molecular interaction with soils**

| Repelled to Weakly Retained | Moderately Retained | Strongly Retained |
|---|---|---|
| $NO_3^-$, $SO_4^{2-}$, $SeO_4^{2-}$ | $H_3BO_3$, $H_2BO_3^-$, $F^-$ | $H_2PO_4^-$, $HPO_4^{2-}$, $H_2S$, $HS^-$ |
| $HCO_3^-$, $CO_3^{2-}$, $ClO_4^-$ | $CrO_4^{2-}$ | $H_2AsO_4^-$, $HAsO_4^{2-}$, $MoO_4^{2-}$ |
| $Cl^-$, $Br^-$, $I^-$ | | |

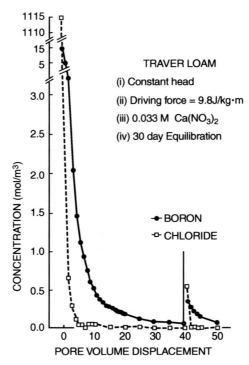

**FIGURE 9.1.** Boron and Cl⁻ concentrations in successive pore volume displacements (PVD) of the Traver loam soil. The solid vertical line at PVD = 40 indicates an intervening 30-day, saturated storage period. F. J. Peryea, F. T. Dingham and J. D. Rhoades. 1985. *Soil Sci. Soc. Am. J.* **49**:840.

and the borate concentration was 0.3 mM. The shape of the chloride and borate curves were the same as the first leaching sequence: Cl⁻ decreased rapidly again and borate decreased more slowly. The relative amount of borate that was released to the leaching solution during the 30-day incubation period was much greater than that of chloride.

Chloride was not retained by the soils. The chloride increase in the second leaching sequence was due to Cl⁻ diffusion from pores that were stagnant during the first leaching sequence. The borate increase, on the other hand, was due to a redistribution of available borate during the 30-day interval. Diffusion would account for only a tiny fraction of the second borate peak because the aqueous borate concentration was so much less than the aqueous chloride concentration. The second borate peak was due to slow release of borate ions from retention sites in the soil, a redistribution of borate to reestablish the "equilibrium" between strongly adsorbed and weakly adsorbed borate that was disturbed when the first leaching sequence depleted the weakly held fraction. The strongly adsorbed sites may have been deeper in the soil's weathered surface layer.

Strongly retained ions such as phosphate are released to the soil solution in the same way as borate but at lower concentrations and at slower rates. Weakly retained ions, in contrast, reach an equilibrium or steady-state concentration between the soil surface and the soil solution very quickly. When added to soil suspensions in the laboratory, phosphate in the soil solution decreases rapidly at first and continues to decrease over periods of weeks in the laboratory and weeks to months in the field. The laboratory reaction goes faster because the mixing and contact between soil particles and the soil solution is more complete.

Weak anion attraction and repulsion (anion exchange) in soils is primarily electrostatic and is similar to cation exchange. Anion exchange is rapid and reversible and the anion attraction is weak; chemical bonding is slower and stronger. Soils in Europe and North America are predominantly weakly to moderately weathered soils of neutral pH and have appreciable organic matter. In these soils the cation exchange capacity CEC greatly exceeds the anion exchange capacity (AEC). This preoccupation with cation exchange goes back to the initial studies by Thompson and Way in England in 1850. Soil chemistry has reflected this geographical bias. Anion exchange is important in Australia, New Zealand, and South Africa, where some soils are strongly weathered and have low organic matter contents and low pH; it occurs in European and North American soils, too.

In strongly weathered soils of low pH and low soil organic matter, and in soils derived from volcanic parent material, the AEC can equal or exceed the CEC (Table 9.2). The predominance of AEC or CEC can change from one stratum to the next in the same soil as the pH and composition of the soil strata change. Many of these high-AEC soils are coincidentally in the Southern Hemisphere, but anion exchange is also significant in acidic and highly weathered soils of the southeastern United States, in European forest soils, and in Japan.

One reason for the lesser interest in anion exchange may be that sulfate is the only macro-essential ion for plants that is retained to a significant extent as an exchangeable anion. Cation exchange, in contrast, covers four major cations: Ca, Mg, K, and Na. Each category in Table 9.1 covers a range of retention. Among the "weakly retained" anions, sulfate and probably selenate are retained the strongest by soils, because of their divalent charge. Nitrate, chloride, and perchlorate are retained the weakest. Nitrate and chloride are indeed considered to be repelled, rather than retained, in predominantly negatively charged soils because their retention is so weak. As in cation exchange, anion retention depends on the size and charge of the hydrated ion and on the ability of the ion to covalently bond with the soil surface.

The soil's retention of water-soluble cations and anions depends largely on colloid and ion charge. The aluminosilicate layer–lattice minerals tend to dominate the clay fraction of temperate and arid region soils. These minerals are predominantly negatively charged, so their physical adsorption of cations exceeds their adsorption of anions. Many anions of interest, however, are weakly soluble because they form strong chemical bonds with the cations in soil clays. These bonds can overcome the electrostatic repulsion of the negative charge and lead to strong soil retention.

Anion retention received little attention in North America until research on toxic wastes and on anionic pesticides demonstrated the importance of anion retention.

**Table 9.2. Charge characteristics of various soil orders**

| Soil Order | Horizon | pH$_{Salt}$ | Charge cmol$_c$ kg$^{-1}$ | |
|---|---|---|---|---|
| | | | Negative | Positive |
| Oxisol (Morais et al., 1976) | A | 3.0 | 3.9 | 3.2 |
| | A | 7.8 | 11.6 | 1.4 |
| | B | 3.0 | 2.5 | 5.1 |
| | B | 8.2 | 5.3 | 2.0 |
| Ultisol (Morais et al., 1976) | A | 3.0 | 1.0 | 1.1 |
| | A | 8.5 | 5.0 | 0.8 |
| | B | 3.0 | 2.5 | 2.4 |
| | B | 8.5 | 3.8 | 1.7 |
| Alfisol (Morais et al., 1976) | A | 2.9 | 3.8 | 2.0 |
| | A | 8.5 | 14.0 | −1.6 |
| | B | 2.9 | 2.4 | 5.5 |
| | B | 8.5 | 9.5 | 1.9 |
| Andisol (Sumner et al., 1993) | A | 4.6 | 3.8 | 2.6 |
| | B | 5.1 | 1.9 | 4.2 |
| Oxisol (Sumner, 1963b) | B | 3.5 | 2.3 | 5.1 |
| | B | 8.2 | 8.0 | 1.0 |

M. E. Sumner. 1998. *Future Prospects for Soil Chemistry*, Soil Science Society of America, Madison WI. Special Publication Number 55.

Strongly weathered soils contain Al and Fe(III) hydroxyoxides whose negative charge is low and whose positive charge can be relatively high, especially at low pH. When acidic, these *"variable charge"* soils can retain more anions than cations. The amorphous weathering products of volcanic soils of Japan also exhibit variable charge.

Molecular retention nowadays mostly refers to the widespread interest in the retention of organic pesticide molecules by soils. These molecules are mostly electrically neutral overall but have functional groups ($PO_4^{3-}$, $Cl^-$, $NO_2^-$, etc.) that interact with organic and inorganic soil solids. Positively and negatively charged molecules behave somewhat like simple cations or anions.

The soil's retention of uncharged molecules is largely independent of colloidal charge. For these substances, the soil is an inorganic matrix whose retention is based on the molecule's tendency to distribute between the gaseous, aqueous, and solid organic and inorganic soil phases. For organic molecules, this tendency depends on relative volatility, molecular weight, chemical composition, physical structure, solubility in the soil solution, the soil's organic matter content, and to some extent the soil's surface area. If the molecule also contains functional groups of inorganic nature, such as R—CO, R—COOH, R—CHO, R—PO$_4$, and R—NH$_2$, soil retention increases. The R—Cl unit of chlorinated hydrocarbons generally adds little to retention except by increasing the molecule's mass and nonvolatility.

The absorption of acid-forming gases ($SO_2$, $NO$, $NO_2$, $HF$, $HCl$) from the air increases with the molecule's water solubility and reactivity and increases with the soil's pH and base status. Weak bases such as $NH_3$ are retained more strongly by acid soils.

*Soluble silica* is ubiquitous in soil solutions, commonly at concentrations of about $10^{-4}$ M, or 2–5 mg Si $L^{-1}$, and is present as $SiOH_4$ rather than as an anion. It is less accurately described as silicic acid $H_4SiO_4$ because it is a very weak acid, $pK \sim 10^{-10}$. Soluble silica already saturates the soil sorption sites so little is removed from solutions flowing through soil. Indeed, the course of soil weathering is the slow release of soluble silica to the soil solution. The chemistry of silica is dominated by very slow reaction rates and by the presence of many forms of silica and aluminosilicate minerals of varying aqueous solubility. The solubility of quartz ($SiO_2$), chalcedony, chert, and other forms of $SiO_2$ is about 3 mg Si $L^{-1}$. Amorphous and hydrated opal $SiO_2 \cdot H_2O$ is soluble to the extent of 100 mg $L^{-1}$. The range of equilibrium aluminosilicate solubility is broader, but equilibrium and silicate solubility are rather incompatible terms. Particle size is at least as important a determinant of the soluble silica concentration in the soil solution.

The anions of concern to agriculture include $Cl^-$, $HCO_3^-$, $NO_3^-$, $SO_4^{2-}$, $HPO_4^{2-}$, $H_2PO_4^-$, $OH^-$, and $F^-$. In addition, some micronutrients ($H_2BO_3^-$, $MoO_4^{2-}$, and $HAsO_4^{2-}$) and heavy metals ($CrO_4^{2-}$) exist as anions in soils, as do some pesticides, such as the dissociated phenoxyacetic acids (2,4,5-T and 2,4-D). Molecular species of interest include $NH_3$, undissociated weak acids such as $H_3BO_3$ and $H_4SiO_4$, and the undissociated forms of many pesticides (DDT, 2,4,5-T, ant 2,4-D.) The study of anionic and molecular retention by soils has been the subject of increasing research in recent years.

The approach in this chapter is to describe various retention mechanisms and to cite examples of their involvement in the retention of specific anionic and molecular species. Several of the mechanisms are general and apply to many of the species listed above. Adsorption isotherms are also discussed because of their widespread use to describe anion and molecular retention by soils.

## 9.1  ANION EXCHANGE

Anions are attracted by positively charged sites on surfaces and repelled by negative charges. Layer silicates in the clay fraction of soils are mostly negatively charged so that anions tend to be slightly repelled electrostatically. Soils, however, contain solids, including the layer silicates, that also develop positive charges (often simultaneously though in different locations). An anion approaching soil solids may thus be simultaneously repelled by negatively charged aluminosilicate surfaces and attracted to positive charges on clay edges, hydrous oxides, and allophane.

If a dilute, neutral solution of KCl is added to dry montmorillonite, the equilibrium $Cl^-$ concentration in the bulk soil solution will be greater than the $Cl^-$ concentration in the solution originally added to the clay. This phenomenon is observed whenever a salt solution is added to a dry colloid having no adsorbing capacity for the anion at the prevailing pH. The process is called *anion repulsion*, or less accurately *negative adsorption*, and is due to anion repulsion from the diffuse double layer (DDL) surrounding charged colloid surfaces. An alternative explanation for the increased $Cl^-$ concentration in the bulk solution is hydration of the montmorillonite with $H_2O$, leaving less water for the salt.

Factors affecting anion repulsion include (1) anion charge and concentration, (2) species of exchangeable cation, (3) pH, (4) presence of other anions, and (5) nature and charge of the colloid surface. Ions commonly exhibiting net anion repulsion include $Cl^-$, $NO_3^-$, and $SO_4^{2-}$. Anion repulsion, as moles repelled per unit area of solid surface, increases with anion charge (valence). If the negative charge of a soil colloid surface remains constant and if no other reactions take place, anions of higher charge are repelled more than anions of lower charge. Mattson found that anion repulsion in a Na-montmorillonite suspension increased in the order: $Cl^- \approx NO_3^- < SO_4^{2-} < Fe(CN)_6^{4-}$. In soils that are dominated by Ca or other polyvalent cations, chemical reactions with the cations often change this purely electrostatic order. Increasing the anion concentration also increases the number of anions repelled, although the volume of the DDL from which the anions are excluded, the *exclusion volume*, decreases.

Anything that affects the DDL also affects anion repulsion. Thus, the $Cl^-$ exclusion volume of layer silicate suspensions increases in the order $Ba^{2+} < Ca^{2+} < K^+ < Na^+$. The multiply charged and more tightly adsorbed cations better neutralize the negative charge and produce a more condensed double layer, so that a smaller number of anions is excluded and the exclusion volume is less. Lowering the pH decreases the soil's net negative charge and increases the positive charge, so anion repulsion decreases with soil pH.

Anion repulsion also decreases when anions can be adsorbed by positively charged sites on soil colloids. Pretreatment of the colloids with highly charged and tightly adsorbed anions such as phosphate can mask the positive charges. These adsorbed anions present a negative surface to anions added later. Anion repulsion is then greater than in the absence of the tightly adsorbed anions.

The greater the negative charge of the soil solids, the greater the anion repulsion. Montmorillonitic soils thus exhibit greater anion repulsion than do kaolinitic soils at all pH values, and especially at low pH, where kaolinite can develop a positive charge. Anion repulsion can have important consequences during solute transport through soils. When anions are excluded from some of the volume of water surrounding soil particles, the anions can travel through the soil as a concentration bulge at the water front. The anions thus appear to travel faster through the soil than the water carrying them.

Anions approaching positively charged sites on layer silicate or hydrous oxide minerals are attracted electrostatically in the same manner as cations are attracted to negatively charged soil colloids. The effects of ion concentration, valence, and complementary ion on the distribution of exchangeable anions are similar to the effects described for cations (Chapter 8). Electrostatically retained anions are said to be *nonspecifically adsorbed*. Figure 9.2a illustrates the nonspecific adsorption of $Cl^-$. The dotted line shows electrostatic attraction of a positively charged mineral

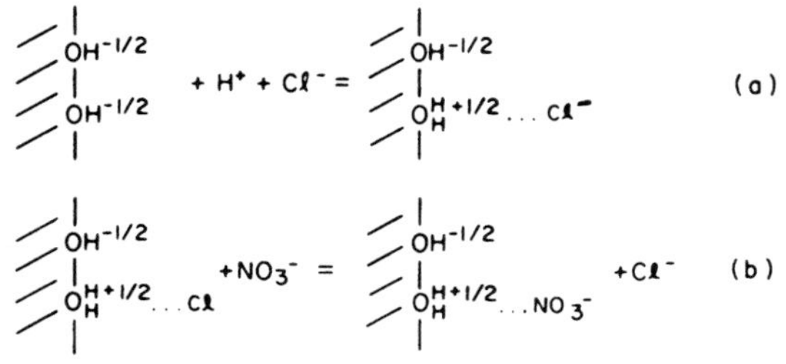

**FIGURE 9.2.** Nonspecific anion reactions at a solid/solution interface: (a) adsorption, (b) anion exchange. (After F. J. Hingston, R. J. Atkinson, A. M. Posner, and J. P. Quirk. 1967. *Nature* **215**:1459–1461.)

surface site for the anion. The positive charge in this case is the result of surface protonation, which increases with soil acidity. Figure 9.2b shows the exchange of one nonspecifically adsorbed anion ($NO_3^-$) for another ($Cl^-$). Exchange equations similar to those developed for cation exchange describe such reactions because non-specifically adsorbed anions are in the solution adjacent to the solid surface and are readily exchangeable.

The $Cl^-$, $NO_3^-$, and $SO_4^{2-}$ anions are considered to be nonspecifically adsorbed. Table 9.1 shows typical data for $Cl^-$ and $SO_4^{2-}$ adsorption by soils. The capacity of soils to adsorb anions increases with increasing acidity and is much greater for the kaolinitic soil, which has significant pH-dependent charge. At all pH values, the divalent $SO_4^{2-}$ ion is adsorbed to a greater extent than the monovalent $Cl^-$ ion, as would be expected on the basis of electrostatic attraction forces alone.

For the montmorillonitic soil, where pH-dependent charge and thus positive charge are of minor importance, $Cl^-$ is adsorbed only slightly at low pH and not at all in the slightly acid to neutral pH range. Such data are typical of nonspecif-ically adsorbed anions. Even for kaolinite, and for soils containing considerable pH-dependent charge, anion adsorption is negligible at pH > 7. The generally negative charge of pH > 7 soils repels nonspecifically adsorbed anions.

Chloride, nitrate, and sulfate are common and important anions in most soils and have been studied extensively. Chloride, in particular, is often used as an indicator of $NO_3^-$ mobility in soils, since $Cl^-$ is not subject to the complicating biological reactions characteristic of $NO_3^-$. In most other respects, $Cl^-$ behaves similarly to $NO_3^-$.

## 9.2  STRONG ANION RETENTION

Anions strongly retained by soils include $PO_4^{3-}$, $AsO_4^{3-}$, $MoO_4^{2-}$, $CrO_4^{2-}$, and $F^-$. These anions are essential microelements for plants and animals and are present in trace concentrations in the solutions of native soils. Because the amounts and tenacity of soil retention of these ions is so much greater than Cl, $NO_3$, and others, this retention has been misnamed as *specific adsorption*. These anions are simply water insoluble in the presence of the typical cations and colloids in soils. The state of these anions in the soil solution is a matter of great economic and environmental concern. Phosphate deficiency of agricultural crops is an ongoing global problem. A newer aspect is that these anions are being added to soils in fertilizers, agricultural wastes, fly ash from coal combustion, and municipal and industrial wastes. Initially, the additions increase the soil solution concentrations and plant availabilities of these anions. Since concentrations may reach levels that are appreciably greater than native levels, land disposal of such wastes has created public fear.

Some concerns about the safety of waste disposal on soils may be exaggerated. Within days to several weeks, the plant availability and movement of many ions decrease sharply and are nearly indistinguishable from the native concentrations, if the wastes are well mixed with the soil. The native concentrations of all ions vary widely from soil to soil yet their concentrations in plants and groundwater are low. If the wastes are distributed widely and the soil is given some time to react with the wastes, the probability of contaminated food or water is very low. Concentrating wastes in "hazardous waste landfills" where we try to isolate wastes from the environment is well intentioned, but prevents soil mixing and may be, in the long run, a counterproductive method of dealing with wastes.

The agricultural contribution to lake and stream contamination probably comes mostly from surface runoff of fertilized fields and from feed lots rather than from actual drainage water. Mixing, dilution, and time can mitigate soil contamination problems.

Anion removal from the soil solution is fast initially, but slows thereafter as the ions diffuse into the weathered and porous aqueous–solid phase on the surfaces of soil particles, increasing the diffusion pathlength. The diffusion also slows because the aqueous concentration, the driving force of the diffusion, is much less. Diffusion continues longer in the field than in the laboratory because the diffusion pathlengths from phosphate fertilizer granules to soil particles are much longer, and the water films on particle surfaces are thin. Phosphate retention in stirred suspensions in the laboratory reaches a steady state after several days. Phosphate from fertilizer granules in the field can release phosphate to plants for weeks before the phosphate is strongly adsorbed by the soil.

A hydrous oxide system is *amphoteric*; that is, its surface charge varies from negative to neutral to positive, depending on the pH of the aqueous solution. An electrostatic approach explains the exchange properties of hydrous oxides for chloride, sulfate, and other water-soluble anions. These surfaces can also interact chemically and strongly with weakly water-soluble anions. This gives Fe oxide- and Al oxide-dominated soils a much greater adsorption capacity for these anions than that predicted from electroneutrality alone, that is, greater than the quantity of adsorbed anions required to neutralize the surface positive charge. Indeed, iron oxides and other oxides scavenge (remove) arsenate, phosphate, molybdate, and other anions from solution with high efficiency.

Oxygen ions on a hydrous oxide surface can be replaced by oxyacid anions, such as phosphate, and by fluoride, which can enter into sixfold coordination with Al or Fe ions. This is known as *ligand exchange*, or *anion penetration*, for it takes place

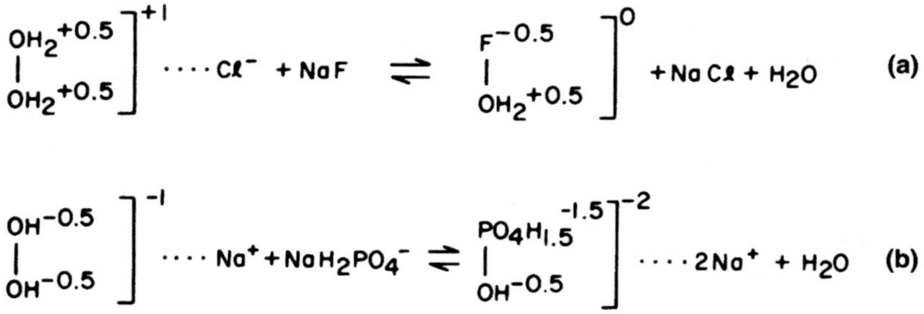

$$\left[\begin{array}{c}OH_2{}^{+0.5}\\ |\\ OH_2{}^{+0.5}\end{array}\right]^{+1}\cdots Cl^- + NaF \rightleftharpoons \left[\begin{array}{c}F^{-0.5}\\ |\\ OH_2{}^{+0.5}\end{array}\right]^{0} + NaCl + H_2O \qquad \textbf{(a)}$$

$$\left[\begin{array}{c}OH^{-0.5}\\ |\\ OH^{-0.5}\end{array}\right]^{-1}\cdots Na^+ + NaH_2PO_4^- \rightleftharpoons \left[\begin{array}{c}PO_4H_{1.5}{}^{-1.5}\\ |\\ OH^{-0.5}\end{array}\right]^{-2}\cdots 2Na^+ + H_2O \qquad \textbf{(b)}$$

**FIGURE 9.3.** Specific anion reactions at a solid/solution surface: (a) Neutralization of positive charge, and (b) ionization of a proton of an adsorbed acid anion. (After F. J. Hingston, R. J. Atkinson, A. M. Posner, and J. P. Quirk. 1963. *Trans. 9th Int. Cong. Soil Sci.* 1:669–677.)

within the crystal and renders the surfaces of oxides more negative. The negative charge arises when part of the liberated hydroxyl ions are neutralized by the formation of water (Fig. 9.3). Ligand exchange can occur on surfaces initially carrying a net negative, positive, or neutral charge. This contrasts with nonspecific anion adsorption, which occurs only when the surface carries a net positive charge. Ligand exchange may explain why weak-acid anions show maximum adsorption at pH values about equal to their p$K$ values (Fig. 9.4). At pH $=$ p$K$, both the amounts of

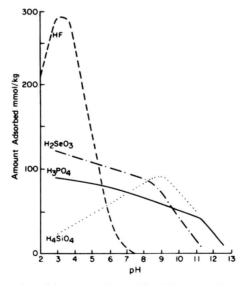

**FIGURE 9.4.** The adsorption of three oxyacids and fluoride on geothite as a function of pH. HF, p$K = 3.45$; $H_2SeO_3$, p$K_2 = 8.35$; $H_3PO_4$. p$K_1 = 2.12$, p$K_2 = 7.21$, p$K_3 = 12.67$; $H_4SiO_4$, p$K_1 = 9.66$. (After F. J. Hingston, R. J. Atkinson, A. M. Posner, and J. P. Quirk. 1968. *Trans. 9th Int. Cong. Soil Sci.* 1:669–677.)

anion (dissociated acid) available for ligand exchange and the amounts of undissociated acid capable of neutralizing liberated $OH^-$ are greatest.

Fluoride ion is moderately adsorbed by soil minerals. Figure 9.4 shows fluoride adsorption by goethite (FeOOH). Fluoride adsorption conforms to the ligand exchange theory and is probably favored by the close similarity in size of $F^-$ and $OH^-$ ions. In acid soils at equal anion concentrations, $F^-$ adsorption predominates over that of other common anions. This makes $F^-$ an effective desorbing agent for previously adsorbed anions.

### 9.2.1  Phosphate Reactions in Soils

Phosphate is probably the most important example of specifically adsorbed anions. Many soils fix large quantities of phosphate by converting readily soluble phosphate to forms less available to plants. In terms of ligand exchange or anion penetration theory, phosphate adsorption on oxide surfaces can be explained by Fig. 9.5. Phosphate replaces singly coordinated ("A-type") OH groups and then reorganizes into a very stable binuclear bridge between cations.

In the laboratory, phosphate adsorption by layer silicates is rapid for a few hours and then continues more slowly for weeks. The initial rapid reaction can be envisioned as a combination of nonspecific adsorption and ligand exchange on mineral edges. The slower reaction probably consists of a complex combination of mineral dissolution and precipitation of added phosphate with exchangeable cations or cations within the lattices.

Low and Black showed that phosphate retention by kaolinite increased with time and phosphate concentration. Silica concentrations in the bulk solution increased simultaneously. The reaction was thought to be a two-stage reaction:

$$Al_2Si_2O_5(OH)_4 + 4H^+ + 2H_2O = 2Al(OH)^{2+} + 2Si(OH)_4 \qquad (9.1)$$
$$\underset{\text{kaolinite}}{}$$

$$Al(OH)^{2+} + HPO_4^{2-} = AlPO_4 \cdot H_2O \qquad (9.2)$$
$$\underset{\text{variscite}}{}$$

with the Al phosphate precipitating on the surface or phosphate tetrahedra substituting for silicon tetrahedra. The more generally accepted explanation nowadays is the dissolution of $Al^{3+}$ as kaolinite breaks down, followed by the precipitation of

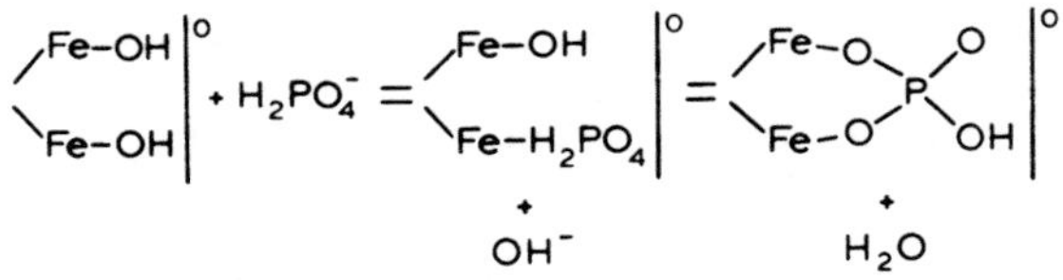

**FIGURE 9.5.** Representation of $H_2PO_4^-$ penetration into an iron oxide surface and subsequent formation of a stable binuclear bridge. (After J. C. Ryden, J. K. Syers, and F. R. Harris. 1973. *Adv. Agron.* **25**:1–45.)

Al phosphate. When hectorite (a 2:1 layer silicate in which $Mg^{2+}$ is the dominant octahedral cation rather than $Al^{3+}$) is substituted for kaolinite, much less phosphate is retained. This points out the importance of Al to phosphate retention. In acid soils, phosphate retention is often closely related to the amounts of extractable $Fe^{3+}$ and Al.

Phosphate forms weakly soluble $Fe^{3+}$ and $Al^{3+}$ compounds at low pH, more soluble $Ca^{2+}$ and $Mg^{2+}$ compounds at pH values near neutrality, and difficultly soluble $Ca^{2+}$ compounds at higher pH. Lindsay and Moreno developed a solubility diagram for phosphate in a system containing variscite ($AlPO_4 \cdot H_2O$), strengite ($FePO_4 \cdot H_2O$), fluoroapatite ($Ca_{10}(PO_4)_6F_2$), hydroxyapatite ($Ca_{10}(PO_4)_6(OH)_2$), octocalcium phosphate ($Ca_4H(PO_4)_3$), and dicalcium phosphate dihydrate ($CaHPO_4 \cdot 2H_2O$). The solubility diagram (Fig. 9.6) describes equilibrium phosphate precipitation reactions at various pH values, but slow kinetics have prevented the quantitative application of solubility data to soils. In addition, assumptions have to be made about the activities and the solid phases that control the activities of $Ca^{2+}$, $Fe^{3+}$, and $Al^{3+}$ in the soil solution. These assumptions limit the diagram to equilibrium, which can be approached but not reached in soils. Despite this shortcoming, Fig. 9.6 illustrates the changes in phosphate minerals as phosphate fertilizers slowly transform to less-soluble states in soils.

Figure 9.6 illustrates the relative stabilities of several phosphate compounds in soils of various pH values. Soil solution compositions can be plotted on the diagram by measuring soil pH and soluble phosphate concentrations. Data above a compound's isotherm represent supersaturation with respect to the solid, indicating that the compound will precipitate. Data below the isotherm indicate undersaturation of phosphate in the soil solution with respect to that compound, so that the solid, if

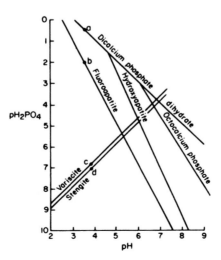

**FIGURE 9.6.** Solubility diagram for phosphorus compounds at 25° C and $5 \times 10^{-3}$ M calcium. (After W. L. Lindsay and E. C. Moreno. 1960. *Soil Sci. Soc. Am. Proc.* **24**:177–182.)

present, would dissolve. Intersections of two isotherms represent solutions in equilibrium with both solids.

Consider a soil of pH 4 to which soluble phosphate fertilizer is added, resulting in a phosphate potential, or $pH_2PO_4$ (= negative logarithm of the $H_2PO_4^-$ activity), of 0.5. This is point $a$, which falls on the dicalcium phosphate dihydrate isotherm and is at equilibrium with that solid phase. Point $a$ is above the isotherms for fluoroapatite, variscite, and strengite, indicating supersaturation of the aqueous solution with respect to these compounds. Fluoroapatite might tend to precipitate first, reducing phosphate levels in solution (and increasing the value of $pH_2PO_4$). Assuming constant soil pH, $pH_2PO_4$ will tend to increase to the equilibrium line of fluoroapatite ($pH_2PO_4 = 2.0$, point $b$). When $pH_2PO_4$ is greater than 0.5, the soil solution is undersaturated with respect to dicalcium phosphate dihydrate and it will dissolve, increasing the phosphate in solution once more. The phosphate concentration can be any value between $a$ and $b$, depending on the rate of $CaHPO_4$ dissolution versus that of $Ca_5F(PO_4)_3$ precipitation. This precipitation–dissolution reaction will continue until all of the dicalcium phosphate dihydrate dissolves. The soil solution is then represented by point $b$, in equilibrium with fluoroapatite. The least soluble (most stable) compounds indicated on the diagram at acid pH are variscite and strengite. The soil solution at point $b$ is highly supersaturated with respect to both of these compounds. As a result they should begin to precipitate immediately. The diagram predicts that both dicalcium phosphate dihydrate and fluoroapatite will eventually disappear to form variscite and/or strengite (points $c$ and $d$). Either transformation results in a substantial reduction in soluble phosphate concentrations compared to those of the initially fertilized soil (point $a$). Large quantities of phosphate should thus be fixed as insoluble $Fe^{3+}$ and Al phosphates in acid soils.

The diagram also indicates that phosphate should precipitate in basic soils as one of several Ca phosphates, the least soluble of which are hydroxy- and fluoroapatite. Variscite and strengite are too soluble to exist under basic conditions, and they should not form in basic soils. Both variscite and strengite, in fact, would be good phosphate fertilizers in alkaline soils because of their solubility in basic soils, if they were applied as finely ground materials. Calcium phosphate ore ("rock phosphate," mostly hydroxy- and fluoroapatite) is similarly effective in acid soils. Rock phosphate is treated with sulfuric acid to make "superphosphate," nominally $CaHPO_4$; treatment with phosphoric acid yields "triple superphosphate," nominally $Ca(H_2PO_4)_2$. Both superphosphate and triple superphosphate are more soluble than rock phosphate and make phosphate more immediately available when added to soils at any pH.

Figure 9.6 explains the observations that (1) phosphate is fixed in large amounts as iron and aluminum phosphates in acid soils, where $Fe^{3+}$ and $Al^{3+}$ activities are high; (2) calcium fixes phosphate similarly in basic soils, where $Ca^{2+}$ activity is high; and (3) maximum amounts of phosphate are available at slightly acid to neutral pH where the solubilities of Fe, Al, and Ca phosphates are highest simultaneously.

For most soils, the various mechanisms responsible for phosphate retention are nearly impossible to separate. Phosphate is retained by a multistage process, probably involving several of the mechanisms described above as well as other unknown reactions. Even carefully designed experiments are often confounded by reactions other than those intended to be studied. Precipitation is especially difficult to eliminate as a mechanism; all of the specific adsorption mechanisms can be viewed as modified precipitation reactions.

The mechanisms of phosphate retention by soil organic matter are not known, but are believed to be important in maintaining plant-adequate levels of phosphate in the soil solution. Inositol hexaphosphate and possibly other organic phosphate compounds apparently are retained in soils by precipitation reactions. Many common, water-soluble, organic phosphate compounds become nonextractable to water at almost the same rate as, and to the same extent as, dissolved inorganic phosphates. Thus, although organic phosphate is reported to leach from soils, a large proportion of it appears to move attached to particulate matter rather than as dissolved phosphate. Retention mechanisms for organic phosphate include (1) sorption through orthophosphate groups to Fe and Al oxides by mechanisms similar to those for inorganic phosphate, and (2) sorption by interaction of the organic portion of the phosphate ester with organic or inorganic soil components.

At current rates of fertilizer usage, we have about a 200-year supply of good-grade phosphate ore worldwide. The importance of phosphate fertilization to agriculture and the relatively short supply of phosphate ore have led to many, as yet unsuccessful, attempts to increase the low availability of the large amounts of phosphate present in almost all soils. These attempts include leaching the soil with silicate, which might replace phosphate that is strongly retained by Fe, Al, and aluminosilicates; creating polymeric phosphate fertilizers (pyro- and metaphosphates) instead of the normal ortho (monomeric) forms; and breeding plant varieties that can better utilize soil phosphate. To overcome the high cost of shipping phosphate per unit P and of spreading it on rugged terrain, New Zealanders went so far as to propose spreading elemental P from airplanes. White P, however, is a dangerous explosive and the safe, polymeric red and black P forms are too insoluble and unreactive in soils. Unlocking native phosphate would be an important step in achieving sustainable agriculture.

The "*breakthrough curves*" of Fig. 9.7 summarize the net effects of repulsion and specific adsorption on the relative adsorption of anions by soils. Solutions containing the anions at initial concentration $C_0$ were added to soil columns. The effluent concentration is $C$. The volume of water is expanded as pore volumes added to the

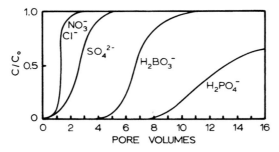

**FIGURE 9.7.** Representative breakthrough curves of anions weakly, moderately, and strongly retained by soils.

soil column. The $Cl^-$ and $NO_3^-$ solutions flowed through the soil columns almost as quickly as the water. The other anions were delayed because of soil adsorption. Sulfate and phosphate retention increased with iron and aluminum hydroxyoxide and allophone contents. The long-term capacity of most soils to adsorb phosphate is orders of magnitude greater than the amounts of phosphate added as fertilizer.

## 9.3   MOLECULAR RETENTION

A solute in water need not be initially charged to be retained by soils. Molecules in the soil solution can become charged and then be adsorbed as cations or anions. They may also remain nonionic and adsorb as a consequence of polarity that produces localized charge within the molecule.

Molecules such as $NH_3$, amino acids, and protein can protonate (add $H^+$) in acid solutions and be adsorbed as cations on negatively charged soil solids:

$$B + H^+ = BH^+ \tag{9.3}$$

where B is a weakly basic molecule. The tendency of a molecule to protonate is characterized by its $pK_a$:

$$K_a = \frac{(H^+)(B)}{(BH^+)} \qquad pK_a = pH + \log \frac{(BH^+)}{(B)} \tag{9.4}$$

The greater the $pK_a$ of a basic molecule, the greater is its tendency to protonate. Important molecules that protonate include the $s$-triazine and $s$-triazole herbicides and ammonia. Their $pK_a$ values are given in Table 9.3.

When soil pH > pK, weak-acid anions are adsorbed by positively charged sites on Fe and Al oxides or layer silicate edges. The weak acids (high $pK_a$ values) include

**Table 9.3.  $pK_a$ values of some molecular species**

| Species | $pK_a{}^{a, b}$ | Reaction |
|---|---|---|
| $s$-Trazine (atrazine) | 1.68 | $BH^+ = B + H^+$ |
| $s$-Trazole (amitrole) | 4.17 | $BH^+ = B + H^+$ |
| 2,4-D | 2.80 | $R\text{—}COOH = R\text{—}COO^- + H^+$ |
| 2,4,5-T | 3.46 | $R\text{—}COOH = R\text{—}COO^- + H^+$ |
| $H_2CO_3$ | 6.37 | $H_2CO_3 = HCO_3^- + H^+$ |
| $H_3BO_3$ | 9.14 | $H_3BO_3 = H_2BO_3^- + H^+$ |
| $NH_3$ | 9.26 | $NH_4^+ + OH^- = NH_3 + HOH$ |
| $H_4SiO_4$ | 9.66 | $H_4SiO_4 = H_3SiO_4^- + H^+$ |

[a] Organic $pK_a$ reproduced from *Pesticides in Soil and Water*, 1974, p. 47.

[b] Inorganic $pK_a$ reprinted with permission from *Handbook Chemistry and Physics*, 50th ed., 1969–1970. Copyright, The Chemical Rubber Co., CRC Press, Inc.

$H_3BO_3$ and $H_4SiO_4$ (or $Si(OH)_4$). These acids remain uncharged in most agricultural soils and do not form anionic or cationic bonds. Important weak acids that dissociate to form anionic bonds with soil include the phenoxyacetic acids (2,4-D and 2,4,5-T) and carbonic acid ($H_2CO_3$). Their $pK_a$ values are also given in Table 9.2.

Molecules that do not protonate or deprotonate to become charged species can still be adsorbed on soil by hydrogen bonding and van der Waals attraction. The *hydrogen bond* is a dipole–dipole interaction in which $H^+$ bridges between two electronegative atoms. The hydrogen is held by a weak electrostatic bond to one electronegative atom and by a stronger covalent bond to the other. The functional groups of the soil's solid phase that are capable of hydrogen bonding include the $O^{2-}$ on silicate surfaces, edge hydroxyls, and the carboxyl, hydroxyl, and amino groups of organic matter. Individual hydrogen bonds are relatively weak, but many polar molecules (particularly pesticides) have numerous sites capable of hydrogen bonding with soils, especially with soil organic matter. The summation of many hydrogen bonds results in strong retention of, for example, carbaryl and carbamate insecticides.

Many organic molecules, although uncharged and without apparent hydrogen bonding, are nonetheless strongly retained by soils. The intense interest in this phenomenon stems from possibility of movement of pesticides and other organic molecules in soils to groundwater. A less obvious phenomenon is the soil's adsorption of organic molecules from the atmosphere. Uncharged molecules have been adsorbed from the atmosphere and produced in the soil by organic decay since the earth was formed, yet the groundwater and atmosphere are remarkably free of them. This is partly due to the strong retention of organic molecules by soils, a second reason is the active degradation of organic substances by soil microorganisms.

Soil retention of uncharged molecules is often described as *van der Waals attraction*, which is a way of saying that the retention mechanism is unknown or poorly understood. In the 19th century, van der Waals modified the ideal gas law to account for the attraction between gas molecules, without knowing the nature of attraction. Charge-induced dipole interactions and dipole-induced dipole interactions are the forces thought to be involved. The van der Waals attractions are weak and short-ranged. They are additive, and each atom of the molecule and its adsorbent contribute to the total bond energy. Such forces operate in all adsorbent–adsorbate relationships but appear to be the principal forces of adsorption for nonpolar molecules such as DDT and $N_2$. The electrostatic forces of charged species overshadow van der Waals attractive forces.

Molecular retention involves no charges and therefore requires no strict 1–1 exchange between the soil and the soil solution. Ion retention requires exchange to maintain charge neutrality. The amount of molecular retention, however, is limited by the number of exposed sorption sites, or by the amount of sorbing surface and material, in the soil.

Percolating solutions containing organic molecules pass through an intricate network of soil pores. Organic molecules tend to be nonpolar and to prefer an environment less polar than that of the highly polar water. If some other less polar phase is present, such as soil surfaces and especially SOM surfaces, the organic molecules are in effect forced out of the aqueous phase onto organic-coated soil surfaces. The SOM also attracts organic molecules by providing a phase into which they can "dissolve" or form a solid solution. That action helps to purify contaminated water or gas flowing through the soil. The soil's adsorption capacity is small for organic molecules but is continually renewed by microbial decay of the adsorbed molecules.

The separation of organic molecules out of the soil solution onto the solid phase is called *partitioning*. The ratio of a molecule's concentrations in the water and SOM phases is a constant, the partition coefficient $K_D$:

$$K_D = \frac{\text{concentration}_{I(\text{soil})}}{\text{concentration}_{II(\text{soil solution})}} \tag{9.5}$$

and is a measure of the relative solubility of the molecule in both phases. For substances that are only slightly soluble in water, the values of $K_D$ are very large. Dibenzothiophene $((C_6H_5)_2C_4H_4S)$ is weakly water soluble and its soil/water $K_D$ value is 11 000. Passing dibenzothiophene-contaminated water through soil containing organic matter greatly depletes the water of this compound.

The extent to which an organic compound partitions out of water onto soil is determined by physical–chemical properties of both the soil and the compound. The soil's organic matter content is the single best characteristic for estimating the amount of soil adsorption of pesticides and other organic molecules. The partition, or sorption, coefficient of the organic molecule $K_{OC}$ (equal to $K_D/\text{SOM}$) is rather independent of soil type. This suggests that SOM is the principal soil component responsible for pesticide sorption and that the role of SOM is similar in different soils.

The $K_{OC}$ value is correlated to physical–chemical properties of the organic molecules. One such easily determined property is the partition coefficient $K_{OW}$ of a molecule between octanol and water replicates fairly well the partitioning between soil and the soil solution. The correlation of $K_{OW}$ to $K_{OC}$ in soils is

$$\log K_{OC} = -0.99 \log K_{OW} - 0.34 \tag{9.6}$$

The $K_{OC}$ is a first approximation of a pesticide's mobility in soil from readily available pesticide and soil properties.

Partitioning by this *hydrophobic adsorption* explains why soils retain organic molecules, but direct soil–organic interaction occurs also. Dry soils retain organic molecules more strongly and in greater amounts than do wet soils. Differences between organic molecules with respect to soil retention become more obvious when the competition with water for soil surfaces is absent. For a gas passing through a dry soil, the soil/gas partition coefficient of methane is about the same as that for dinitrogen ($N_2$) and helium. Methane flows as easily through soil as the unreactive $N_2$ and He molecules. The partition coefficient increases exponentially with molecular weight to about $10^5$ for gaseous octane (Fig. 9.8). Unsaturated double bonds and aromatic ring structures increase retention slightly, and alcohol, aldehyde, and acid functional groups in the gas molecule increase its soil retention greatly. Presumably, nitrogen, phosphate, and sulfur functional groups also increase retention. Those dis-

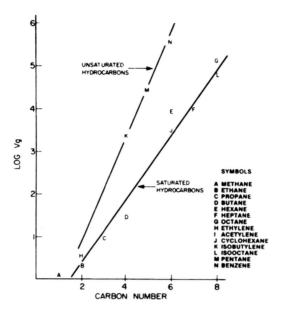

**FIGURE 9.8.** Retention of various hydrocarbons by a dry soil at 15° C. $V_g$, the retention volume, is closely related to the soil/gas partition coefficient $K_D$. (From H. L. Bohn et al. 1980. *J. Environ. Qual.* **4**:563.)

tinctions, however, are less obvious in the presence of water because it competes with the gases for adsorption sites. For more polar and water-soluble organic molecules, direct adsorption by inorganic soil surfaces and dissolution into the soil solution are also important.

The adsorption of a particular uncharged species can rarely be identified with only one mechanism, although the dominant mechanism can often be inferred. Thus, care should be taken in extrapolating data from, for example, a weakly basic herbicide to other weakly basic herbicides. The individual properties of the molecule, such as (1) chemical character, shape, and configuration, (2) acidity or basicity, (3) water solubility, (4) charge distribution, (5) polarity, (6) size, and (7) polarizability, all influence molecular adsorption by soil.

As with cations and anions, soil interaction with molecules happens only if the substance contacts soil particles. Surface spreading or burial is insufficient; the wastes must be mixed with the soil to react with it. The sensational Love Canal case, for example, involved the burial of organic liquids in 55-gallon (215-L) drums stacked in shallow trenches underground. Had the organic liquids been mixed and allowed to interact with the soil, the leakage and movement after the thin steel drums corroded might never have happened, and it certainly would have been less severe.

## 9.4   ADSORPTION ISOTHERMS

Adsorption isotherms describe solute adsorption by solids at constant temperature and pressure. An adsorption isotherm shows the amount of adsorbate sorbed as a function of its equilibrium concentration. A variety of isotherm shapes are possible, depending on the affinity of the adsorbent for the adsorbate (Fig. 9.9).

   To generate adsorption data, a known amount of adsorbate in aqueous solution is mixed with a known amount of *adsorbent*. At equilibrium, the amount of adsorbate removed from solution is assumed to be adsorbed. Secondary reactions (such as precipitation) must be eliminated or corrected for. Precipitation is indicated in some cases by a rapid increase in apparent adsorption (disappearance from solution) with a small change in solution concentration. Three equations are commonly used to describe adsorption: the Langmuir, Freundlich, and Brunauer–Emmett–Teller (BET) equations.

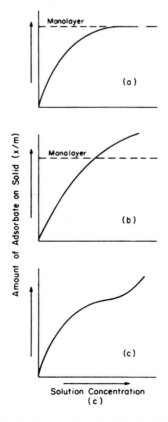

**FIGURE 9.9.** Typical adsorption isotherms described by the (a) Langmuir, (b) Freundlich, and (c) BET equations.

The Langmuir equation was initially derived for the adsorption of gases by solids. The derivation was based on three assumptions: (1) a constant energy of adsorption that is independent of the extent of surface coverage (i.e., a homogeneous surface); (2) adsorption on specific sites, with no interaction between adsorbate molecules; and (3) maximum adsorption equal to a complete monomolecular layer on all reactive adsorbent surfaces (Fig. 9.8a). A common form of the Langmuir equation is

$$\frac{x}{m} = \frac{KCb}{1 + KC} \tag{9.7}$$

where $x/m$ is the weight of adsorbate per unit weight of adsorbent, $K$ is a constant related to the binding strength, $b$ is the maximum amount of adsorbate that can be adsorbed (i.e., a complete monomolecular layer), and $C$ is the adsorbate concentration. Rearranging Eq. 9.7 yields the more convenient linear form

$$\frac{C}{x/m} = \frac{1}{Kb} + \frac{1}{b} \tag{9.8}$$

If adsorption conforms to the Langmuir model, plotting $C/(x/m)$ versus $C$ yields a straight line with a slope $1/b$ and intercept $1/Kb$. The Langmuir constant $K$ is the quotient of the slope $1/b$ and intercept $1/Kb$.

Equation 9.7 assumes constant free energy of adsorption on the surface, a situation that rarely occurs in nature. Instead, the energy of adsorption tends to decrease with increasing surface coverage. The interaction with already-adsorbed molecules increases with increasing surface coverage. The net effect is that the two phenomena tend to compensate for each other, yielding a relatively constant energy of adsorption. In systems where the energy of adsorption is not constant, the Langmuir equation may still describe adsorption over a portion of the adsorption range, since the variation in energy of adsorption over such a range can be small if only one type of bonding site or mechanism predominates.

A true adsorption maximum, however, is rarely observed. Precipitation reactions can exhibit Langmuir-type behavior. If only a limited quantity of a solute that precipitates is present, a Langmuir isotherm can result as the solute increases, that is, a "sorption maximum" occurs. This behavior is found at low solute concentrations, where no precipitation occurs until the solute's solubility product is reached.

An advantage of using the Langmuir equation for describing adsorption is that it defines an adsorption limit on a given array of sites that meet the Langmuir model's criteria. This limit has been used to estimate the adsorption capacity of soils for phosphate and various herbicides. Comparing such capacities can also suggest adsorption mechanisms. Unfortunately, the adsorption maxima usually do not occur, instead adsorption continues but at ever-decreasing amounts.

If data fail to conform to the Langmuir equation, the Freundlich equation often fits the data successfully:

$$\frac{x}{m} = KC^{1/n} \tag{9.9}$$

where $K$ and $n$ are empirical constants and the other terms are defined above. The equation was originally empirical, without a theoretical foundation. It implies that the energy of adsorption decreases logarithmically as the fraction of covered surface increases, similar to the solid solution ideas. The Freundlich equation can be derived theoretically by assuming that the decrease in energy of adsorption with increasing surface coverage is due to surface heterogeneity. The degree of heterogeneity is unknown in most adsorption studies, and both the Langmuir and Freundlich equations are better thought of as empirical curve fitting of adsorption data, rather than describing the actual mechanism of adsorption.

The linear form of the Freundlich equation is

$$\log \frac{x}{m} = \frac{1}{n} \log C + \log K \tag{9.10}$$

The frequent good fit of adsorption data to the Freundlich equation is influenced by the insensitivity of log–log plots and by the flexibility afforded curve fitting by the two empirical constants $K$ and $n$. This flexibility does not guarantee accuracy, however, if the data are extrapolated beyond the experimental range. The Freundlich equation has the further limitation that it does not predict a maximum adsorption capacity, however mythical the adsorption maximum may be. Despite its shortcomings, the Freundlich equation is a common adsorption equation and is included in several models for predicting pesticide behavior in soil.

## APPENDIX 9.1   MULTISITE AND MULTILAYER ADSORPTION

A number of recent studies involving the adsorption of solutes from solution by mineral surfaces have resulted in data suggesting *multiple-site adsorption*. That is, several different arrays of sites are postulated, each of which fulfills the requirements of the Langmuir model. For example, Fig. 9.10 shows data for phosphate adsorption on

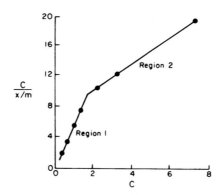

**FIGURE 9.10.** Phosphorus adsorption data plotted according to the Langmuir equation. (After J. K. Syers, M. G. Browman, G. W. Smillie, and R. B. Corey. 1973. *Soil Sci. Soc. Am. Proc.* **37**:358–363.)

soil plotted according to the traditional linear Langmuir form. The data suggest two sets of sites, each with its own binding strength and adsorption maximum. Alternatively, the data suggest two mechanisms of adsorption on similar sites. The adsorption curve is resolved by dividing the curve into several straight-line components. For Fig. 9.10, only two straight lines are needed. The form of the Langmuir equation for two-site adsorption is

$$\frac{x}{m} = b_1 - \frac{x/m_1}{K_1 C} + b_2 - \frac{x/m_2}{K_2 C} \tag{9.11}$$

where subscripts 1 and 2 refer to regions (or mechanisms) 1 and 2. The adsorption maximum for the soil is the sum $b_1$ plus $b_2$. This form of the Langmuir equation can be used to describe adsorption sites for species whose total adsorption appears to be the sum of both a high-energy and a low-energy component. Such a distribution of adsorption energies is reasonably well established for soil phosphate. To fit adsorption data more closely to the Langmuir model, even three- and four-site models have been invoked.

In addition to multisite adsorption, many gases and vapors adsorbed by solids do not produce a typical monolayer-type adsorption isotherm (Fig. 9.9a), but rather produce an isotherm indicating multilayer adsorption (Fig. 9.9c). An equation that treats multilayer adsorption is the BET equation, named after developers Brunauer, Emmett, and Teller. Multilayer adsorption is characteristic of physical or van der Waals attraction. It often proceeds with no apparent limit, since multilayer adsorption merges directly into capillary condensation as the vapor pressure of the adsorbate approaches its saturation value.

The BET equation has been used to determine the surface area of solids from gas adsorption data. The equation not only predicts the shape of the adsorption isotherm, but also gives the volume of gas $V_m$ required to form a monolayer. The BET equation has the form

$$A = \frac{V_m}{V_0} N_a A_m \tag{9.12}$$

where $A$ is the surface area of the adsorbent, $V_m$ is the volume of gas adsorbed, $V_0$ is the molar volume of adsorbate gas (22.4 liters at 25° C), $N_a$ is Avogadro's number, and $A_m$ is the cross-sectional area of the adsorbate molecule. Surface area determinations are often based on $N_2$ adsorption by the solid at −195° C. For $N_2$, $A_m = 16.2 \times 10^{-20}$ m$^2$. The linear form of the BET equation is

$$\frac{P}{V(P_0 - P)} = \frac{1}{V_m C} + \frac{(C - 1)P}{V_m C P_0} \tag{9.13}$$

where $P$ is the equilibrium pressure at which a volume $V$ of gas is adsorbed, $P_0$ is the saturation pressure of the gas, and $C$ is a constant related to the heat of adsorption of the gas on the solid. If a plot of $P/V(P_0 - P)$ is a straight line, the effective surface area of the solid can be calculated after $C$ has been determined, either from the slope of the line $(C - 1)V_m C$, or from the intercept, $1/V_m C$.

The BET equation has been applied to ion adsorption from soil solutions, although the extended Langmuir equation (Eq. 9.11) would seem to apply as well. The BET equation has also been used to study the adsorption of pesticides having relatively high vapor pressures.

An alternative explanation of retention by adsorption is to consider the substance being held by solid-state mixing on the surfaces of soil particles (see Chapter 3). The advantage to the solid-state mixing concept is that the quantitative interpretation is based on the same thermodynamics as ions in aqueous solutions. It provides a plausible explanation of why the retention energy progressively decreases as the adsorbate concentration increases, why soils retain ions at different energies, and why soil particles retain substances much more strongly than their own pure minerals while releasing other ions as the soil weathers.

Retention by solid solution cannot be differentiated from adsorption on the basis of experiment. They are simply different explanations for the same phenomenon. The idea of adsorption sites can be reconciled by considering that those areas are where solid-state mixing, because of the solid's surface structure and composition, is most likely.

## BIBLIOGRAPHY

Edzwald, J. K., D. C. Toensing, and M. C. Leung. 1976. Phosphate adsorption reactions with clay minerals. *Environ. Sci. Tech.* **20**:485–490.

Guenzi, W. E. (ed.). 1974. *Pesticides in Soil and Water*. Soil Science Society of America, Madison, WI.

Mott, C. J. B. 1981. Anion and ligand exchange. In *The Chemistry of Soil Processes* (D. J. Greenland and M. H. B. Hayes, eds.). Wiley, New York, Ch. 5.

## QUESTIONS AND PROBLEMS

1. Distinguish between specific and nonspecific reactions of anions soils. Give examples of anions that tend to be specifically and nonspecifically reactive in soils.

2. What are the forces acting on an anion as it approaches a layer silicate? How will these forces vary from soil to soil? What is the dominant for acting on anions in most agricultural soils?

3. How do the following factors affect anion repulsion:
   (a) Anion charge
   (b) Anion concentration
   (c) Exchangeable cation
   (d) Soil pH
   (e) Other anions

4. Are all anions adsorbed alike in soils? If not, explain the differences, giving examples of each reaction type.

5. What reactions are responsible for the fixation of phosphate in acid and basic soils?

6. How are molecular species such as $N_2$ and $NH_3$ retained by soils?

7. Certain mechanisms are active in the retention of all species (cationic, anionic, and molecular), while other mechanisms are active in the retention of only certain species. Explain, giving examples of species retained predominantly by each specific mechanism.

8. Refer to Fig. 9.5. Describe the sequence of events if a soil of pH 7.5 were fertilized to a $pH_2PO_4$ level of 2.

9. Given the data below, determine if the adsorption of 2,4,5-T conforms to the Langmuir or the Freundlich models and determine the appropriate adsorption parameters $(K, n, b)$. You may need to restrict your attention to a limited concentration range.

| Initial Solution Concentration (mg $L^{-1}$) | Final Solution Concentration (mg $L^{-1}$) | Volume of Solution (mL) | Weight of Soil (g) |
|---|---|---|---|
| 5 | 3 | 10 | 5 |
| 10 | 6 | 10 | 5 |
| 25 | 15 | 10 | 5 |
| 50 | 30 | 10 | 5 |
| 100 | 70 | 10 | 5 |

10. Explain in your own words the peaks and inflection points of Fig. 9.3.

11. Maximum phosphate availability in soils tends to occur around pH 6 to 6.5. Explain why in terms of Fig. 9.5.

12. Based on Table 9.3, predict the relative mobility of (a) $s$-triazole, (b) $H_3BO_3$, and (c) 2,4-D in pH 4.5, 7.0, and 8.5 soils.

# 10

# ACID SOILS

Rainfall over a large portion of the earth's surface exceeds evapotranspiration for much of the year, and soil leaching results. The leaching gradually removes soluble salts, more readily soluble soil minerals, and bases (nonacidic cations such as $Ca^{2+}$, $Mg^{2+}$, $K^+$, and $Na^+$). Consequently, the leached surface soil becomes slightly to moderately acid, although the subsoil may remain neutral or alkaline. As weathering proceeds, even acidic components are leached from the soil. At this stage, the surface soil pH, and ultimately the pH of the entire profile, once more approaches neutrality. Only iron and aluminium oxides, and some of the trace metal oxides that are also highly resistant to weathering, remain in the soil from the original parent material.

Local highly acidic conditions can also arise when mine wastes containing iron pyrite ($FeS_2$) and other sulfides are exposed to the air. The sulfide oxidizes to $H_2SO_4$ and $Fe(OH)_3$. Acidities of pH 2 or lower are not uncommon in these soil solutions. Extremely acid soils also result from drainage of marine floodplains containing high-sulfide sediments (•cat claysŽ or Acid Sulfate soils), which oxidize to $H_2SO_4$ upon exposure.

Crop fertilization can also produce substantial soil acidity. Continued use of ammonia fertilizers can lead to acidic soil conditions by the microbially mediated reaction

$$NH_4^+ + 2O_2 = NO_3^- + 2H^+ + H_2O \tag{10.1}$$

Less acidity is generated from $NH_4NO_3$ per unit of nitrogen than from $(NH_4)_2SO_4$, because only half of the nitrogen in $NH_4NO_3$ can be further oxidized. The $H_3PO_4$ released by dissolving phosphate fertilizer granules can lead to pH values near the granule as low as pH 1.5. The $H_3PO_4$ is rapidly neutralized by soils, but the acidic reaction products may remain to influence soil properties. Despite considerable publicity about acid rain, the rate of soil acidification due to acid rain (containing $H_2SO_4$

and $HNO_3$ as a result of human activities) is normally severalfold lower than that from the use of ammoniacal and phosphate fertilizers. In nonagricultural soils, acid rain is a relatively greater factor in soil acidification, partly because these soils are not limed to overcome acidity, as are agricultural soils.

Finally, acidity may be produced by plant residues or organic wastes decomposing under somewhat reducing conditions into organic acids. This is of particular importance in many forest soils. The organic acids account in part for the dissolution and movement of Fe, Al, and Mn through the soil beneath many forest litter layers. Chelation or complexation by soluble organic molecules also contributes to cation transport through the soil under such conditions.

The chemical behavior and properties of acid soils and the diagnosis and amelioration of their adverse effects are the main subjects of this chapter. The problem of plant growth in acid soils is treated only lightly. Excellent books by Black and by Pearson and Adams treat this aspect of the subject in more detail.

The soil chemistry literature of the 1930s, 1940s, and early 1950s contains reports of numerous studies on the properties of hydrogen-saturated soils and clays. One aim was to predict the amount of lime needed to counteract soil acidity. A common approach was to titrate the soil potentiometrically (as a function of pH) with a base. Figure 10.1 shows typical curves for the potentiometric titration of acids, and of titrating an acid montmorillonite suspension, with a strong base such as NaOH. Curve 1 is typical of strong acids such as HCl, and is also typical of a freshly prepared acidic clay suspension. The pH remains relatively constant until nearly all of the acid is neutralized. Then the pH rises rapidly until it is determined by the concentration of the added base.

Curves 2 and 3 are NaOH titration curves of polyprotonated weak acids, such as $H_3PO_4$, of titrating an acid clay suspension prepared slowly by dialysis, of clay suspensions prepared rapidly by newer methods and then aged for a few days, and

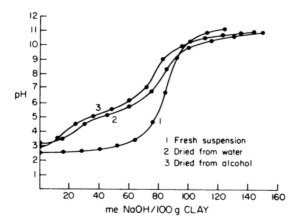

**FIGURE 10.1.** Potentiometric titration of montmorillonite suspensions after treatment with NaCl–HCl solution and H-resin. (From D. G. Aldrich and J. R. Buchanan. 1958. *Soil Sci. Soc. Am. Proc.* **22**:281–286.)

of field soils. Here the pH increases more or less continuously as base is added, with an occasional plateau corresponding to weak-acid groups having p$K$ values near the plateau pH. The intermediate plateaus in this figure, for example, correspond to weak-acid p$K$ values of 5.5 to 6.

In 1947, Chernov of the Soviet Union summarized the results of many studies about the nature and properties of acid soils and clays. He recognized that hydrogen-saturated minerals were highly unstable, and that they rapidly broke down to release Al, Mg, and Fe from within their lattices. He suggested that most hydrogen-saturated soil materials were in reality saturated primarily with $Al^{3+}$ and $Fe^{3+}$. His work was supported in the United States during the early 1950s by Jenny, Coleman, and others. They showed that weak-acid properties commonly attributed to "hydrogen" clays (e.g., curves 2 and 3 in Fig. 10.1) were in reality the result of partial or complete saturation of exchange sites with weakly acidic Al ions. For these two curves, 40 to 50% of the cation exchange capacity (CEC) was actually occupied by exchangeable Al. Hydrogen clays that were analyzed immediately after rapid preparation behaved much more like strong acids. For curve 1 in Fig. 10.1, for instance, less than 10% of the clay's CEC was Al saturated. Results such as these suggested that acid soil clays were behaving like Al and Fe(III) clays rather than H clays.

## 10.1 INSTABILITY OF HYDROGEN SOILS AND CLAYS

Hydrogen soils and clays prepared by strong-acid leaching or dialysis decompose rapidly to Al- and Fe(III)-saturated materials. The half-life for the temperature-dependent decomposition is only a few hours for many minerals (Table 10.1). Hydrogen-saturated Utah bentonite (smectite) was half converted to the corresponding Al-saturated form after 18 hours at 30° C, and three-fourths converted after 36 hours. Corresponding times at 60° C were 2.1 and 4.2 hours. Wyoming (Volclay) bentonite was stable nearly three times longer than the Utah bentonite when H saturated, and kaolinite was stable nearly seven times longer. The reaction can be slowed markedly by storage at temperatures near freezing, but truly H-saturated soils or clays must still be studied as rapidly as possible.

**Table 10.1. Rates of decomposition of hydrogen-saturated layer silicates, showing the time for one-half of the exchangeable hydrogen to the loss at various temperatures[a]**

| Layer Silicate | Rates of Decomposition (Half-time) (min) | | | | |
| --- | --- | --- | --- | --- | --- |
| | 30° C | 50° C | 60° C | 70° C | 80° C |
| Montmorillonite (Utah Bentonite) | 1080 | 260 | 125 | 60 | 32 |
| Montmorillonite (Volclay bentonite) | — | — | 340 | — | 86 |
| Kaolinite | — | — | 850 | — | 194 |

[a] N. T. Coleman and D. Craig. 1961. *Soil Sci.* **91**:14–18.

Vermiculite is particularly susceptible to decomposition under acid conditions. This susceptibility is one explanation for the relatively low vermiculite contents of slightly acid surface soils from arid and semiarid regions, when compared with corresponding subsoils. Another explanation for low vermiculite contents of surface soils is K cycling by plants, with the subsequent conversion of vermiculite particles to their collapsed (micaceous) equivalents.

## 10.2  HYDROLYZED ALUMINIUM IONS

The chemical behavior of acid soils and minerals is intimately linked to the aqueous solution chemistry of aluminium. Aluminium hydrolyzes to monomeric and polymeric hydroxyaluminium complexes made up of $Al(OH)^{2+}$ and $Al(OH)_2^+$. Ultimately Al precipitates as solid-phase gibbsite $(Al(OH)_3)$ when the solubility product of this mineral is exceeded. The hydrolysis reactions of the monomers are

$$Al(H_2O)_6^{3+} + H_2O = Al(OH)(H_2O)_5^{2+} + H_3O^+$$

$$Al(OH)(H_2O)_5^{2+} + H_2O = Al(OH)_2(H_2O)_4^+ + H_3O^+ \qquad (10.2)$$

$$Al(OH)_2(H_2O_4)^+ + H_2O = Al(OH)_3(H_2O)_3^0 + H_3O^+$$

$$Al(OH)_3(H_2O_3)^0 + H_2O = Al(OH)_4(H_2O)_2^- + H_3O^+$$

Each reaction is driven to the right by the consumption of $H^+$ (hydronium, $H_3O^+$) ions by reacting with hydroxyl ions. Successive hydrolysis reactions are associated with increasing pH. The distribution of Al ions with pH is shown in Fig. 10.2. For convenience, the $H_2O$ ligands have been omitted from the formulas. Such diagrams

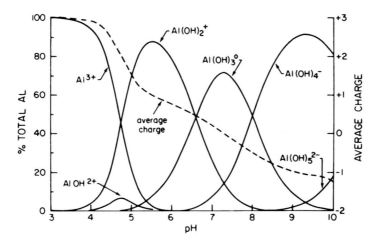

**FIGURE 10.2.** Relative distribution and average charge of the soluble aluminum species as a function of pH, ionicstrength = 0.1 M. (From G. Marion et al. 1976. *Soil Sci.* **121**:76–82.)

identify the Al hydrolysis species at various pH values and show their relative contribution to total soluble Al. The $Al(OH)^{2+}$ ion is of minor importance and exists over only a narrow pH range. The $Al^{3+}$ ion is predominant below pH 4.7, $Al(OH)_2^+$ between pH 4.7 and 6.5, $Al(OH)_3^0$ between pH 6.5 and 8, and $Al(OH)_4^-$ above pH 8. The $Al(OH)_5^{2-}$ ion occurs only at pH values above those common to soils. Solid-phase $Al(OH)_3$ precipitates throughout the pH range covered, whenever its solubility product is exceeded. The total concentration of soluble aluminium is strongly pH dependent and is minimal at about pH 7.

The hydrolysis reactions of Eq. 10.2 liberate $H^+$ and lower the solution pH unless $OH^-$ is present. This stepwise production of hydrogen ions is similar to that which occurs during the dissociation of polyprotonated acids. It is the primary reason why early workers attributed weak-acid properties to acid soils and clays.

Monomeric hexaquoaluminium $(Al(H_2O)_6^{3+})$ is exchangeable, although its trivalent charge results in a strong retention or preference by many soil colloids. The hydrolyzed Al ions rapidly polymerize to form large, multicharged units. The polymerization is enhanced by soil colloid surfaces. Hydroxyl groups are shared by adjacent Al ions to produce polymers of the general formula $(Al(OH)_x(H_2O)_{6-x}^{(3-x)+})_n$, where $n$ is the average number of Al ions per polymer. These multicharged polymers polymerize (age) further with time. The polymers are strongly retained by soil colloid surfaces and behave as if they are virtually nonexchangeable. This is probably because sufficient numbers of exchanging ions are rarely present at any time or place to exchange them. An exception to the general rule of nonexchangeability for hydroxy aluminium polymers occurs when previously formed polymers are adsorbed on expanded montmorillonite interlayers. In this rare circumstance, the polymer may attach to the montmorillonite surface in only a few places, and hence remain exchangeable when exposed to sufficient numbers of replacing cations.

Retention of positively charged and virtually nonexchangeable hydroxy aluminium polymers lowers the net negative charge of soil colloids. Thus, formation of hydroxy aluminium polymers on the surface of inorganic soil colloids decreases the cation exchange capacity (CEC) of the colloids. Raising the pH decreases the positive charge on the polymers (Eq. 10.2 and Fig. 10.2) and increases the CEC of the mixture. This is an important source of *pH-dependent charge* for inorganic soil colloids. Conversely, lowering the pH of soils containing large quantities of adsorbed hydroxy aluminium polymers decreases the soil CEC, by increasing the positive charge on the polymers. In some cases it may result in zero to positive charge.

Iron hydrolysis is similar to that of aluminium. The $pK$ of the first step of Fe(III) hydrolysis,

$$Fe(H_2O)_6^{3+} + H_2O = Fe(OH)(H_2O)_5^{2+} + H_3O \qquad (10.3)$$

is close to 3, while that for $Al(H_2O)_6^{3+}$ is 5. The Fe(III) ion is a stronger acid and its acidity is buffered by the Al hydrolysis reactions. Most of the soil's large reserve of Al would have to react before pH could decrease to the point where Fe hydrolysis could control soil pH. So $Al^{3+}$ is the primary ion of concern in acid soils.

Hydroxy aluminium and hydroxy iron polymers also can adsorb anions with concurrent release of hydroxyl ions. The pH increase due to this anion exchange can be masked, however, by the simultaneous hydrolysis of desorbed aluminium ions (Eq. 10.2). Adsorption of multicharged anions can also decrease the net positive charge on hydroxy aluminium or hydroxy iron polymers, and thus increase the net negative charge of the soil–polymer mixture. The anion adsorption capacity of soils decreases with increasing pH and becomes virtually zero for all anions except phosphate and arsenate at pH values greater than 5.5 or 6.

Hydroxy Al and hydroxy Fe polymers can be held between the lattices of expanding soil minerals, preventing collapse of these lattices as water is removed during drying or freezing. Swelling of dried soil materials may also be restricted, since water is less able to enter between mineral sheets once the minerals have collapsed. As little as 1/16 coverage by a hydroxy aluminium or hydroxy iron monolayer can stabilize soil minerals against shrinking and swelling. The resultant minerals are termed *intergrades*: chlorite–vermiculite and chlorite–montmorillonite intergrades are particularly common. Although intergrade minerals commonly are described as having chlorite-like structures, the interlayer generally consists of either hydroxy aluminium or hydroxy iron compounds, rather than the hydroxy magnesium of most geologic chlorites. The polymeric materials in acid soils and clays also tend to exist as intermittent islands between the interlayers of expanding minerals, rather than as the continuous sheets typical of traditional chlorites.

Once present as interlayer material, hydroxy aluminium or hydroxy iron polymers can be removed only by fairly complete neutralization of their charge to form uncharged $Al(OH)_3$ and $FeOOH$ or by acidification to produce monomeric (and hence exchangeable) $Al^{3+}$ or $Fe^{3+}$. Upon neutralization to form the aluminium or iron hydroxides, which is complete at about pH 8 for aluminium, the hydroxides and the negatively charged silicate layer lattices are no longer electrostatically attracted. Furthermore, particles of intergrade minerals are generally smaller than the free gibbsite ($Al(OH)_3$) and goethite ($FeOOH$) particles of soils. The higher surface energy of the smaller particles causes them to dissolve from mineral surfaces and to reprecipitate as part of the larger particles. At still higher pH values, the dominant soluble species are negatively charged $Al(OH)_4^-$, $Fe(OH)_4^-$, and $Fe(OH)_5^{2-}$. Hence, they are repelled by negatively charged layer silicates.

## 10.3  CLASSIFICATION AND DETERMINATION OF SOIL ACIDITY

Various approaches have been used to classify the components of soil acidity. As a carryover from the titration curves used to characterize soil acidity in earlier studies, a common category is *titratable acidity* or *total acidity*. This is the quantity of a strong base ($NaOH$ or $Ca(OH)_2$) required to raise soil pH to a predetermined level. Time, method of stirring, and period between additions of base must be specified because the neutralization of soil acidity is highly dependent on reaction conditions. The values are also meaningless unless the initial and final pH values are specified, because more base is consumed if the reaction is carried out over a wider pH range.

For example, titration curve 3 in Fig. 10.1 required approximately 200 mmol hydroxyl $kg^{-1}$ of clay to raise the pH from 4 to 5, and another 200 mmol hydroxyl $kg^{-1}$ of clay to raise the pH from 5 to 6.

Common endpoints of such titrations are pH 7 or pH 8.2, although soils in the field are rarely limed above pH 6 or 6.5. The value 8.2 was chosen historically because it approximates the pH of soil containing free $CaCO_3$ in equilibrium with the normal $CO_2$ content (0.0003 mol fraction) of the atmosphere. This pH also corresponds closely with the pH of complete neutralization of soil hydroxy aluminium compounds. The pH 8.2 is conveniently maintained by Mehlich's $BaCl_2$-triethanolamine extraction technique.

The titration process, if carried out so slowly that the reaction is fairly complete following each addition of base, does not distinguish between exchangeable and virtually nonexchangeable components. Hence, titratable acidity is only a measure of the total acidity neutralized during the experimental technique employed. The titratable or total acidity is nonetheless useful for determining the lime requirement of acid soils.

Further classification of acid soils includes the distinction between exchangeable and nonexchangeable acidity. *Exchangeable acidity* is that exchanged by an unbuffered neutral salt solution, such as 1 M KCl or NaCl. *Nonexchangeable acidity* also has been more ponderously termed "titratable but nonexchangeable acidity." It includes hydroxyl-consuming reactions such as neutralization of hydroxy aluminium polymers on soil surfaces:

$$X—Al(OH)^{2+} + OH^- = X—Al(OH)_2^+ \qquad (10.4)$$

neutralization of protons from weakly acidic organic functional groups:

$$R—COOH + OH^- = R—COO^- + H_2O \qquad (10.5)$$

(creating additional pH-dependent charge or CEC); and displacement of adsorbed anions:

$$X—Al(OH)(H_2PO_4) + OH^- = X—Al(OH)_2 + H_2PO_4^- \qquad (10.6)$$

Exchangeable acidity consists of monomeric aluminium and exchangeable hydrogen. The exchangeable aluminium and exchangeable hydrogen can be roughly divided by soil pH, for exchangeable hydrogen normally is present in measurable quantities only at pH values less than 4. Hence, exchangeable hydrogen is of concern only for extremely acidic materials, such as mine spoils or acid sulfate soils from marine floodplains, organic acids from decomposing soil organic matter, and from acid rain. Exchangeable aluminium normally occurs in significant amounts only at soil pH values less than about 5.5. In the range of pH 5.5 to 7, hydroxy aluminium polymers predominate among acidic soil components, exchangeable acidity is virtually absent, and only nonexchangeable and titratable acidity are present in measurable quantities. Significant quantities of such acidity from weakly acidic R—COOH and

R—OH groups of soil organic matter, and from incompletely neutralized hydroxy aluminium polymers can be present in soils at pH > 7.

Although exchangeable acidity is essentially absent above pH 5.5, some direct proton exchange by weaker carboxylic groups and most of the phenolic groups on SOM, as well as some of the weakly acidic protons on soil mineral edges, may still occur above this pH. $H^+$ production from such groups is relatively minor in most agricultural soils. Some ambiguity also exists when the CEC of organic colloids is neutralized by difficultly exchangeable $Al^{3+}$ and $Fe^{3+}$. These ions may react during relatively rapid titrations as titratable but nonexchangeable acidity.

More precise separation of acidity into its components requires a separate aluminium (and possibly iron) determination on the neutral salt extract, or a conductimetric titration (Fig. 10.3). The electrical conductance changes rapidly with the degree of neutralization when relatively mobile ions, such as hydrogen or hydroxyl, predominate in solution. It changes little when immobile ions, such as aluminium or hydroxy aluminium, dominate. In Fig. 10.3, the conductance (reciprocal of resistance) decreases as the $H^+$ concentration decreases during the change from pH 3 to 5. The resistant then remains unchanged, near pH 5, as the added $OH^-$ precipitates aluminium. When all the $Al^{3+}$ ions have been titrated to $Al(OH)_3$, the conductance increases again as more $OH^-$ is added. The similarity of such a titration curve to the titration curves of acid soils is one type of evidence that led to the conclusion that acid soils are aluminium soils rather than true weak acids.

The complexity of soil acidity emphasizes the need to specify the final pH of a titration if the components of soil acidity are to be classified meaningfully. Classification also should include specifying the initial pH, which can indicate the most probable forms of acidity present.

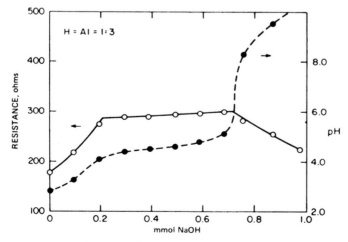

**FIGURE 10.3.** Potentiometric (●) and resistance (○) curves for titrating mixtures of HCl plus AlCl₃ with NaOH. (From P. F. Low. 1955. *Soil Sci. Soc. Am. Proc.* **19**:135–139.)

## 10.4 SOIL PH MEASUREMENTS

Soil pH measurements can be ambiguous. Two factors that affect soil pH measurements are the soil–solution ratio and the salt concentration. Increasing either factor normally decreases the measured soil pH because H and Al cations on or near soil colloid surfaces can be displaced by exchange with soluble cations. Once displaced into solution, the Al ions can hydrolyze (Eq. 10.2) and further lower the pH. Preferential retention of hydroxy aluminium polymers by soil colloids drives the hydrolysis reactions further toward completion and leads to lower pH. Increasing the neutral salt concentration to 0.1 or 1 M can lower the measured soil pH as much as 0.5 to 1.5 units, compared to soil pH measured in distilled water suspensions.

Because of the cation distribution in the diffuse double layer (DDL), and possibly because of higher concentrations of hydrogen near weakly ionized organic groups and mineral edges, the hydrogen ion concentration near soil colloid surfaces appears to be 100 to 1000 times greater than in the bulk solution. Such concentrations have been substantiated for dry clays by measuring the infrared spectra of adsorbed organic acids, and for wet clays by the pH-dependent reaction rates of adsorbed enzymes. The greater acidity near soil colloid surfaces should be kept in mind whenever adsorption mechanisms are being postulated for weakly acidic organic molecules, including many of the common pesticides.

In soil suspensions at low salt concentrations, extraneous (*junction*) *potentials* can affect pH readings (Appendix 10.1). The most plausible explanation for junction potentials is that K ions in the KCl bridge of the reference electrode diffuse more rapidly, and Cl ions less rapidly, when negatively charged soil particles are near the bridge. One answer is to place the reference electrode in the clear supernatant solution and the glass electrode in the settled clay suspension (where $H^+$ is concentrated) to obtain valid soil pH measurements. Junction potentials are essentially eliminated at salt concentrations greater than 0.01 M.

## 10.5 PERCENT BASE SATURATION

A soil parameter of historical importance is the *percent base saturation*, defined as

$$\text{percent base saturation} = \sum(\text{exchangeable Ca, Mg, Na, K}) \times \frac{100}{\text{CEC}} \quad (10.7)$$

at pH 7 or 8.2. The pH used for this measurement and for CEC determinations must be specified whenever this concept is used. As an example, consider a soil of pH 5 that has 5 mmol(+) of exchangeable bases (Ca, Mg, K, and Na), 1 mmol(+) of exchangeable acidity, and a CEC of 80 at pH 7 and of 100 at pH 8.2. The percent base saturation of this soil is 62% based on the CEC at pH 7, 50% based on the CEC at pH 8.2, and 83% at the native soil pH. If organically chelated or polymeric forms of aluminium are present, the basic cations that are displaced may vary somewhat with the pH of the extracting solution. This is probably because of competition for exchange sites from small, pH-dependent quantities of displaced aluminium.

In the early literature dealing with soil acidity, soils were characterized by their percent base saturation at specified pH levels. Soils with low percent base saturation values were considered to be dominated by kaolinite and hydrous oxide minerals, but soils of high percent base saturation were considered to be dominated by 2:1-type minerals, such as montmorillonite, vermiculite, chlorite, and the micas. Base saturation is a criterion of soil taxonomy in the U.S. soil classification scheme. Fifty percent base saturation (based on soil CEC at pH 7) is one criterion for distinguishing between mollic epipedons (dark, high organic horizons) and their umbric (low organic) counterparts.

Unfortunately, percent base saturation is as much a measure of the pH-dependent charge of soils as it is of the actual percentage of cation exchange sites occupied by exchangeable bases. The denominator includes any additional charge (CEC) generated SOM and hydrous oxide–mineral complexes between the initial soil pH and the reference pH (7 or 8.2). Since neither exchangeable aluminium nor exchangeable hydrogen is appreciable above pH 5.5, the CEC above this pH should be 100% base saturated. However, soils in the pH range 5.5 to 7 or 8.2 generally still have measured base saturations well below 100%. Such base saturation values are particularly low for minerals that have a high proportion of pH-dependent charge, such as kaolinite.

Although imprecise, the percent base saturation is still useful for soil genesis and classification purposes and for empirical liming recommendations. For example, each unit rise in soil pH was related to a 20 to 30% increase in base saturation for field soils in Virginia. From the standpoint of soil chemical properties and reactions, however, the base saturation is more correctly an acidity index or liming index. In addition, the degree of nonbase saturation is more meaningful if separated into exchangeable acidity and pH-dependent charge.

## 10.6   LIME REQUIREMENT

A major problem of managing acid soils is to estimate the quantity of lime required to raise the soil pH to a certain level. As shown in Table 10.2, plant species vary considerably in their response to soil pH. Such data must be interpreted carefully. In this case, the nonlegumes benefited from nitrogen fixed by legumes in the rotation. Much of the pH response may actually be the pH response of nitrogen fixation by the legume–*Rhizobium* pair.

The most theoretically satisfying way to estimate the *lime requirement* of acid soils is to measure the quantity of base required to raise soil pH to a specified level. To be realistic the titration must be slow enough for the added base to react completely with the soil. Both exchangeable and titratable acidity will be neutralized during the titration.

Ten $mmol(+)\,kg^{-1}\,OH^-$ consumed during the titration is equivalent to 4.5 tonnes pure $CaCO_3$ (ha-30 cm)$^{-1}$ of field soil. The $mol(+)$ equivalent weight of $CaCO_3$ is 100 in this case because only one $OH^-$ is produced per $CaCO_3$ molecule, in the normal pH range of limed acid soils. Although many people still regard the primary effect of lime to be the provision of adequate soil calcium, its main value is really to

**Table 10.2. Yield of crops grown in corn, small grain, legume, or timothy rotation at different soil pH values**[a]

| | Average Relative Yield at pH Indicated | | | | |
| Crop | 4.7 | 5.0 | 5.7 | 6.8 | 7.5 |
| --- | --- | --- | --- | --- | --- |
| Sweet clover | 0 | 2 | 49 | 89 | 100 |
| Alfalfa | 2 | 9 | 42 | 100 | 100 |
| Red Clover | 12 | 21 | 53 | 98 | 100 |
| Alsike clover | 13 | 27 | 72 | 100 | 95 |
| Mammoth clover | 16 | 29 | 69 | 100 | 99 |
| Timothy | 31 | 47 | 66 | 100 | 95 |
| Barley | 0 | 23 | 80 | 95 | 100 |
| Corn | 34 | 73 | 80 | 100 | 93 |
| Wheat | 68 | 76 | 89 | 100 | 99 |
| Oats | 77 | 93 | 99 | 98 | 100 |

[a]From Ohio Agric. Expt. Sta. 1938. *Ohio Agric. Expt. Sta. Special Circular 53.*

provide hydroxyl ions:

$$CaCO_3 + H_2O = Ca^{2+} + HCO_3^- + OH^- \qquad (10.8)$$

Increased quantities of soluble and exchangeable Ca and Mg are by-products of liming, though their greater amounts may be beneficial to plants, such as legumes, having high Ca requirements.

A hydroxyl ion is also consumed during the displacement of adsorbed anions as the soil pH is raised. This effect is not normally a major one, but contributes to field lime requirements of several metric tons per hectare for some highly acid Piedmont soils from the southeastern United States.

Field liming reactions are generally incomplete, because of incomplete mixing, and require considerable time. The reaction rate varies inversely with pH, limestone particle size, and solubility of the liming agent. Hence, the laboratory lime requirement value is often further multiplied by a conversion factor to better estimate the amount of lime needed to achieve a given field pH.

The titration of individual soil samples is impractical for soil-testing purposes, because of the time and experimental precision required. Such titration is also highly dependent on the time allowed for each increment of base to react with the soil (Fig. 10.4). The usual procedure is to add a pH buffer solution to the soil, measure the amount of buffer consumed or the resulting pH of the soil–buffer suspension, and calibrate results with field lime requirements for similar soils from the same geographical area.

If the soil is leached with buffer solution until the soil pH equals that of the buffer, titrating the remaining buffer capacity of the solution measures the soil acidity that must be neutralized to produce a soil pH equal to that of the buffer. A more rapid method is add buffer solution to soil without attempting to bring the final pH of the

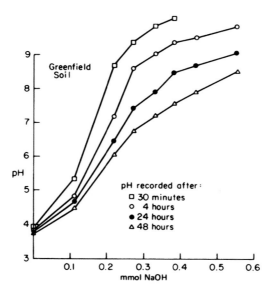

**FIGURE 10.4.** NaOH titration curves for vermiculitic Greenfield soil as influenced by time of the titration interval. (From A. L. Page et al. 1965. *Soil Sci. Soc. Am. Proc.* **29**:246–250.)

mixture to the initial pH of the buffer. The pH of the soil–buffer suspension indicates the degree of soil acidity present and, after field calibration, indicates the quantity of lime required to raise the soil pH to a certain level. For example, a pH change of 0.1 unit from the initial buffer pH might correspond to 1.0 Mg limestone ha$^{-1}$. If a soil has an initial pH of 5.5, the buffer solution has an initial pH of 6.8, and the final mixture has a pH of 6.3, then the lime requirement for this soil would be 1.0 × 5, or 5.0 metric tons ha$^{-1}$. The pH values of the final mixtures are calibrated against those of similar samples that have been leached with buffer solutions, or that have been titrated to specified pH levels, to obtain more precise estimates of the lime requirement.

Estimates of soil texture and measurements of initial soil pH for similar soils from a rather homogeneous geographical area can provide a simpler but less precise estimate of soil lime requirements. Such techniques must be calibrated against one of the more precise lime requirement methods to accurately estimate the amount of lime required. Different limed-soil pH values each require a separate calibration curve.

To achieve maximum crop production, soil pH must be raised to the optimum level for the crop in question. Little is gained by raising the pH to still higher levels. The growth increase from each successive increment of lime diminishes, but the cost of adding the increment remains the same. Although complete equilibration with lime may not occur until pH 8 or 8.2, acid soils are rarely limed above pH 6 or 6.5.

Plant growth in strongly weathered soils can be hampered by acidic subsoils. Surface application and mixing by plowing and discing are ineffective in treating the subsoils. Lime diffuses no more than 10 cm even 10 years after application. Adding gypsum and lime at the same amounts (ca. 5–10 Mg ha$^{-1}$ each) on the surface improves plant growth and considerably improves root growth and penetration of the treatment into the subsoil. Gypsum ($CaSO_4 \cdot 2H_2O$) is more soluble than $CaCO_3$. The sulfate anion is thought to penetrate to the subsoil, saturate the positively charged clays, raise pH by 0.4 units, and reduce Al toxicity.

## 10.7  ALUMINIUM AND MANGANESE TOXICITY

Many plants grow poorly in acid soils. Early workers supposed that this was a consequence either of hydrogen ion toxicity or of Ca and Mg deficiencies. The soil acidity must be greater than about pH 3, however, before the H$^+$ concentration itself is toxic to most plant species. Although the components of acidity are emphasized in acid soils, the major exchangeable cations are Ca, Mg, and to a lesser extent K in soils of pH > 4.5 to 5.

Plant growth problems associated with poor root penetration into acid subsoils are frequently associated with high plant availability of Al or M, which are toxic to most plants. Aluminium restricts or stops root growth at solution concentrations as low as 1 mg L$^{-1}$. Plants tolerate higher levels of soluble manganese, but reducing conditions in flooded or periodically inundated acid soils can result in soluble manganese concentrations as high as 100 mg L$^{-1}$.

## 10.8  PH AND MACRONUTRIENTS

The effects of low pH on plant growth are generally caused by increases of toxic ions, or decreases of essential ions, in the soil solution. Such effects can also arise from nutritional imbalances because the concentrations can increase or decrease as soil acidity changes.

Although effects of pH on plant nutrient levels in soils are complicated and interrelated, some generalizations are possible. Plants able to utilize ammonium forms of nitrogen have a considerable advantage in acid soils, because nitrification (microbial oxidation of ammonium to nitrate) is slow below pH 5.5. Ammonium ions may accumulate in acid forest soils, because the microbes that mineralize organic nitrogen to ammonia are less dependent on soil pH than are the nitrifying organisms.

Another important facet of nitrogen availability in acid soils is the pH dependence of ammonium–ion fixation between the lattices of expanding layer–silicate minerals. Such fixation generally decreases with increasing soil pH. Although the mechanism for this pH effect is incompletely understood, the decrease may be due to "islands" of hydroxy aluminium and hydroxy iron polymers, which prevent the complete collapse of mineral lattices and hence decrease $NH_4^+$ fixation.

The availability of soil phosphate is highly pH dependent and, as with nitrogen, is only partially understood. The main mechanism for phosphate fixation (decreased availability) under acid conditions appears to be the precipitation of highly insoluble iron and aluminium phosphates. Phosphate availability also tends to decrease at high soil pH, because of precipitation on insoluble calcium phosphate compounds. The pH range of greatest phosphate availability is about 6 to 7 for most agricultural soils.

Liming acid soils can increase or decrease potassium availability. Decreased K availability can be attributed to increased K fixation in limed soils, similar to ammonium fixation. Liming can increase K availability where the soil in its native state may have insufficient nutrient cations for plant growth. This would be typical of sandy soils or of highly weathered tropical soils. Insufficient exchange sites may be present in such soils to retain K and other nutrient cations against leaching. The increased soil CEC upon liming retains greater quantities of fertilizer K within the root zone and also retains it longer.

The change in ion-exchange specificity with pH (Chapter 8) can also cause opposing trends of K availability in limed soils. Increased availability can be attributed to greater quantities of K in the soil solution, because of Ca replacement of K in the DDL of the soil's colloids. Decreased K availability after liming can be due to greater quantities of K leaching from limed soils.


## 10.9  PH AND MICRONUTRIENTS

The micronutrients of major interest to soil chemistry because of plant deficiencies are boron, manganese, iron, cobalt, copper, zinc, and molybdenum. Other ions— chromium, nickel, cadmium, mercury, and lead—behave similarly in soils but the problems are usually plant toxicity. The availability of most of the micronutrient and toxic ions increases with increasing soil acidity. Those present as anions— molybdenum, chromium, and boron—differ in that their availability generally decreases with increasing acidity.

Acid soils generally provide sufficient micronutrients, occasionally even toxic amounts, to plants or to animals grazing on those plants. Because of the small quantities of micronutrients required for plant growth, adequate amounts can be taken up from small portions of the root zone, if such regions are sufficiently acidic. In basic soils the acidity from fertilizers or from small quantities of elemental sulfur or sulfuric acid added to a portion of the root zone may provide adequate micronutrients to plants.

Molybdenum is unique among the micronutrients because it is less available to plants at low pH. Occasionally, the harmful effect of soil acidity on leguminous plants seems to be caused by Mo deficiency rather than by Al toxicity (Table 10.3). In this case of a soil in the foothills of the Cascade Mountains in Oregon, the normal fertility program of P, K, B, and S gave only low to moderate alfalfa yields. Liming at a rate of 5000 kg ha$^{-1}$ gave high yields in all cases, as did only 0.5 kg ha$^{-1}$ of Mo on the unlimed soil. Molybdenum is required for nitrogen fixation by legumes.

In addition to pH effects on the availability of individual ions, various nutrients often interact with respect to their effects on plant growth. Some such interactions may arise from similarities in uptake mechanisms for different nutrients, whereas others may arise from precipitation or immobilization of ions near the plant root or

**Table 10.3. Effect of lime and molybedenum on the yield of alfalfa hay in the Willamette Valley of Oregon**[a]

| Treatment | Average Alfalfa Hay Yield $(kg\ ha^{-1})$ | | |
|---|---|---|---|
| | Melbourne Soil | Aiken Soil #1 | Aiken Soil #2 |
| PKBS[b] | 1280 | 4840 | 6010 |
| PKBS lime (5000 kg ha$^{-1}$) | 7060 | 9030 | 9810 |
| PKBS Mo (0.5 kg ha$^{-1}$) | 5980 | 8100 | 10020 |
| PKBS lime + Mo | 7310 | 9670 | 9790 |

[a] From T. L. Jackson et al. 1967. Reproduced from *Soil Acidity and Liming.* ASA Monograph No. 12, p. 267, by permission of the American Society of Agronomy.

[b] P = phosphorus, K = potassium, B = boron, and S = sulfur, at recommended rates for alfalfa in western Oregon.

within the plant root itself. Chemically similar ions also may compete for absorption sites on the plant root surface.

## 10.10 MANAGEMENT

Managing acid soils requires that crop tolerance to soil acidity be weighed against the cost of liming. Availability of lime, transportation charges, and the necessity and cost of grinding the limestone all influence the quantities that can be applied economically. A rule of thumb is to apply sufficient lime initially to raise soil pH to the desired range, and then to provide 2 to 5 metric tons lime ha$^{-1}$ every 3 to 5 years to maintain soil pH in that range. Sometimes substituting a more acid-tolerant crop may be more economic than liming.

Soil samples from the surface 20 to 30 cm of soil should be collected every 2 to 3 years to determine soil pH and to assist with predictions of additional liming or fertilization needs. Such sampling is normally done anyway on well-managed croplands to maintain adequate levels of available soil P and K. Hence, sampling costs should not be assigned entirely to managing soil acidity. Attaining the desired soil pH may require 6 to 8 months after lime application and the pH may change appreciably for as long as 18 months thereafter. Adequate soil water is necessary to permit hydroxyl and calcium ion diffusion and to carry out the associated liming reactions.

When the more slowly reacting dolomitic limestone is used, soil pH may increase for as much as five years after liming. In general, more finely ground liming materials cost more but react faster and more thoroughly with the soil. Finer particles also can be dispersed more evenly throughout the soil than can smaller numbers of large particles.

## APPENDIX 10.1    PH AND ION ACTIVITY MEASUREMENTS

The measurement of pH is the most common chemical measurement in soil, biology, and aqueous solutions. In addition, electrodes similar in principle and sensitive to $Na^+$, $K^+$, $Ca^{2+}$, $Mg^{2+}$, $Cl^-$, $NO_3^-$, $CN^-$, $F^-$, $S^{2-}$, and other ions are available. Such electrodes are in increasing use and respond roughly to the activity of the ion in question.

In a strict thermodynamic sense, single-ion activities are not measurable. Adherence to strict thermodynamics, however, can sometimes be unnecessarily limiting. Ion-sensitive electrodes do respond to changes in the concentrations of ions in solution, but they probably do not respond exactly to ion activity. Electrodes can also measure spurious potentials under unfavorable conditions. With reasonable care, most of these unwanted potentials can be minimized or eliminated.

The unique property of ion-sensitive electrodes is a membrane between the test solution and the electrode sensor that develops an electrical potential, or voltage, in response to a change in the concentration of a single ion. The pH electrode, for example, is shown schematically in Fig. 10.5. Other ion-sensitive electrodes differ in the composition of the membrane and in the salts necessary to develop the potential. In the pH electrode, a silver wire coated with AgCl dips into an HCl solution. The HCl solution is separated from the test solution by a membrane of special glass, usually a lithium silicate. Differences in $H^+$ activity across this glass membrane cause a difference in electrical potential, which can be measured by a sensitive potentiometer.

The electrode potentials developed by this electrode are the membrane potential plus the potential of the Ag—AgCl—HCl reaction inside the electrode:

$$AgCl + e^- = Ag + Cl^- \tag{10.9}$$

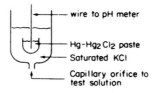

**FIGURE 10.5.** Diagram of the pH and calomel reference electrodes.

This reaction is reversible and, since the activities of Ag and AgCl can be taken as unity and the $Cl^-$ activity is fixed by the constant HCl concentration, the Ag—AgCl potential ($E^0 = 0.222$ V at 25° C) is constant. This potential is accounted for when the electrode is standardized against a standard pH buffer solution.

The electrode develops a second potential across the membrane separating the standard HCl from the test solution. The tiny current flow required by the pH meter causes ion exchange at the inner and outer surfaces of the glass membrane and causes diffusion of ions across the glass membrane. The electrical current is of the order of nano- or picoamperes and diffusion of trace quantities of $Na^+$ in the glass apparently carries the current. The potential of the pH electrode is

$$E = E_{ref} - 0.059 \log \frac{a_{H^+, \text{ test}}}{a_{H^+, \text{ std}}} \tag{10.10}$$

where $E$ is the measured potential and is converted to pH units by the scale of the pH meter. $E_{ref}$ includes all other (and hopefully constant) potentials which are nullified by standardizing the system with a standard pH buffer solution. Despite this complexity, the potential across the glass membrane can be closely calibrated to the approximate value of the $H^+$ activity.

A second or *reference electrode* is necessary to complete the electrical circuit. The reference electrode is sometimes welded to the pH electrode so that the pair look like a single electrode. Reference electrodes are too often taken for granted; their spurious potentials are a common source of error in soil pH measurements. A typical reference electrode is also sketched in Fig. 10.5. The wire dipping into the liquid mercury makes electrical contact with the pH meter, and current flows from the electrode to the solution phase through the reversible reaction ($E^0 = 0.268$ V at 25° C):

$$Hg_2Cl_2 + 2e^- = 2Hg + 2Cl^- \tag{10.11}$$

The $Cl^-$ activity is fixed by the KCl concentration (usually saturated KCl). This potential is also compensated when the pH electrode system is standardized in a standard pH buffer. KCl diffusion through the orifice makes the electrical contact between the reference cell and the test solution. This KCl connection forms a "salt bridge" between the test solution and the reference electrode.

A junction or diffusion potential always develops when two dissimilar substances or solutions of different composition come in contact. An example in this case is the saturated KCl of the reference electrode as it diffuses into the test solution. The potential is minimal when the rates of $K^+$ and $Cl^-$ diffusion into the test solution are equal. Indeed, KCl was chosen as the reference cell electrolyte because of the similarity of $K^+$ and $Cl^-$ diffusion rates in water.

When immersed in solution, the reference electrode in Fig. 10.5 usually fulfills its role of simply completing the electrical circuit. In a colloidal suspension, however, the colloid may cause $K^+$ and $Cl^-$ to diffuse at different rates. Because of attraction or repulsion by the charged colloid, one ion moves ahead of the other. Ion separation at the junction between the electrode solution and the suspension produces a charge separation or electrical potential, the liquid–liquid junction potential ($E_j$). Accurate pH measurements require a negligible $E_j$, because such potentials are unpredictable. The potentiometer of the pH meter measures all potentials of the circuit and cannot distinguish between the $H^+$ potential at the glass membrane and the spurious $E_j$ values. The value of $E_j$ at the interface between saturated KCl and a colloidal suspension of very low salt concentration can be as high as 240 mV, equivalent to more than 4 pH units. Such extreme $E_j$ values are unlikely in soil suspensions, because salt concentrations even in highly leached soils are normally at least several mmoles per liter. Values of $E_j$ greater than 30 mV, an error equivalent to 0.5 pH unit, are probably uncommon for soils. Measuring pH in salt solutions of 0.01 M or greater virtually eliminates $E_j$.

Another simple way to minimize $E_j$ in soil pH measurements is to allow the tip of the reference electrode to contact only the supernatant solution above the colloidal phase. The rates of $K^+$ and $Cl^-$ diffusion are then unaffected by the colloid. The glass electrode, on the other hand, can be placed either in the supernatant solution or in the colloidal suspension. The $H^+$ activity is the same in both phases, and the glass electrode is unaffected by the presence of the colloid.

The electrodes of a pH measurement circuit can be shown as

$$Ag \mid AgCl \mid standard\ HCl \mid test\ solution \parallel saturated\ KCl \mid Hg_2Cl_2 \mid Hg$$
$$\underset{E_{membrane}}{} \qquad \underset{E_j}{}$$

where each bar represents a phase boundary and where the double bar represents the liquid–liquid junction. Other ion-sensitive electrodes differ primarily in the composition of the membrane.

The glass membrane of the pH electrode has proved to be by far the most successful ion-sensitive membrane. The glass has a uniform response to a wide range of $H^+$ activities (or concentrations), requires little maintenance, is resistant to contamination, is structurally strong, and is insensitive to interfering ions. The extent of interference is denoted by the selectivity ratio, which is the concentration ratio of the test ion to interfering ion at which the interfering ion exerts a significant potential at the membrane. The selectivity ratio of the pH glass membrane for $H^+$ over $Li^+$, the most serious interfering ion, is about $10^9$. That is, $Li^+$ would cause a significant pH error in a pH 9 solution containing 0.1 M or more $Li^+$. The selectivity ratio for $Na^+$, the next most serious interference, is about $10^{13}$.

Other ion-sensitive electrodes are not nearly as effective in screening out interfering ions. Selectivity ratios are as low as 1, meaning that the electrode is as sensitive to the interfering ion as to the test ion. Such measurements are valid only when the concentrations of interfering ions are considerably lower than that of the test ion. Ion-sensitive membranes are being continually improved and hold considerable promise for soil chemical analysis.

## BIBLIOGRAPHY

Chernov, V. A. 1947. *On the Nature of Soil Acidity.* Academy of Sciences, Moscow U.S.S.R. 170 pp. (Translated and published by the Soil Science Society of America, Madison, WI.)

Coleman, N. T., and G. W. Thomas. 1967. The basic chemistry of soil acidity. *Agronomy* **12**:1–41. Excellent review article on acid soil chemistry.

## QUESTIONS AND PROBLEMS

1. Calculate the relative acidifying tendencies of 100 kg ha$^{-1}$ of nitrogen as $(NH_4)_2SO_4$, $NH_4NO_3$, and $NH_3$.

2. Based on the data of Table 10.1, calculate the times required for Utah bentonite, Volclay bentonite, and kaolinite to convert from $H^+$ form to >99% $Al^{3+}$ form.

3. A mineral subsoil of initial pH 4.8 and CEC 76 mmoles(+) kg$^{-1}$ is titrated with $OH^-$ to pH 6.5. Sketch the variation in CEC with pH that you would expect during this process.

4. For the soil of Problem 3, discuss the probable composition of the exchange complex at
   (a) pH 4.8
   (b) pH 5.7
   (c) pH 6.5

5. An acidic pesticide of p$K$ 5 is being applied to soils of a given region. Would leaching be greatest for soils of
   (a) pH > 5?
   (b) pH 6 to 8?
   (c) pH < 5?
   Why?

6. A soil of pH 5.5 retains 60 mmol(+) kg of exchangeable bases and has a CEC at pH 7 of 80 mmol(+) kg$^{-1}$:
   (a) What is its percent base solution?
   (b) What is its approximate CEC at pH 5.5?

7. A soil has a pH of 5.2, retains 70 mmol(+) kg$^{-1}$ of exchangeable bases, 10 mmol(+) kg$^{-1}$ of exchangeable $Al^{3+}$, and 30 mmol(−) kg$^{-1}$ of phosphorus at pH 5.2, and has CEC and phosphorus-retention capacities at pH 7 of 100(+) and 15 mmol(−) kg$^{-1}$, respectively:
   (a) What is its percent base saturation?
   (b) What is its percent exchangeable acidity?
   (c) What is its amount of titratable but nonexchangeable acidity?

8. For the soil of Problem 7, what is the approximate field lime requirement if the pH is to be raised to 6.5?

9. Based on the data of Fig. 10.4, how much effect would variation in titration time from 0.5 to 48 hours have on the lime requirement of pH 5 Greenfield soil, if a final pH of 6.5 were sought and if 5 g of soil were used for the titration shown?

10. Based on the data of Table 10.2, tabulate the approximate pH values required to produce 50, 75, and 90% of maximum yield for each of the crops listed.

11. Calculate the titration curve of a 0.01 M HCl solution titrated with 0.01 M NaOH to verify the strong acid curve of Fig. 10.1.

12. Based on Fig. 10.4, how are laboratory titration data related to lime requirement values in the field?

13. With a glass electrode $H^+/Li^+$ selectivity ratio of $10^9$, calculate the pH error if 1 M LiCl were used to extract exchangeable cations at measured pH values of 6, 8, and 10.

# 11

# SALT-AFFECTED SOILS

Salt-affected soils are common in arid and semiarid regions, where annual precipitation is insufficient to meet the evapotranspiration needs of plants. As a result, salts are not leached from the soil. Instead, they accumulate in amounts or types detrimental to plant growth. Salt problems are not restricted to arid or semiarid regions, however. They can develop in subhumid and humid regions under appropriate conditions. Basic principles of soil chemistry directly apply to the study and management of salt-affected soils.

## 11.1   DISTRIBUTION AND ORIGIN

Salt-affected soils often occur within irrigated lands. In the United States, 5 million ha of irrigated land are estimated to be salt-affected, mostly in the 17 western states. A recent survey indicates that as much as one-third of all irrigated lands in the world (or approximately 70 million ha) may be plagued by salt problems. When salt problems of nonirrigated semiarid and humid regions, greenhouse crops, mine spoils, and waste disposal areas are added to these figures, the dimensions of the problem are truly impressive. Though one might think that naturally saline areas would be better left unfarmed, the typically favorable year-round climates of many such areas, the desire to develop all of a farm for crop production, and the expense of installing and maintaining a water conveyance system can dictate the reclamation of many saline areas.

The three main natural sources of soil salinity are mineral weathering, atmospheric precipitation, and fossil salts (those remaining from former marine or lacustrine environments). The human activities that add salts to soil include irrigation and saline industrial wastes. Seawater encroachment can also harm soils.

The ultimate source of all soil salts is the exposed rocks and minerals of the earth's crust, from which salts have been released during chemical and physical weathering. In humid areas, soluble salts are carried down through the soil profile by percolating rainwater and ultimately are transported to the ocean or to inland seas. In arid regions, leaching is generally more localized. Salts tend to accumulate because of the relative scarcity of rainfall, high evaporation and plant transpiration rates, or landlocked topography.

Without leaching, in situ weathering of primary minerals would eventually allow soluble salts to accumulate to hazardous levels, but this degree of accumulation is rare. Salts are released during weathering. Mafic mineral (dark, Mg- and Fe-rich) minerals, for example, are common in arid-region soils. If present in sufficient quantities, they can increase the salt concentration of slowly percolating waters by as much as 3 to 5 mmol(+) $L^{-1}$. In arid regions, the occasional rains that cause the weathering are usually sufficient to flush out most of the salts.

Weathering minerals rarely dissolve congruently (in strict proportion to their composition). Instead, they release their most soluble components first. A mineral high in Ca and Mg may therefore initially release significant amounts of Na and K to the percolating solution. The water weathering the minerals is usually of sufficient quantity to carry the soluble salts thus created to the sea, to a landlocked lake, to a nearby saline seep, or at least to the average annual depth of wetting of the soil.

So-called fossil salts can introduce large amounts of salinity into soil and ground water. This was dramatized in the 1960s by the Wellton–Mohawk irrigation project of Arizona, where saline groundwaters were discharged into the Gila River after irrigation raised the groundwater level in a valley underlain by saline deposits. The drainage water mixed with the Colorado River and significantly increased the river's salinity. Downstream farmers in Mexicali, Mexico, were understandably angered when the more saline water damaged their irrigated crops. Fossil salts dissolving in percolating waters contribute materially to the salinity added to the Colorado River from several irrigation projects along its upper reaches.

Fossil salts can also be dissolved when water-storage or water-transmission structures are placed over saline sediments. The Lake Mead reservoir behind Hoover Dam in southern Nevada overlies deposits of gypsiferous sediments. Dissolution of this gypsum substantially increases the salinity of the Colorado River during its passage through the reservoir.

Appreciable salt can also be deposited in some areas from the atmosphere. Rain droplets form around tiny condensation nuclei such as salt or dust particles. The total salt concentration of rainfall may be as high as 50 to 200 mg $L^{-1}$ near the seacoast, but rapidly decreases to only a few mg $L^{-1}$ in the continental interior. The exact pattern of the decrease depends on local topography and weather patterns. Changes in composition of the rainfall also occur. The salts in rain near the seacoast are high in Na, Cl, and Mg. Inland precipitation is dominated by Ca and Mg sulfates and bicarbonates.

The quantities of salt added from the atmosphere to arid and semiarid regions may amount to only a few kilograms per hektar per year, but the amounts introduced over periods of tens to thousands of years can be substantial. The vegetation of such areas normally has reached a balance with incoming precipitation, so salts tend to accumulate below the surface at the average depth of soil wetting. They can then be flushed from the soil at relatively high concentrations when a period of particularly high rainfall occurs, or when human activities tend to change the annual water balance. Such changes have contributed to the *saline seeps* now common in certain controlled-brush and overgrazed portions of Australia, and in summer-fallowed wheat areas of eastern Montana and the Dakotas. The conditions causing saline seeps and saltpans on an Australian landscape are shown in Fig. 11.1. Salt accumulates at the seep on the slope because of arrested soil drainage. The rates of salt input and evapotranspiration at the seeps are greater than the rates of leaching or down slope runoff.

The saltpan in the basin of Fig. 11.1 exemplifies a more common occurrence of soil salinity. Soils in low-lying areas, even in arid regions, may have high water tables. Water from groundwater tables within a few meters of the surface can move by capillarity to the soil surface, where it evaporates and leaves behind its salts. Figure 11.2 shows an example of salt distribution above a water table 90 cm below the soil surface. The soil salinity concentration is expressed as electrical conductivity, the common method of measurement.

Large-scale examples of water collection and evaporation are the Great Salt Lake of Utah, a remnant of ancient landlocked Lake Bonneville that once covered much of the western United States; the Caspian Sea in Asia; Lake Chad in Africa; and Lake Ayre in Australia. More common examples are the fringes of salt accumulation along arid-region rivers and drainageways and small playas in many arid regions. Salt accumulation on a local level also includes "slick spot" patches of sodic (sodium-rich) soil, which can be accompanied by marked soil morphological changes as a result of repeated clay swelling and migration.

Many present-day salt-affected soils result from human activities. Salts commonly are transported from areas of overirrigation to accumulate in poorly drained areas. As drainage waters or irrigation return flows (drainage waters) evaporate, high concentrations of salts may remain. An example is the Salton Sea of southern California, which formed initially after a break in the dikes of the Colorado River during the early 1900s. It has become highly saline during subsequent accumulation and evaporation of irrigation return flows from the nearby Coachella and Imperial Valleys. Salts also accumulate in underirrigated fields, particularly if relatively saline irrigation waters are used. The salt concentration of the soil solution increases steadily as water is removed during plant growth. Proper irrigation management includes periodic irrigation with water in excess of plant needs, to leach accumulated salts from the plant root zone. Since water in these regions is scarce and expensive, minimizing the amount of this leach water is important.

Such human activities as oil-field development, waste-spreading operations, and fertilization can also add sizeable quantities of soluble salts to soils. Development of tidal or formerly marine areas can lead to salinity from saltwater intrusion whenever freshwater is insufficient to keep out seawater. The seaward flow of freshwater is often decreased by pumping from wells and diversion of streams for irrigation. In the Netherlands, treated municipal wastewater is pumped into the ground to prevent

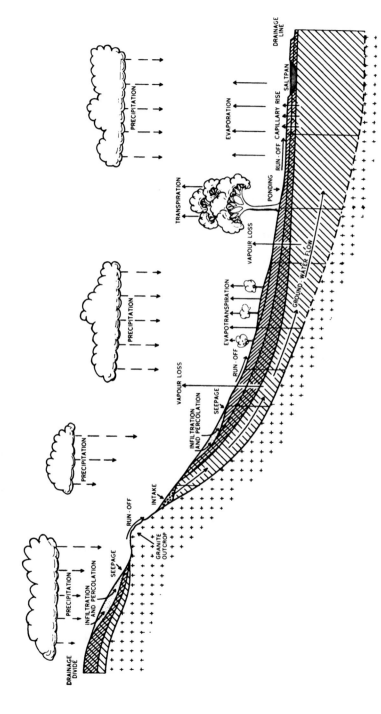

**FIGURE 11.1.** Hydrologic cycle in the Belka Valley of Australia, showing the geomorphology that causes salt seeps and a saltpan in arid regions. (Adapted from E. Bettenay et al. 1964. *Aust. J. Soil Res.* **2**:187–210.)

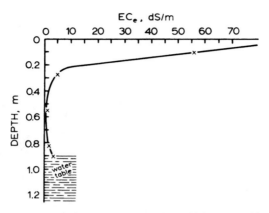

**FIGURE 11.2.** Typical salinity profile in soil exposed to a high water table. (From R. S. Ayers and D. W. Westcot. 1976. *Water Quality for Agriculture.* Food and Agriculture Organization of the UN, Rome.)

seawater intrusion along the coast. The soil pores prevent any mixing of the freshwater with saltwater so the freshwater is an effective dam against seawater movement inland.

## 11.2  IRRIGATION WATER QUALITY

The major cause of soil salinization is unsatisfactory irrigation and drainage. Various systems have been proposed to classify the quality of irrigation and drainage waters. Irrigation involves applying water to the soil surface, displacing unused water downward through the soil during subsequent irrigations, and eventual emergence of drainage waters from bottom of the plant root zone. Some water is lost during evaporation at the soil surface, and the plant removes considerably more water during transpiration. Although plants absorb some salts, both evaporation and transpiration increase the residual concentration of dissolved salts, so the salt concentration of the soil solution increases with soil depth.

Typical salt distributions in irrigated soil profiles are shown in Fig. 11.3. As the proportion of irrigated water passing through the root zone (the *leaching fraction*) increases, so does the depth of soil that has essentially the same salt concentration as the irrigation water. As the leaching fraction increases, salt accumulation is pushed down to lower depths.

As the salt concentration of the soil increases, so does the potential for salinity effects on plant growth. Early appraisals of the salinity of irrigation waters were in terms of *total dissolved solids* (TDS). The TDS were determined by evaporating a known volume of water to dryness. The presence of hygroscopic water in the resultant salt mixtures made the values for TDS strongly dependent on the drying conditions. The concentration of salts in most irrigation waters is less than 1000 mg

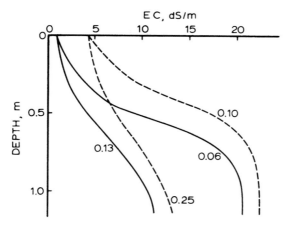

**FIGURE 11.3.** The steady-state profile of soil salinity, expressed as the electrical conductivity of the saturation extract, in lysimeters. The irrigation water has EC values of 2 (solid lines) and 4 (dashed lines) dS m$^{-1}$. Numbers on the figure are the respective leaching fractions.

L$^{-1}$ TDS. Over half of the waters used for irrigation in the western United States have TDS values less than 500 mg L$^{-1}$; less than 10% have values greater than 1500 mg L$^{-1}$. Groundwaters used for irrigation are usually higher in TDS than surface waters. Some groundwaters have been used successfully for irrigation despite TDS values approaching 5000 mg L$^{-1}$.

More recently salinity has been measured in terms of the *electrical conductivity* (EC) of a solution. In addition to overcoming some of the ambiguities of TDS measurements, the EC measurement is quicker and sufficiently accurate for most purposes. To determine the EC, the solution is placed between two electrodes of constant geometry, including constant distance of separation. When an electrical potential is imposed, the electrical current varies directly with the total concentration of dissolved salts. The current is inversely proportional to the solution's resistance and can be measured with a resistance bridge. Conductance is the reciprocal of resistance and has units of reciprocal ohms or siemens (formerly mhos). The EC of the saturation extract of the soil measures the salinity of the soil.

The measured conductance is a result of the solution's salt concentration and the electrode geometry in the measuring cell. The effects of electrode geometry are embodied in the cell constant, which is related to the distance between electrodes and their cross-sectional area. The cell constant is measured by calibration with KCl solutions of known concentration. The conductivity of KCl solutions is available in published tables. For example, calibration might yield a cell constant of 2.0 cm$^{-1}$. A test solution that measures 2000 $\Omega$ resistance (conductance of 1/2000 $\Omega^{-1}$ or 0.0005 siemens) in this cell has a conductivity of 1.0 dS m$^{-1}$, or 1.0 mmho cm$^{-1}$.

The former unit of electrical conductivity was millimhos per centimeter (mmho cm$^{-1}$). In SI units, the unit of conductivity is siemens (1 S = 1 mho, so that 1 dS m$^{-1}$ = 1 mmho cm$^{-1}$). When dealing with rainwater or with river water of low salinity, results were also reported as micromhos per centimeter ($\mu$mho cm$^{-1}$). A water with an EC of 0.2 mmho cm$^{-1}$ has an EC of 200 $\mu$mho cm$^{-1}$ or 0.2 dS m$^{-1}$.

Now, in situ soil EC measurements are being made by (1) sensors embedded in porous ceramic, thus maintaining solution contact with the electrodes; (2) groups of electrodes (commonly four) placed across the soil surface to measure the salinity of underlying soils; or (3) mobile electromagnetic "wands," which can be carried across the landscape to give similar information.

Several empirical relationships have been developed for converting one type of water quality analysis to another. For solutions in the EC range from 0.1 to 5 dS m$^{-1}$,

$$\text{Sum of cations or anions (mmol(+ or $-$) L}^{-1}) \approx \text{EC (dS m}^{-1}) \times 10 \qquad (11.1)$$

and

$$\text{TDS (mg L}^{-1}) \approx (\text{EC (dS m}^{-1}) \times 640 \qquad (11.2)$$

For soil extracts in the EC range from 3 to 30 dS m$^{-1}$,

$$\text{OP (bars)} \approx \text{EC (dS m}^{-1}) \times (-0.36) \qquad (11.3)$$

where OP is the *osmotic potential*, or the negative of the osmotic pressure of the water. The osmotic pressure or osmotic potential most directly measures the effects of salinity on plant growth. An irrigation water containing 3 mmol(+) L$^{-1}$ Ca$^{2+}$, 2 mmol(+) L$^{-1}$ Mg$^{2+}$, and 3 mmol(+) L$^{-1}$ Na$^+$ has 8 mmol(+) L$^{-1}$ total cations, an EC of approximately 0.8 dS m$^{-1}$, a TDS value of approximately 510 mg L$^{-1}$, and an OP of approximately $-0.3$ bars or $-30$ kPa.

Whenever complete chemical analyses are provided for soil extracts or irrigation waters, the sum of major cations (mmol(+) L$^{-1}$) should approximately equal the sum of all major anions (mmol($-$) L$^{-1}$). Repeated exact agreement, however, indicates that one ion is being determined by difference. This is usually sulfate for recent analyses, or sodium for older analyses. Also, reported concentrations of carbonate should be negligible at solution pH >9. In the water-supply literature, Ca plus Mg concentrations are reported as *hardness*, the chemically equivalent quantity of CaCO$_3$ in milligrams per liter. Concentrations of bicarbonate plus carbonate may be reported as *alkalinity*, the equivalent acid-neutralizing capacity of the water.

### 11.2.1  Sodium Hazard

Another important measurement of water quality is its relative amount of sodium (*sodicity*). Irrigation waters with a high sodium content tend to produce soils with high exchangeable sodium levels. Such soils crust badly and swell and disperse, greatly decreasing the soil's hydraulic conductivity, or water permeability. Clay particles disperse and plug the soil–water flow channels, as does swelling of clay particles. Decreased permeability interferes with the drainage required for salinity control and with the water supply and aeration required for plant growth.

Early estimates of sodicity were based on sodium content. Because of the strong preference of most soil particles for divalent cations over monovalent cations, however, waters with high Na contents may still produce relatively low exchangeable Na levels in soils, if the Ca + Mg concentration is appreciable.

Cation exchange equations contain the ratio of the monovalent cation concentrations to the square root of the divalent cation concentration, or the square of this ratio. The equation may involve ion activities rather than concentrations, and may include corrections for ion pairs. For most field practice, however, the ratios of total ion concentrations alone are sufficient.

Workers at the U.S. Salinity Laboratory proposed the *sodium adsorption ratio* (SAR) to characterize the sodium status of irrigation waters and soil solutions:

$$SAR = \frac{[Na^+]}{\left[[Ca^{2+} + Mg^{2+}]/2\right]^{1/2}} \tag{11.4}$$

where the brackets indicate that the concentrations are in millimoles(+) per liter. The Ca plus Mg term is divided by two because most ion-exchange equations express concentrations as moles per liter or mmoles per liter, rather than as millimoles(+) per liter. Combining Ca and Mg is not strictly correct but seems to cause little loss of accuracy. The combination is necessary because many early water analyses combined Ca with Mg, and it is justified because these two divalent cations behave similarly during cation exchange. Recent work has distinguished between the "true" SAR (involving ion activities), and the "practical" SAR (SARp) involving the ratio of concentrations.

The exchangeable sodium status of soils can be predicted quite well from the SAR and a Gapon-type exchange equation:

$$ESR = \frac{[NaX]}{[CaX + MgX]} = \frac{K_G[Na^+]}{\left[[Ca^{2+} + Mg^{2+}]/2\right]^{1/2}} = K_GSAR \tag{11.5}$$

where ESR is the *exchangeable sodium ratio* of the soil, X is the soil, the exchangeable ion concentrations are in millimoles(+) per kilogram, and $K_G$ is the Gapon exchange constant. The range of $K_G$ is commonly 0.010 to 0.015 (L mmol)$^{-1/2}$. The values of ESR of the soils and ESP/100 (*exchangeable sodium percentage*) of the irrigation water are approximately equal for many irrigated soils at ESP values below 25 or 30% . The exact relation between the two parameters is

$$ESP = \frac{100\ ESR}{1 + ESR} \tag{11.6}$$

Water having an EC of 1 dS m$^{-1}$ and a sodium percentage ([mmol(+) L$^{-1}$ of Na$^+$]/[mmol(+) L$^{-1}$ of total cations] $\times 100$) of 92% has an SAR of approximately 15.

### 11.2.2 Bicarbonate Hazard

Another property related to the sodium hazard of irrigation waters is the bicarbonate concentration. Bicarbonate toxicities associated with some waters generally arise from deficiencies of iron or other micronutrients caused by the resultant high pH. Precipitation of calcium carbonate from such waters,

$$Ca^{2+} + 2HCO_3^- = CaCO_3 + H_2O + CO_2 \qquad (11.7)$$

lowers the concentration of dissolved Ca, increases the SAR, and increases the exchangeable-sodium level of the soil. The $CaCO_3$ precipitation can be accounted for by the adjusted SAR:

$$\text{Adjusted SAR} = \text{SAR} \times [1 + (8.4 - pH_c)] \qquad (11.8)$$

where $pH_c$ is as defined and discussed in Appendix 11.1. The concept of an adjusted SAR has found widespread applicability. Figure 11.4 shows the relation of the adjusted SAR of the irrigation water to the SAR of the saturation extracts for a group of Middle Eastern soils after three years of irrigation.

Early workers used the *residual sodium carbonate* (RSC) to predict the tendency of calcium carbonate to precipitate from high-bicarbonate waters and thus create a sodium hazard. The RSC was defined as

$$\text{RSC} = [HCO_3^- + CO_3^{2-}] - [Ca^{2+} + Mg^{2+}] \qquad (11.9)$$

with all values in millimoles($\pm$) per liter. Waters of RSC greater than 2.5 were considered hazardous under all conditions. RSC values between 1.25 and 2.50 were

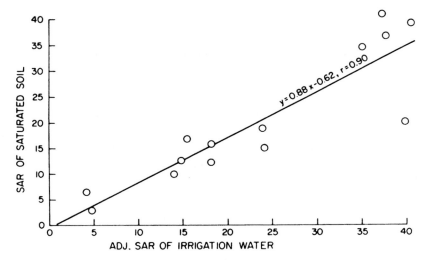

**FIGURE 11.4.** Influences of adjusted SAR of the irrigation water upon the SAR of saturation extracts from Pakistani soils after three years of cropping. (From R. S. Ayers and D. W. Westcot. 1976. *Water Quality for Agriculture.* Food and Agriculture Organization of the UN, Rome.)

considered potentially hazardous, and waters with RSC values less than 1.25 were considered safe. These predictions worked reasonably well.

The main disadvantage of the RSC was that it treated all bicarbonate in the water as if it would precipitate. This was incorrect, for the amount of bicarbonate that precipitates depends on the degree to which salts are concentrated by evapotranspiration in the plant root zone. As an extreme example, if no water evapotranspired, all the bicarbonate would pass through the soil unchanged. Conversely, if all of the water evapotranspired, all of the bicarbonate would precipitate. Hence, the quantity of bicarbonate precipitating depends on the proportion of water percolating through the soil, or the leaching fraction.

### 11.2.3  Other Toxic Solutes

Irrigation waters also contain potentially toxic ions such as boron, lithium, sodium, and chloride. The boron concentrations of irrigation waters are particularly important, because many crops are susceptible to even extremely low concentrations of this element. The differences between deficient and toxic B concentrations are only a few milligrams per liter. Sodium and chloride ions also are hazardous to fruit and berry crops and to other woody plants. Their ranges of hazardous concentration are considerably higher than for boron. In addition to absorption through roots, toxic ions also can be taken up by foliage. Sprinkling water high in sodium or chloride on the leaves of horticultural plants and vegetables, fruits, and berry crops can cause plant damage as the water evaporates. Although Li is potentially a problem in the Coachella Valley of California, the management that controls B, Na, and Cl also prevents Li toxicity.

## 11.3  CHARACTERIZING SALT-AFFECTED SOILS

The sodium status of soils is generally best described by the soil's exchangeable sodium percentage. Measuring ESP, however, is tedious and subject to error. The concentration of "soluble" sodium in the bulk solution must be measured and subtracted from the total quantity of sodium extracted to obtain the exchangeable sodium. Soluble sodium can be measured in the saturation extract, but anion exclusion can produce excessive soluble sodium concentrations in extracts from high-clay soils. This results in low ESP values. Incomplete removal of the index salt solution during the wash step of CEC determinations can lead to high CEC values and therefore to low ESP estimates. Hydrolysis during removal of the index salt solution, trapping of $NH_4^+$ from the index solution between soil mineral lattices, and calcium carbonate or gypsum dissolution in the index or replacement solutions can all lead to low CEC values and hence to high ESP estimates.

Still another special problem in CEC and ESP determinations occurs for soils of high pH containing significant amounts of the slightly soluble zeolite minerals. Zeolites such as analcime and leucine contain replaceable monovalent cations in their crystal lattices. These structural cations are readily displaced by other monovalent cations, but not by divalent cations. If a monovalent cation is used as the index or replacement cation, the amounts of sodium or ammonium extracted are erroneously high, for many of the extracted cations would not be available for normal exchange by divalent or trivalent ions. This problem should be suspected whenever soils of high pH have unusually high ESP–SAR relationships. In such cases, estimating the true ESP from the SAR of the saturation extract may be more accurate than measuring the ESP directly.

As a result of these potential errors in soil ESP determinations, and because of the generally good relationship between SAR of the soil solution and ESP of the soil, the SAR of the saturation extract is normally a satisfactory index to the exchangeable-sodium status of salt-affected soils. Since the saturation extract is already required to determine the EC, using the SAR requires only that a few additional chemical determinations be made on this extract. In fact, when the quantity of solution or the cost of analyses is limiting, the SAR can be estimated from the EC and either the sodium or the calcium plus magnesium concentration alone. The EC reflects the total cation concentration, and saline-soil extracts typically contain few cations other than sodium, calcium, and magnesium. A solution having an EC of 0.6 dS m$^{-1}$ and a sodium concentration of 3 mmol(+) L$^{-1}$ would have a total salt concentration of approximately 6 mmol(+) L$^{-1}$, a calcium plus magnesium concentration of approximately 3 mmol(+) L$^{-1}$, and an SAR of approximately $3/(3/2)^{1/2} = 2.5$.

The traditional classification of salt-affected soils in the United States has been based on the soluble salt (EC) concentrations of extracted soil solutions and on the exchangeable sodium percentage of the associated soil. The dividing line between saline and nonsaline soils was established at 4 dS m$^{-1}$ for water extracts from saturated soil pastes. Salt-sensitive plants, however, can be affected in soil whose saturation extracts have ECs of 2 to 4 dS m$^{-1}$. The Terminology Committee of the Soil Science Society of America has recommended lowering the boundary between saline and nonsaline soils to 2 dS m$^{-1}$ in the saturation extract.

The traditional and recently proposed classification categories for salt-affected soils are given in Table 11.1. *Saline* (white alkali) soils are those in which plant growth is reduced by excess soluble salts. These soils can be converted to normal soils by leaching the excess salts from the plant root zone. The pH of saline soils

**Table 11.1. Traditional and proposed classifications of salt-affected soils[a]**

|  | Normal Soils | Saline Soils | Sodic Soils | Saline–Sodic Soils |
|---|---|---|---|---|
| Traditional classification | EC < 4 dS m$^{-1}$ | EC > 4 dS m$^{-1}$ | ESP > 15% | EC > 4 dS m$^{-1}$ |
|  | ESP < 15% |  |  | ESP > 15% |
| Proposed classification | EC < 2 dS m$^{-1}$ | EC > 2 dS m$^{-1}$ | SAR > 15 | EC > 2 dS m$^{-1}$ |
|  | SAR < 15 |  |  | SAR > 15 |

[a] From Terminology Committee. 1973. *Glossary of Soil Science Terms*, Soil Science Society of America, Madison, WI.

generally is less than 8.5, and they are normally well flocculated (i.e., as permeable as might be expected from soil texture alone). Plants growing on such soils may appear stunted and have thickened leaves and a dark green color. Substantial reductions in plant growth can occur without appreciable changes in plant appearance.

Soils containing both high soluble-salt and high exchangeable-sodium levels are called *saline–sodic*. Such soils also reduce plant growth because of their high soluble-salt content. Because the soluble salts prevent hydrolysis, the pH of saline–sodic soils is typically less than 8.5. The main hazard occurs when these soils are leached to remove salts. Leaching removes the salts faster than it removes exchangeable Na, causing conversion to sodic soils. This can severely reduce soil permeability or hydraulic conductivity and affect plant–water relations and the ability to leach for salinity control.

*Sodic* (black alkali) soils are a particularly difficult management problem. The water permeability of these soils to water is very slow. The pH of sodic soils is commonly greater than 9 or 9.5, and the clay and organic fractions are dispersed. Dispersed organic matter accumulates at the surface of poorly drained areas as water evaporates and imparts a black color to the surface, hence the name "black alkali." Sodic soils are found in many parts of the western United States. In some locations they occur in small patches, "slick spots," less than 0.5 ha in extent. Such patches occupy slight depressions, which become accentuated as surface soil particles disperse and are blown away by wind erosion. The percolation of insufficient water to satisfy plants and to control salinity is the main problem associated with sodic soils. In addition, their relatively low soluble-salt concentrations and high pH values can result in direct Na toxicities to the most sensitive plants.

## 11.4   EFFECTS OF SALTS ON SOILS AND PLANTS

The main effect of soluble salts on plants is osmotic—plants must expend large amounts of energy to absorb water from the soil solution, energy that would otherwise be used for plant growth and crop yield. The plant root contains a semipermeable membrane permitting water to pass but rejecting most of the salt. Thus, water is osmotically more difficult to extract from increasingly saline solutions. Plants growing on saline media can somewhat increase their internal osmotic concentrations by producing organic acids or by absorbing salts. This process is called *osmotic adjustment*. The effect of salinity on the plant appears primarily to be energy diversion from growth processes to maintain the osmotic differential between the interior of the plant and the soil solution. One of the first processes from which this energy is diverted is cell elongation. Leaf tissue cells continue to divide but do not elongate. The occurrence of more cells per unit leaf area accounts for the typically dark green color of osmotically stressed plants.

The relative growth of plants in the presence of salinity has been termed their *salt tolerance*. Earlier data were summarized by separating plants into several salt-tolerant groups. Subsequent listings were in terms of relative plant growth at various salinity levels (EC) of the soil's saturation extract (Table 11.2). Some recent listings

**Table 11.2. Salt tolerance of plants[a]**

| Crop | EC (dS m$^{-1}$ at 25°C) at Which Yield Will Be Decreased by[b] | | |
|---|---|---|---|
| | 10% | 25% | 50% |
| *Forage Crops* | | | |
| Bermudagrass[c] (*Cynodon dactylon* (L.) Pers.) | 13 | 16 | 18 |
| Tall wheatgrass (*Agropyron elongatum* (Host) Beauv.) | 11 | 15 | 18 |
| Crested wheatgrass (*Agropyron desertorum* (Fisch. ex Link) Schult.) | 6 | 11 | 18 |
| Tall fescue (*Festuca arundinacea* Schreb) | 7 | 10.5 | 14.5 |
| Barley, hay[d] (*Hordeum vulgare* L.) | 8 | 11 | 13.5 |
| Perennial ryegrass (*Lolium perenne* L.) | 8 | 10 | 13 |
| Harding grass (*Phalaris stenoptera* Hack) | 8 | 10 | 13 |
| Narrow-leaf birdsfoot trefoil (*Lotus tenuifolius* (L.) Reich) | 6 | 8 | 10 |
| Beardless wild rye (*Elymus triticoides* Buckley) | 4 | 7 | 11 |
| Alfalfa (*Medicago sativa* L.) | 3 | 5 | 8 |
| Orchardgrass (*Dactylis glomerata* L.) | 2.5 | 4.5 | 8 |
| Meadow foxtail (*Alopecurus pratensis* L.) | 2 | 3.5 | 6.5 |
| Alsike and red clovers (*Trifolium hybridum* L. and *T. pratense* L.) | 2 | 2.5 | 4 |
| *Field Crops* | | | |
| Barley, grain[d] (*Hordeum vulgare* L.) | 12 | 16 | 18 |
| Sugarbeet[e] (*Hordeum vulgare* L.) | 10 | 13 | 16 |
| Cotton (*Gossypium hirsutum* L.) | 10 | 12 | 16 |
| Safflower (*Carthamus tinctorius* L.) | 8 | 11 | 12 |
| Wheat[d] (*Triticum aestivum* L.) | 7 | 10 | 14 |
| Sorghum (*Sorghum vulgare* Pers.) | 6 | 9 | 12 |
| Soybean (*Glycine max* (L.) Merr.) | 5.5 | 7 | 9 |
| Sesbania[d] (*Sesbania macrocarpa* Muhl.) | 4 | 5.5 | 9 |
| Sugarcane (*Saccharum officinarum* L.) | 3 | 5 | 8.5 |
| Rice, paddy[d] (*Oryza sativa* L.) | 5 | 6 | 8 |
| Corn (*Zea mays* L.) | 5 | 6 | 7 |
| Broadbean (*Vicia faba* L.) | 3.5 | 4.5 | 6.5 |
| Flax (*Linum usitatissimum* L.) | 3 | 4.5 | 6.5 |
| Field bean (*Phaseolus vulgaris* L.) | 1.5 | 2 | 3 |

*(continued)*

have been given instead in terms of EC at the point of initial yield decline and in terms of percent yield decrease per unit increase in salinity beyond this threshold. Most yield data were obtained from uniformly salinized field plots having nearly constant salinity with depth. Actual distributions under field conditions more closely resemble those from the plots in Fig. 11.3, where the plant can extract most of its water from the least-salinized portion of the profile.

As evidenced by the footnotes to Table 11.2, some plants are particularly sensitive to salinity during the germination or seedling stages when a restricted root zone makes the plant extremely vulnerable to osmotic stress. Seedbed shape is often modified for such crops to minimize salt accumulation in the vicinity of young seedlings (Figure 11.5). Alternate-furrow irrigation (where only one side of the crop row is irrigated at any one time) can also be used to flush salts past the young seedling if

**Table 11.2. (Continued)**

| Crop | EC (dS m$^{-1}$ at 25°C) at Which Yield Will Be Decreased by[b] | | |
| --- | --- | --- | --- |
| | 10% | 25% | 50% |
| *Vegetable Crops* | | | |
| Beets[e] (*Beta vulgaris* L.) | 8 | 10 | 12 |
| Spinach (*Spinacia oleracea* L.) | 5.5 | 7 | 8 |
| Tomato (*Lycopersium esculentum* Mill.) | 4 | 6.5 | 8 |
| Broccoli (*Brassica oleracea* var. italica L.) | 4 | 6 | 8 |
| Cabbage (*Brassica oleracea* var. capitata L.) | 2.5 | 4 | 7 |
| Potato (*Solanum tuberosum* L.) | 2.5 | 4 | 6 |
| Sweet corn (*Zea mays* L.) | 2.5 | 4 | 6 |
| Sweet potato (*Ipomoea batatas* (L.) Lam.) | 2.5 | 3.5 | 6 |
| Lettuce (*Lactuca sativa* L.) | 2 | 3 | 5 |
| Bell pepper (*Capsicum annum* L.) | 2 | 3 | 5 |
| Onion (*Allium cepa* L.) | 2 | 3.5 | 4 |
| Carrot (*Dancus carota* L.) | 1.5 | 2.5 | 4 |
| Green bean (*Phaseolus vulgaris* L.) | 1.5 | 2 | 3.5 |

[a] From L. Bernstein. 1964. *Salt Tolerance of Plants*. U.S. Department of Agriculture Information Bulletin 283.

[b] In gypsiferous soils, EC values causing equivalent yield reductions will be about 2 dS m$^{-1}$ greater.

[c] Average for different varieties. Suwanne and Coastal bermudagrass are about 20% more tolerant, and Common and Greenfield are about 20% less tolerant, than the average. For most crops, varietal differences are relatively insignificant.

[d] Less tolerant during the seeding stage. Salinity at this stage should not exceed an EC of 4 to 5 dS m$^{-1}$.

[e] Sensitive during germination, when salinity should not exceed 3 dS m$^{-1}$.

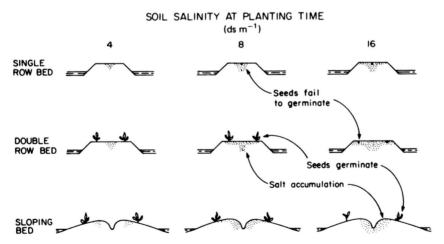

**FIGURE 11.5.** Effects of bed shape on seedling emergence at various salinity levels. (L. Bernstein et al. 1955. *U.S. Dept. Agric. Bull. ARS-41-4*, 16 pp.)

single rows are used. If double-row beds are used, alternate-furrow irrigation can flush salts to the vicinity of the bed edge opposite the irrigated furrow and hence stress the seedlings near this edge. Drip irrigation, though generally flushing salts to the periphery of the wetted soil volume, can also lead to serious salinity problems when high rates of fertilizer are added through the drip lines, upon replanting, or whenever rainfall flushes accumulated salts toward previously nonstressed plant roots.

In addition to the general osmotic effects summarized in Table 11.2, many plants are sensitive to specific ions in irrigation waters or soil solutions. Boron toxicity is probably the most common. Table 11.3 lists some plants according to their sensitivity to the B concentration of irrigation water. Boron is more difficult to control than is salinity in general because it leaches more slowly than more soluble salts.

Direct sensitivity to exchangeable or soluble sodium is more apparent at low salt levels, and therefore is difficult to differentiate from the effects of sodium on soil permeability. For plants that are extremely sensitive to sodium, as little as 5% exchangeable sodium may lead to toxic accumulations of sodium in leaf tissues (Table 11.4).

Chloride toxicity appears to be similar to Na toxicity. Excessive accumulations in tissues near plant tips, the end of the plant transpiration stream, lead to necrosis, death of leaf tips and margins, and eventual death of the plant. Some plants are able to screen out such ions through their root membranes. In addition, different rootstocks may possess varying abilities to exclude sodium or chloride from above-ground parts (Table 11.5). Some grape rootstocks exhibit up to 30-fold differences in their abilities to exclude chloride ions. Selection of a rootstock that screens out ions may prevent toxic accumulations in plant tops.

A third mechanism for salt injury to plants is *nutritional imbalances*. An example is the bicarbonate toxicities reported for some saline environments. These result primarily from reduced Fe availability at the high pH common in high-bicarbonate soils, rather than from the bicarbonate ions themselves. The nutritional needs of plants may also vary with the types of salts present. For example, high Na levels could lead to Ca and Mg deficiencies. The high pH levels of sodic soils can accentuate deficiencies of many of the microelements. High soil pH levels also might lead to high concentrations of soluble aluminium, such as the aluminate ($Al(OH)_4^-$) species. Salt tolerance also can vary with soil fertility, and especially when inadequate fertility limits yields. Nutritional effects of salinity on plants are poorly understood at present, however. Many of the supposed consequences are still largely speculative.

## 11.5  SALT BALANCE AND THE LEACHING REQUIREMENT

Management of salt-affected soils once centered around maintaining the *salt balance*. This concept dictates that the quantity of salt leaving an area be equal to, or greater than, the quantity of salt entering the area. The concern was justified by the difficulty in maintaining long-term agriculture for many irrigated areas of the world, such as the Tigris and Euphrates valleys of Iraq, where farming has taken place for several millenia. Some irrigation projects, however, appear able to operate indefi-

**Table 11.3. Boron tolerance of plants in relation to boron concentration of irrigation water[a]**

| Tolerant[b] | Semitolerant | Sensitive |
|---|---|---|
| *4.0 mg L$^{-1}$ of Boron* | *2.0 mg L$^{-1}$ of Boron* | *1.0 mg L$^{-1}$ of Boron* |
| Athel (*Tamarix aphylla*) | Sunflower (native) (*Helianthus annuus*) | Pecan (*Carya pecan*) |
| Asparagus (*Asparagus officinalis*) | Potato (*Solanum tuberosum*) | Walnut (black and Persian or English)(*Juglans* spp.) |
| Palm (*Phoenix canariensis*) | Cotton (Acala and Pima) (*Gossypium* spp.) | Jerusalem artichoke (*Helianthus tuberosus*) |
| Date palm (*Phoenix dactylifera*) | Tomato (*Lycopersicum esculentum*) | Navy bean (*Phaseolus vulgaris*) |
| Sugar beet (*Beta vulgaris*) | Sweetpea (*Lathyrus odoratus*) | American elm (*Ulmus americana*) |
| Mangel (*Beta vulgaris*) | Radish (*Raphanus sativus*) | Plum (*Prunus domestica*) |
| Garden beet (*Beta vulgaris*) | Field pea (*Pisum sativum*) | Pear (*Pyrus communis*) |
| Alfalfa (*Medicago sativa*) | Ragged-robin rose (*Rosa*) | Apple (*Pyrus malus*) |
| Gladiolus (*Gladiolus* spp.) | Olive (*Olea europaea*) | Grape (Sultanina and Malaga) (*Vitis vinifera*) |
| Broadbean (*Vicia faba*) | Barley (*Hordeum vulgare*) | Kadota fig (*Ficus carica*) |
| Onion (*Allium cepa*) | Wheat (*Triticum vulgare*) | Persimmon (*Diospyros* spp.) |
| Turnip (*Brassica rapa*) | Corn (*Zea mays*) | Cherry (*Prunus avium*) |
| Cabbage (*Brassica oleracea var. capitata*) | Milo (*Sorghum vulgare*) | Peach (*Prunus persica*) |
| Lettuce (*Lactuca sativa*) | Oat (*Avena sativa*) | Apricot (*Prunus armentaca*) |
| Carrot (*Daucus carota*) | Zinnia (*Zinnia elegans*) | Thornless blackberry (*Rubus* spp.) |
| | Pumpkin (*Cucurbita pepo*) | Orange (*Citrus sineosis*) |
| | Bell pepper (*Capsicum frutescens*) | Avocado (*Persea americana*) |
| | Sweet potato (*Ipomoea batatas*) | Grapefruit (*Citrus paradisi*) |
| | Lima bean (*Phaseolus lunatus*) | |
| *2.0 mg L$^{-1}$ of Boron* | *1.0 mg L$^{-1}$ of Boron* | *0.3 mg L$^{-1}$ of Boron* |

[a]From L. V. Wilcox and W. H. Durum. 1967. *Irrigation of Agricultural Lands*. American Society of Agronomy Monograph 11, Madison, WI.

[b]In each group the plants named first are considered to be more tolerant. Those named last are more sensitive. The figures at the top and bottom of each column represent limiting boron concentrations of the irrigation water.

**Table 11.4. Tolerance of various crops to percentage of exchangeable sodium in soils[a]**

| Tolerance to ESP[b] and Range at Which Affected | Crop | Growth Response Under Field Conditions |
|---|---|---|
| Extremely sensitive (ESP = 2–10) | Deciduous fruits Nuts Citrus Avocado | Sodium toxicity symptoms even at low ESP values |
| Sensitive (ESP = 10–20) | Beans | Stunted growth at low ESP values even though the physical condition of the soil may be good |
| Moderately tolerant (ESP = 20–40) | Clover Oats Tall fescue Rice Dallisgrass | Stunted growth due to both nutritional factors and adverse soil conditions |
| Tolerant (ESP = 40–60) | Wheat Cotton Alfalfa Barley Tomatoes Beets | Stunted growth usually due to adverse physical conditions of soil |
| Most tolerant (ESP = more than 60) | Crested and Fairway wheatgrass Tall wheatgrass Rhoades grass | Stunted growth usually due to adverse physical conditions of soil |

[a]From G. A. Pearson. 1960. *U.S. Dept. Agric. Bull. 216.*
[b]ESP = exchangeable sodium percentage.

nitely at a negative salt balance (more salt entering than leaving) with few adverse effects on soils or plants. The key in such cases is the amount of salt precipitating (and hence inactivated with respect to plants) in the soil. Normal plant growth can continue, provided that the quantities of salt precipitated do not lead to sodic soil conditions or to nutritional imbalances.

The most common approach to salinity management is to maintain a prescribed *leaching requirement* (LR), defined as

$$\text{LR} = \frac{D_{dw}}{D_{iw}} = \frac{EC_{iw}}{EC_{dw}} \tag{11.10}$$

where $EC_{dw}$ and $EC_{iw}$ are the electrical conductivities (salt concentrations) of the drainage and irrigation waters, and $D_{iw}$ and $D_{dw}$ are the amounts of irrigation and drainage water. The relationship is based on the assumptions that a salt balance exists (i.e., that $EC_{iw} D_{iw} = EC_{dw} D_{dw}$) and that the plant is a perfect semipermeable mem-

**Table 11.5. Chloride tolerances of fruit varieties and rootstocks**[a]

| Crop | Rootstock or Variety | Limit to Tolerance to Chloride in Soil Saturation Extracts (mmole $L^{-1}$) |
|------|---------------------|-----|
| | *Rootstocks* | |
| Citrus | Rangpur lime, Cleopatra mandarin | 25 |
| Citrus | Rough lemon, tangelo, sour orange | 15 |
| Citrus | Sweet orange, citrange | 10 |
| Stone fruit | Marianna | 25 |
| Stone fruit | Lovell, Shalil | 10 |
| Stone fruit | Yunnan | 7 |
| Avocado | West Indian | 8 |
| Avocado | Mexican | 5 |
| | *Varieties* | |
| Grape | Thompson seedless, Perlette | 25 |
| Grape | Cardinal, Black Rose | 10 |
| Berries[b] | Boysenberry | 10 |
| Berries | Olallie blackberry | 10 |
| Berries | Indian Summer raspberry | 5 |
| Strawberry | Lassen | 8 |
| Strawberry | Shasta | 5 |

[a]From L. Bernstein. 1965. U.S. Department of Agriculture Information Bulletin 292.
[b]Data available for a single variety of each crop only.

brane removing only water from the soil solution and leaving all salts behind. The relationship is inaccurate when substantial salt precipitates in the plant root zone, dissolves from soil minerals, or is taken up by the crop. Despite these constraints, leaching requirement calculations are sufficiently accurate for most crop management purposes.

Workers in Israel have demonstrated that careful management of extremely low leaching requirement values can still maintain adequate salinity control. One source of error in leaching requirement estimates is the substitution of EC of the saturation extract for the EC of the drainage water in the leaching requirement formula. Most salt-tolerance data apply to freely draining soil profiles. The salt concentration at saturation is commonly two to three times more dilute than in the water of a freely draining soil profile ("field capacity"). Hence, using saturation extract data in place of drainage water data gives an EC in the denominator of Eq. 11.10 that is 1/2 to 1/3 too small, and a leaching requirement estimate that is 2 or 3 times too high.

To calculate the leaching requirement one needs an estimate of the allowable EC of the saturation extract, such as can be obtained from existing salt tolerance data

(Table 11.2). Soil texture or some other parameter must then be used to convert the EC value to an estimated EC of the drainage water for the soil–plant system. This value and the EC of the irrigation water can be used to estimate the fraction of leaching water that must be passed through the plant root zone for salinity control. This excess water can then be compared to the soil infiltration rate, to plant tolerance at waterlogged conditions, and to drainage system capacity to see if salinity control is feasible under the chosen set of crop, soil, and water management conditions.

The high leaching requirements that have been recommended in the past have taken into account nonuniform water distribution of many irrigation systems and the very great spatial variability of soil permeability. Excess water leaches through the more permeable parts of the field, while salts can remain in less permeable areas. Large quantities of water are therefore necessary to remove salts from the entire field. More uniform water application is possible with sprinkler irrigation or dead-level surface systems. More accurate estimates of the minimum leaching requirement can then be used.

## 11.6  RECLAMATION

The aims of reclamation are to make $Ca^{2+}$ the major exchangeable ion and to reduce the salt concentration in the soil solution. The main requirement to reclaim salt-affected soils is that sufficient water must pass through the plant root zone to lower the salt concentration to acceptable values. Passing 1 m of leaching water per meter of soil depth under ponded conditions normally removes approximately 80% of the soluble salt from soils (Fig. 11.6). If leaching is under unsaturated conditions, such as with the use of intermittent pending or sprinkler irrigation, this quantity of water may be reduced to as little as 350 to 200 mm of water. Boron removal can require up to three times more water than removal of Na and Cl salts, because B is retained to some extent by soils.

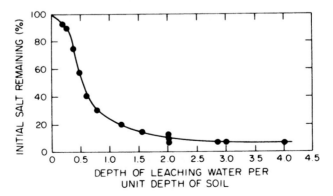

**FIGURE 11.6.** Depth of water per unit depth of soil required to leach a highly saline soil. (From R. C. Reeve et al. 1955. *Hilgardia* **24**:69–91.)

Several techniques have been developed for reclaiming salt-affected soils. Ponding is a traditional method involving the construction of a large dike around the field. A substantial depth of water (commonly 0.3 m or more) is then maintained inside the dike to leach salts from the soil. Such an approach requires drainage facilities capable of removing large quantities of drainage water. The reclamation process is relatively inefficient, because much of the water passes through large soil pores that have already been purged of salts. Salt is removed only slowly from the fine pores of the adjacent soil mass.

A more efficient leaching technique is the basin-furrow method. The soil is nearly leveled and irrigation water is allowed to meander back and forth across the field through adjacent sets of furrows. The water may take as long as a week to meander across the entire field under such conditions, but the quantities of water required are less than for ponded leaching. Furthermore, this technique does not produce sterile strips corresponding to former dike positions, where large quantities of salt accumulate during ponded leaching.

Soluble divalent ions (generally Ca) must be present during the reclamation of sodic soils. A common amendment for such purposes is gypsum ($CaSO_4 \cdot 2H_2O$), added at rates of up to several thousand kilograms per hectare in order to provide Ca as water percolates through the soil. Tests of soil exchangeable-sodium levels should be made every two to three years to estimate the need for reapplication of gypsum.

Another Ca source is the lime in many salt-affected soils. If a soil is only slightly sodic and rather sandy in texture, tillage bringing subsoil lime to the surface before water application may be sufficient to maintain soil permeability during reclamation. Deep plowing to the 0.7- to 0.9-m depth has also proved helpful in redistributing subsurface lime and in opening the soil to maintain adequate water permeability during reclamation. In most instances, however, lime is not sufficiently soluble to serve as an amendment for sodic-soil reclamation. It can be used as a source of soluble calcium only if an acidifying amendment is applied to dissolve the lime before reclamation begins.

Common acidifying amendments for the reclamation of calcareous sodic soils are sulfuric acid and elemental sulfur. Sulfur must be oxidized to sulfuric acid by soil microorganisms before it becomes effective. A lead time of several weeks or months may be required for microbial oxidation before leaching begins. The reaction is

$$2S + 3O_2 + 2H_2O = 2H_2SO_4 \tag{11.11}$$

$$CaCO_3 + H_2SO_4 = CaSO_4 + H_2O + CO_2 \tag{11.12}$$

$$2NaX + CaSO_4 = CaX + Na_2SO_4 \tag{11.13}$$

Gypsum produced by the acid reacting with soil lime behaves like added gypsum during the remainder of the reclamation process.

Still another reclamation procedure for sodic soils is the *high-salt water reclamation* method. Saline–sodic soils will remain permeable as long as soil solution salt concentrations are high enough to flocculate the soil. The soil is leached with successively more dilute water while trying to increase exchangeable Ca by displacing exchangeable Na. Each increment of dilution is small enough to prevent the soil from swelling or dispersing. The initial step of adding high-salt water may increase either the soluble salt concentration or the exchangeable-sodium percentage of the soil. For example, treatments of soil with seawater, which has a salt concentration of 600 mmol(+) $L^{-1}$ and an SAR of approximately 60, generally produces soils with exchangeable-sodium percentages of 40 to 50. A fourfold dilution of the seawater, by mixing one part with three parts of freshwater, produces a salt concentration of 150 mmol $L^{-1}$ and an SAR of 30 (because the SAR changes as the square root upon dilution of salt concentration). A second fourfold dilution step produces a salt concentration of 37.5 mmol $L^{-1}$, an EC of 3.8 dS $m^{-1}$, and an SAR of 15. A third fourfold dilution produces an EC of 9 and an SAR of 7.5. This three-step leaching process would reclaim the soil if drainage facilities were adequate to remove the resultant large quantities of salty drainage water. The high-salt water reclamation method also depends on ready access to high-salt water. Saline ground and surface waters are common in salt-affected areas, and this is usually not a limitation. The ideal environment for high-salt water reclamation is near an ocean or saline lake, where both water access and disposal are available.

## APPENDIX 11.1    THE LANGELIER INDEX

Several workers have characterized the bicarbonate levels of waters with the Langelier Index (LI):

$$LI = (pH_a - pH_c) = pHa - (pK_2' - pK_c') + pC_a + pAlk \qquad (11.14)$$

where $pH_a$ is the measured pH of the soil or water, $pH_c$ is the calculated pH of the irrigation water if equilibrated with $CaCO_3$, $pK_2'$ is the second dissociation constant of $H_2CO_3$, $pK_c'$ is the solubility product of $CaCO_3$, and pCa and pAlk are the negative logarithms of the molar Ca and molar(+) carbonate plus bicarbonate concentrations, respectively. The $pH_c$ can be derived from the reaction

$$HCO_3^- = H^+ + CO_3^{2-} \qquad (11.15)$$

with its dissociation constant

$$K_2 = \frac{(H^+)(CO_3^{2-})}{(HCO_3^-)} \qquad (11.16)$$

In a system at equilibrium with solid-phase $CaCO_3$,

$$K_c = (Ca^{2+})(CO_3^{2-}) \qquad (11.17)$$

because the activity of solid-phase $CaCO_3$ can be taken as unity. Substituting Eq. 11.17 into 11.16 and rearranging gives

$$(H^+) = \frac{K_2(HCO_3^-)(Ca^{2+})}{K_c} \qquad (11.18)$$

or, in terms of the concentrations of $HCO_3^-$ and $Ca^{2+}$,

$$(H+) = \frac{K_2 \gamma_{HCO_3}[HCO_3^-]\gamma_{Ca}[Ca^{2+}]}{K_c} \qquad (11.19)$$

where brackets indicate concentrations and the $\gamma$'s are activity coefficients. Taking negative logarithms gives

$$pH_c = pK_2' - pK_c' + p(\gamma HCO3) + p(\gamma Ca) + p[PCO_3^-] + p[Ca] \qquad (11.20)$$

Combining the two activity coefficients with the dissociation and solubility constants gives

$$pH = (pK_2' - pK_c') + p[HCO_3^-] + p[Ca^{2+}] \qquad (11.21)$$

where the $pK'$ terms are treated as a single, concentration-dependent quantity (such "constants" are truly constant only if expressed in terms of activities). Except at pH $>9$, when appreciable carbonate exists, Eq. 11.21 is equivalent to the $pH_c$ portion of Eq. 11.14.

The fraction of bicarbonate precipitating from an irrigation water is often related linearly to both $pH_c$ and the leaching fraction. With the advent of computers, however, it is more common now to predict the amount of $CaCO_3$ precipitating under a given set of management conditions from irrigation-water chemistry, initial and time-course soil chemistry, the leaching fraction, and measured or inferred $CO_2$ levels of the soil atmosphere.

## BIBLIOGRAPHY

Ayers, R. S., and D. W. Westcot. 1976. *Water Quality for Agriculture*. Food and Agriculture Organization of the UN, Irrigation and Drainage Paper 29, Rome.

Bernstein, L. 1964. *Salt Tolerance of Plants*. United States Department of Agriculture. Information Bulletin 283.

United States Salinity Laboratory Staff. 1954. *Diagnosis and Improvement of Saline and Alkali Soils*. United States Department of Agriculture. Handbook 60.

## QUESTIONS AND PROBLEMS

1. An arid area receives 150 mm of rainfall annually with an average salt concentration of 10 mg $L^{-1}$. If the surface soil from this area contains an average of 30% water at saturation, how many years would be required for sufficient salt to be added from the atmosphere to increase the EC of the saturation extract by 1 dS $m^{-1}$?

2. An irrigation water contains 750 mg $L^{-1}$ soluble salts. If used at an average leaching fraction of 0.15, what would be the average EC of the drainage water leaving the bottom of the crop root zone?

**3.** What is the EC (in dS $m^{-1}$) of a solution having an electrical resistance of 1500 $\Omega$ in a conductivity cell with a cell constant of 5.0 $cm^{-1}$?

**4.** An irrigation water has an EC of 0.8 dS $m^{-1}$ and a sodium concentration of 35 mg $L^{-1}$. Calculate:

   **(a)** Its osmotic potential.

   **(b)** Its SAR.

   **(c)** The equilibrium ESP for soils having a Gapon exchange constant of 0.015 $(L\ mmol^{-1})^{1/2}$.

**5.** An irrigation water contains 7 mmol(+) $L^{-1}$ total cations, 1.5 mmol(+) $L^{-1}$ $Ca^{2+}$, 1 mmol(+) $L^{-1}$ $Mg^{2+}$, and 5 mmol(+) $L^{-1}$ $HCO_3^-$. Calculate the $pH_c$ and the adjusted SAR of this water. To what extent would the water be regarded as hazardous?

**6.** If the irrigation water of Problem 5 has a pH of 7.8, what is the residual sodium carbonate value? To what extent would the water be regarded as hazardous based on this criterion?

**7.** An irrigation water contains 600 mg $L^{-1}$ TDS and has an SAR of 6. If it is applied to a soil containing 40% water at saturation and 20% water at "field capacity," what will be the EC and SAR of the saturation extract of surface soil and the resultant salinity classification category after prolonged irrigation with this water? If the same water-holding characteristics are found at the bottom of the crop root zone, what will be the EC and SAR of the saturation extract for soil from this portion of the profile after prolonged irrigation at a leaching fraction of 20%?

**8.** For a soil having the water-holding characteristics described in Problem 7, what will be the leaching requirement if an irrigation water of EC = 0.8 dS $m^{-1}$ is used to irrigate alfalfa under conditions where no more than a 25% yield reduction due to salinity can be tolerated?

**9.** A 30-cm depth of surface soil contains 28% exchangeable sodium and has a CEC of 150 mmol(+) $kg^{-1}$. How many tonnes per hektar of sulfur will be required to lower its ESP to 5%?

**10.** Explain how overgrazing, conversion from shrubs to grasses, burning, and summer fallowing can lead to "saline seeps."

**11.** Explain why soils containing appreciable amounts of zeolites may give unreliable ESR/SAR relations, and why the SAR is the more accurate index in such soils.

**12.** Explain how the salt tolerance data of Table 11.2 might be related to "real-world" salinity distributions, such as shown in Figs. 11.2, 11.3, and 11.5.

# INDEX

CPSIA information can be obtained at www.ICGtesting.com
Printed in the USA
243223LV00003B/3/A